AF360717

Fluid Interfaces

Fluid Interfaces

Editor

Eduardo Guzmán

MDPI • Basel • Beijing • Wuhan • Barcelona • Belgrade • Manchester • Tokyo • Cluj • Tianjin

Editor
Eduardo Guzmán
Complutense University of Madrid
Spain

Editorial Office
MDPI
St. Alban-Anlage 66
4052 Basel, Switzerland

This is a reprint of articles from the Special Issue published online in the open access journal *Coatings* (ISSN 2079-6412) (available at: https://www.mdpi.com/journal/coatings/special_issues/Fluid).

For citation purposes, cite each article independently as indicated on the article page online and as indicated below:

LastName, A.A.; LastName, B.B.; LastName, C.C. Article Title. *Journal Name* **Year**, *Volume Number*, Page Range.

ISBN 978-3-03943-633-0 (Hbk)
ISBN 978-3-03943-634-7 (PDF)

Contents

About the Editor

Eduardo Guzmán, Associate Professor at the Physico-Chemistry Department and Researcher at the Multi-disciplinary Institute in the Complutense University of Madrid (Spain), received his MSc in Chemistry and in Science and Technology of Colloids and Interfaces, and his PhD in Science at the Complutense University of Madrid (Spain). After his PhD, he worked for a period of four years at the Istituto per l'Energetica e le Interfasi in Genoa (Italy), after which he returned to his alma mater. He has published over 80 paper in JCR journals and 10 chapters in books (https://orcid.org/0000-0002-4682-2734), corresponding to a H-index of 26, and he has co-authored more than 100 contributions to different national and international conferences. His main research interests are in LbL assembly, interfacial rheology, drug delivery, biophysics, cosmetics, and pest control. He has supervised 3 PhD students, 10 students MSc students, and 20 undergraduate students. He has been involved in 2 EU and 6 Spanish funded I+D grants, and has had scientific responsibility for 2 cooperation projects between academia and industry. He is a member of the editorial board of numerous scientific journals, including Coatings (Editor-in-Chief of the Section "Liquid–Fluid Interfaces") and Polymers, and has edited Special Issues in Coatings, Processes, and Advances in Colloid and Interface Science.

Preface to "Fluid Interfaces"

Fluid interfaces are ubiquitous in science and technology, playing a main role in different aspects of industry, nature, or even life. This makes it necessary to understand the most fundamental physicochemical bases underlying the behavior of such interfaces. However, this is tricky, and makes it necessary to explore the impact of molecular and supramolecular species on the flows occurring within the interfacial region and the thickness of the interface. Furthermore, the broken symmetry associated with the formation of a fluid interface can be used as a platform for the assembly of innovative materials with reduced dimensionality, which may have impacts in food science, cosmetics, biology, oil recovery, electronic, drug delivery, detergency, or tissue engineering. This has made of the study of fluid interfaces a multidisciplinary challenge involving researchers from different areas of chemistry, physics, pharmacy, biophysics, medicine, engineering, and materials science.

The papers included in this book present a broad perspective of the current trends on the study of fluid interfaces.

Eduardo Guzmán
Editor

 *coatings*

Editorial

Fluid Interfaces

Eduardo Guzmán [1,2]

[1] Departamento de Química Física, Facultad de Ciencias Químicas, Universidad Complutense de Madrid,
Ciudad Universitaria s/n., 28040 Madrid, Spain; eduardogs@quim.ucm.es; Tel.: +34-91-394-4107

[2] Instituto Pludisciplinar, Universidad Complutense de Madrid, Paseo Juan XXIII 1, 28040 Madrid, Spain

Received: 28 August 2020; Accepted: 24 September 2020; Published: 20 October 2020

Abstract: Fluid interfaces are promising candidates for the design of new functional materials by confining different types of materials, e.g., polymers, surfactants, colloids, or even small molecules, by direct spreading or self-assembly from solutions. The development of such materials requires a deep understanding of the physico-chemical bases underlying the formation of layers at fluid interfaces, as well as the characterization of the structures and properties of such layers. This is of particular importance, because the constraints associated with the assembly of materials at the interface lead to the emergence of equilibrium and dynamic features in the interfacial systems that are far from those found in traditional 3D materials. These new properties are of importance in many scientific and technological fields, such as food science, cosmetics, biology, oil recovery, electronics, drug delivery, detergency, and tissue engineering. Therefore, the understanding of the theoretical and practical aspects involved in the preparation of these interfacial systems is of paramount importance for improving their usage for designing innovative technological solutions.

Keywords: interfaces; confinement; dynamics; materials; applications

---

A fluid interface can be defined as the nanoscopic region of a system containing two fluid phases of different nature, commonly, a liquid combined with a second liquid or vapor, where the separation between two fluid phases occurs. This simple definition excludes many aspects of interest for the daily life of modern society. Fluid interfaces are ubiquitous in science and technology, which has stimulated extensive research activity aiming to disentangle the main physico-chemical bases governing the assembly of molecular and colloidal species in fluids, and to explore the properties of the obtained layers and the potential of the obtained quasi-2D systems for the fabrication of innovative functional materials [1,2]. Examples of the importance of the fluid interfaces appear in different products of interest for food science, e.g., oil–aqueous solution interfaces stabilizing the adsorption of different proteins are found in dietary emulsions such as mayonnaise or milk, and foams stabilized by the adsorption of different types of molecules with surface activity appear in beverages such as beer. Furthermore, interfacial phenomena play a fundamental role in the development of cosmetic formulations, with foams appearing in shampoos and bath gels, affecting consumer sensorial perception of the products (softness, creaminess, etc.) and even cleanliness feeling [3–5]. Interfacial phenomena also play a very important role in many processes of industrial interest, e.g., metal recovery by flotation, the tertiary recovery of oils, interfacial catalysis, gas storage, and biomass conversion [6–8]. In addition, there are many processes of biophysical and biochemical interest, such as endocytosis or the inhalation and transport of colloidal particles through the respiratory tract, in which the dynamic aspects of the behavior of fluid interfaces are involved (see work by Guzmán et al. [9] and Carrascosa-Tejedor et al. [10] in this Special Issue) [11,12]. Therefore, the understanding of the phenomena and applications involving fluid interfaces requires the combination of theoretical and experimental efforts from researchers belonging to a broad range of scientific areas, including chemistry, physics, biophysics, engineering, pharmacy, and cosmetic or materials science. Therefore, the study of fluid interfaces has become a multidisciplinary challenge, with its implications going beyond the understanding of the most fundamental bases governing the behavior of this type of system. This importance is clear from the growing number of publications devoted to the study of fluid interfaces published within the last 20 years (see Figure 1).

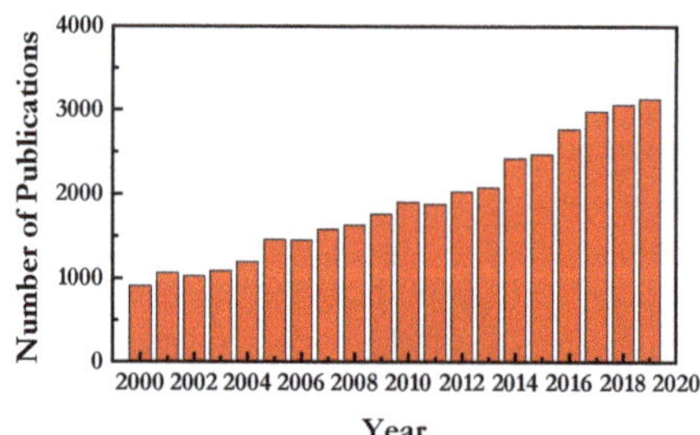

Figure 1. Number of publications per year devoted to the study of fluid interfaces (period 2000–2019) (source: Web of Science, Clarivate Analytics).

This Special Issue is devoted to the fundamental and applied aspects involved in the study of fluid interfaces, with the aim of providing a comprehensive perspective on the current status of the research field. It is expected that the work contained within this Special Issue can help to provide a bridge between the most fundamental knowledge on fluid interfaces and the development of new applications based on it, closing the gap between different approaches.

Conflicts of Interest: The author declares no conflict of interest.

References

1. Shi, S.; Russel, T.P. Nanoparticle Assembly at Liquid–Liquid Interfaces: From the Nanoscale to Mesoscale. *Adv. Mater.* **2018**, *30*, 1800714. [CrossRef] [PubMed]
2. Forth, J.; Kim, P.Y.; Xie, G.; Liu, X.; Helms, B.A.; Russell, T.P. Building Reconfigurable Devices Using Complex Liquid-Fluid Interfaces. *Adv. Mat.* **2019**, *31*, 1806370. [CrossRef] [PubMed]
3. Durian, D.; Raghavan, S. Making a frothy shampoo or beer. *Phys. Today* **2010**, *63*, 62. [CrossRef]
4. Morrison, I.D.; Ross, S. *Colloidal Dispersions: Suspensions, Emulsions, and Foams*; Wiley-Interscience: Hoboken, NJ, USA, 2002.
5. Llamas, S.; Guzmán, E.; Ortega, F.; Badghadli, N.; Cazeneuve, C.; Rubio, R.G.; Luengo, G.S. Adsorption of polyelectrolytes and polyelectrolytes-surfactant mixtures at surfaces: A physico-chemical approach to a cosmetic challenge. *Adv. Colloid Interface Sci.* **2015**, *222*, 461–487. [CrossRef] [PubMed]
6. Nguyen, A.; Schulze, H.J. *Colloidal Science of Flotation*; CRC Press: Boca Raton, FL, USA, 2003.
7. Huang, J.S.; Varadaraj, R. Colloid and interface science in the oil industry. *Curr. Opin. Colloid Interface Sci.* **1999**, *1*, 535–539. [CrossRef]
8. Asuri, P.; Karajanagi, S.S.; Dordick, J.S.; Kane, R.S. Directed Assembly of Carbon Nanotubes at Liquid–Liquid Interfaces: Nanoscale Conveyors for Interfacial Biocatalysis. *J. Am. Chem. Soc.* **2006**, *128*, 1046–1047. [CrossRef] [PubMed]
9. Guzmán, E.; Santini, E.; Ferrari, M.; Liggieri, L.; Ravera, F. Interaction of Particles with Langmuir Monolayers of 1,2-Dipalmitoyl-Sn-Glycero-3-Phosphocholine: A Matter of Chemistry? *Coatings* **2020**, *10*, 469. [CrossRef]
10. Carrascosa-Tejedor, J.; Santamaria, A.; Pereira, D.; Maestro, A. Structure of DPPC Monolayers at the Air/Buffer Interface: A Neutron Reflectometry and Ellipsometry Study. *Coatings* **2020**, *10*, 507. [CrossRef]

11. Guzmán, E.; Orsi, D.; Cristofolini, L.; Liggieri, L.; Ravera, F. Two-Dimensional DPPC Based Emulsion-like Structures Stabilized by Silica Nanoparticles. *Langmuir* **2014**, *30*, 11504–11512. [CrossRef]
12. Guzmán, E.; Santini, E. Lung surfactant-particles at fluid interfaces for toxicity assessments. *Curr. Opin. Colloid Interface Sci.* **2019**, *39*, 24–39. [CrossRef]

Publisher's Note: MDPI stays neutral with regard to jurisdictional claims in published maps and institutional affiliations.

 *coatings* 

Article

Two Different Scenarios for the Equilibration of Polycation—Anionic Solutions at Water–Vapor Interfaces

Eduardo Guzmán [1,2,*], Laura Fernández-Peña [1], Andrew Akanno [1,2], Sara Llamas [1], Francisco Ortega [1,2] and Ramón G. Rubio [1,2]

[1] Departamento de Química Física, Facultad de Ciencias, Universidad Complutense de Madrid, Ciudad Universitaria s/n, 28040 Madrid, Spain
[2] Instituto Pluridisciplinar, Universidad Complutense de Madrid, Paseo Juan XXIII, 1, 28040 Madrid, Spain
* Correspondence: eduardogs@quim.ucm.es; Tel.: +34-91-394-4107

Received: 24 June 2019; Accepted: 11 July 2019; Published: 13 July 2019

Abstract: The assembly in solution of the cationic polymer poly(diallyldimethylammonium chloride) (PDADMAC) and two different anionic surfactants, sodium lauryl ether sulfate (SLES) and sodium N-lauroyl-N-methyltaurate (SLMT), has been studied. Additionally, the adsorption of the formed complexes at the water–vapor interface have been measured to try to shed light on the complex physico-chemical behavior of these systems under conditions close to that used in commercial products. The results show that, independently of the type of surfactant, polyelectrolyte-surfactant interactions lead to the formation of kinetically trapped aggregates in solution. Such aggregates drive the solution to phase separation, even though the complexes should remain undercharged along the whole range of explored compositions. Despite the similarities in the bulk behavior, the equilibration of the interfacial layers formed upon adsorption of kinetically trapped aggregates at the water–vapor interface follows different mechanisms. This was pointed out by surface tension and interfacial dilational rheology measurements, which showed different equilibration mechanisms of the interfacial layer depending on the nature of the surfactant: (i) formation layers with intact aggregates in the PDADMAC-SLMT system, and (ii) dissociation and spreading of kinetically trapped aggregates after their incorporation at the fluid interface for the PDADMAC-SLES one. This evidences the critical impact of the chemical nature of the surfactant in the interfacial properties of these systems. It is expected that this work may contribute to the understanding of the complex interactions involved in this type of system to exploit its behavior for technological purposes.

Keywords: polyelectrolyte; surfactants; kinetically trapped aggregates; interfaces; surface tension; interfacial dilational rheology; adsorption

---

1. Introduction

The study of polyelectrolyte oppositely charged surfactant solutions, either in bulk or close to interfaces (fluid and solid ones), has grown very fast in the last two decades [1], mainly as result of its interest for a broad range of technological and industrial fields, e.g., drug delivery systems, food science, tertiary oil recovery, or cosmetic formulations [1–9]. Most of such applications take advantage of the chemical nature of the compounds involved, structural features of the formed complexes, and the rich phase diagrams appearing in this type of system [10–12].

Despite the extensive research, the description of the physico-chemical behavior of these colloidal systems remains controversial, in part because the self-assembly processes of polyelectrolytes and surfactants bearing opposite charges leads to the formation of non-equilibrium complexes [10,13–16]. They are expected to impact significantly on the properties of the solutions and in their adsorption at the interfaces [7]. This makes it necessary to pay attention to aspects such as the polymer-surfactant mixing protocol, the elapsed time from the preparation of solutions until their study, or the addition of inert electrolytes when comparisons between different studies are performed [17–19]. The role of the above-mentioned aspects in the physico-chemical properties and the phase diagrams of polyelectrolyte-surfactant solutions have been the focus of many studies, which have evidenced the complex behavior of polyelectrolyte-surfactant solutions [6,8,17–19]. It is worth mentioning that the non-equilibrium nature of the complexation process of polymer-surfactant solutions has an extraordinary impact on the interfacial properties of such solutions, as was recently stated by Campbell and Varga [20].

The role of the presence of non-equilibrium aggregates on the adsorption of polymer-surfactant solutions at fluid interfaces was already evidenced by the seminal works of the groups of Campbell and of Meszaros, focused on the analysis of the surface tension of polyelectrolyte-surfactant solutions [7,21–23]. However, it was necessary to use neutron reflectometry, which provides information on the composition and structure of the interfaces to deepen the most fundamental aspects of the physico-chemical behavior of these systems [24–26]. The studies of Penfold's group were a preliminary step toward the understanding of the correlations existing between the aggregation occurring in polyelectrolyte-surfactant solutions and the behavior of these complexes' fluid interfaces [27–31]. However, such works used an extended Gibbs formalism to describe the adsorption at fluid interfaces, i.e., provide a thermodynamic description. This approach was able to account for the non-regular dependences of the surface tension on the bulk concentration (surface tension peaks), even though it neglects the impact of non-equilibrium aspects [32,33]. More recently, Campbell et al. [17,18,34–39], using surface tension measurements and neutron reflectometry combined with ellipsometry, Brewster angle microscopy, and different bulk characterization techniques, tried to link the interfacial properties of the solutions to the bulk phase behavior, paying special attention to the role of the non-equilibrium effects. Their physical picture takes into account the role of the depletion of the interface as a result of the aggregation in the bulk [40], and the enrichment of the interface in virtue of direct interactions of the formed aggregates [19].

Most studies that analyze the behavior of the adsorption of polyelectrolyte-surfactant solutions at fluid interfaces only consider the interfaces as static systems. However, a comprehensive description of their behavior requires taking into consideration the response of such systems against mechanical deformations, i.e., the rheological response of the interfaces [7,41–45]. The understanding of such aspects is essential because most technological applications of interfacial systems, e.g., foam stabilization [42], rely on the response of the interfaces against mechanical perturbations [43]. The seminal studies on the rheological characterization of polyelectrolyte-surfactant layers at the water–vapor interface done by Regismond et al. [26,46] pointed out the strong synergetic effect on the interfacial properties as result of the influence of the bulk complexation process in the interfacial properties. More recent studies by Bhattacharyya et al. [47] and Monteux et al. [48] correlated the interfacial rheological response of polyelectrolyte-surfactant solutions with their ability to stabilize foams. They found that the formation of gel-like layers at the interface hindered destabilization processes such as bubble coalescence and foam drainage. Deepening the understanding of the rheological response of polyelectrolyte-surfactant solutions, Noskov et al. [26,42,43,45,49] showed that the mechanical behavior of the interface is controlled by the heterogeneity of layers, which is reminiscent of the structure of the complexes formed in solution.

It is worth mentioning that most studies in the recent literature deal with solutions containing relative low polymer concentrations, which hold limited interest from an industrial point of view. It is expected that polymer concentration can present an important contribution in both the complexation process and the interfacial properties of polyelectrolyte-surfactant solutions [19,41,43]. Previous studies have shown that, whereas in diluted polyelectrolyte-surfactant solutions, equilibrium between free surfactant molecules and complexes is always present in solution, the role of the free surfactant is rather limited when polymer concentration is increased. For the latter, the binding degree of surfactant molecules to the polymer chain reach values above 90%, which makes it possible to assume that they are mostly complexes that are presented in solution, even for compositions in the vicinity of the onset of the phase separation region [50]. The differences in the complexation phenomena occurring in concentrated and diluted mixtures may significantly affect the interfacial assembly of polymer-surfactant solutions, with concentrated mixtures leading to the formation of interfacial layers, with composition mirroring the composition of the bulk solutions. The latter is far from the scenario found for diluted solutions [50,51].

This work presents a comparative study of the equilibrium and dynamic properties of interfacial layers formed upon adsorption at the water–vapor interface of solution formed by poly(diallyldimethylammonium chloride) (PDADMAC) and two different anionic surfactants: sodium lauryl-ether sulfate (SLES) and sodium N-lauroyl-N-methyltaurate (SLMT). PDADMAC was chosen as the polymer because of its common utilization as a conditioner in cosmetic formulations for hair care and cleansing. Furthermore, SLES and SLMT have been recently included in formulations of shampoos to replace sodium dodecylsulfate (SDS) due to their softness and mildness, which limits skin and mucosa irritation [1].

The main aim of this work is to unravel the different interfacial behavior appearing in polycation-oppositely charged surfactant mixtures. The adsorption at the water–vapor interface is studied by surface tension measurements obtained with different tensiometers. It is worth mentioning that although polyelectrolyte-surfactant may be out of equilibrium, for simplicity we will refer to the effective property measured in this work as surface tension. In addition to the steady state measurements of the surface tension, we will follow the adsorption kinetics of the complexes at the water–vapor interface by the time evolution of the surface tension (dynamic surface tension) and the mechanical performance of the interfaces against dilation using oscillatory barrier experiments in a Langmuir trough [52]. The obtained results will be combined with the information obtained from the study of the self-assembly phenomena taking place in solution. This will provide a comprehensive description of the equilibration processes occurring during the formation of interfacial layers in this type of system. It is expected that the results contained here may help to shed light on the complex physico-chemical behavior of these systems.

2. Materials and Methods

2.1. Chemicals

PDADMAC, with an average molecular weight in the 100–200 kDa range, was purchased as a 20 wt.% aqueous solution from Sigma-Aldrich (Saint Louis, MO, USA), and was used without further purification. SLES was supplied by Kao Chemical Europe S.L. (Barcelona, Spain) as an aqueous solution of surfactant concentration 70 wt.% and was purified by lyophilization followed by recrystallization of the obtained powder using acetone for HPLC (Acros Organics, Hampton, NH, USA) [50]. SLMT was synthetized and purified following the procedures described in a previous study [50]. Scheme 1 shows the molecular formula for PDADMAC and the two surfactants used in this work.

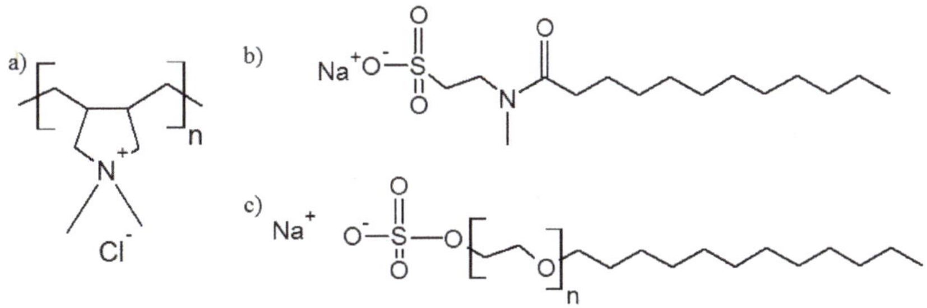

Scheme 1. Molecular formula of the three surfactants used in this work: PDADMAC (**a**), SLMT (**b**) and SLES (**c**).

Ultrapure deionized water used for cleaning and solution preparation was obtained using a multicartridge purification system AquaMAX^{TM}-Ultra 370 Series. (Young Lin, Anyang, Korea). This water presents a resistivity higher than 18 MΩ·cm, and a total organic content lower than 6 ppm. Glacial acetic acid and KCl (purity > 99.9%) purchased from Sigma-Aldrich were used to fix the pH and the ionic strength of solutions, respectively.

2.2. Preparation of Polyelectrolyte-Surfactant Solutions

The preparation of polyelectrolyte-surfactant solutions was performed following a procedure adapted from that proposed by Llamas et al. [53]. Firstly, the required amount of PDADMAC aqueous stock solution (concentration 20 wt.%) for obtaining a solution with polyelectrolyte concentration of 0.5 wt.% was weighted and poured into a flask. Then, KCl up to a final concentration of 40 mM was added into the flask. The last step involved the addition of the surfactant and the final dilution with an acetic acid solution of pH~5.6 to reach the final composition. The addition of surfactant was performed from stock aqueous solutions (pH~5.6) with a concentration one order of magnitude higher than that in the final solution. In this work, polyelectrolyte-surfactant solutions with surfactant concentration, c_s, in the range 10^{-6}–10 mM were studied. Once the solutions were prepared, these were mildly stirred (1000 rpm) for one hour using a magnetic stirrer to ensure the compositional homogenization of the solutions. Samples were left to age for 1 week prior to their use to ensure that no phase separation appeared in samples within the aging period [52].

2.3. Techniques

2.3.1. Turbidity Measurements

The turbidity of the solutions was evaluated from their transmittance at 400 nm, obtained using a UV-Visible spectrophotometer (HP-UV 8452, Hewlett Packard, Palo Alto, CA, USA). The turbidity of the samples was determined by the optical density at 400 nm (OD_{400} = [100 − T(%)]/100, where T is the transmittance). It is worth mentioning that neither the polyelectrolyte nor the surfactant present any absorption band above 350 nm.

2.3.2. Binding Isotherm

The binding isotherm of the anionic surfactant to the polycation PDADMAC was determined by potentiometric titration using a surfactant selective electrode model 6.0507.120 from Metrohm (Herisau, Switzerland). The binding degree of surfactant β was estimated from the potentiometric measurements, as was proposed by Mezei and Meszaros [22]

$$\beta = \frac{c_s^{free}}{c_{monomer}} \tag{1}$$

where c_s^{free} and $c_{monomer}$ are the concentrations of free surfactant in solution and charged monomers of the polyelectrolyte chains, respectively. This method of determining the binding isotherm provides information about the amount of free surfactant remaining in the solution.

2.3.3. Surface Tension Measurements

Surface tension measurements as functions of the surfactant concentration (SLMT or SLES) for pure surfactant and polyelectrolyte-surfactant solutions were performed using different tensiometers. In all the cases, the adsorption was measured until the steady state conditions were reached. Special care was taken to limit the evaporation effects. Each value was obtained as an average of three independent measurements. All experiments were performed at 25.0 ± 0.1 °C. From the results of the experiments, it is possible to define the surface pressure as $\Pi(c_s)= \gamma_0 − \gamma(c_s)$, where γ_0 is the surface tension of the bare water–vapor interface and $\gamma(c_s)$ is the surface tension of the solution–vapor interface. Further details on surface tension experiments can be obtained from a previous study [23].

- Surface force tensiometers. Two different surface force tensiometers were used to measure the equilibrium surface tension: a surface force balance from Nima Technology (Coventry, UK), fitted with a disposable paper plate (Whatman CHR1 chromatography paper) as a contact probe; and a surface force tensiometer Krüss K10 (Hamburg, Germany), using a Pt Wilhelmy plate as a probe.
- Drop profile analysis tensiometer. A home-built drop profile analysis tensiometer in pendant drop configuration allowed determination of the surface tension of the water–vapor interface. This tensiometer enabled evaluation of the time dependence of the surface tension during the adsorption process, thus providing information related to the adsorption kinetics.

2.3.4. Dilational Rheology

A Nima 702 Langmuir balance from Nima Technology equipped with a surface force tensiometer was used to measure the response of the surface tension against sinusoidal changes in the surface area. Thus, it is possible to obtain information about the dilational viscoelatic moduli of the water–vapor interface $\varepsilon^* = \varepsilon' + i\varepsilon''$, with ε' and ε'' being the dilational elastic and viscous moduli, respectively, in the frequency range of 10^{-1}–10^{-2} Hz and at an area deformation amplitude $\Delta u = 0.1$, which was verified to be an appropriate value to ensure results within the linear regime of the layer response [52].

3. Results and Discussion

3.1. PDADMAC-Surfactants Assembly in Solution

The equilibrium condition implies that the chemical potential of all the species in both the bulk and at the interfaces are the same. Therefore, any physical understanding of the latter implies knowledge of the behavior of the different species in the bulk. Figure 1a shows the surfactant-binding isotherms deduced from electromotive force (EMF) measurements. Comparing the curves of EMF obtained for surfactants and PDADMAC-surfactant solutions, it is possible to obtain the binding isotherms for the corresponding surfactant to PDADMAC chains following the approach described by Mezei and Meszaros [50]. The results point out a high degree of binding over the whole range of studied compositions, providing an additional confirmation of the high efficiency of PDADMAC in binding anionic surfactants. Campbell et al. [38] found for PDADMAC-SDS solutions binding degrees of surfactant to PDADMAC close to 0.3 in the vicinity of the isoelectric point (surfactant concentration around 0.2 mM). The extrapolation of such results in similar conditions to those considered in this work, i.e., polymer concentration 50-fold the one used by Campbell et al. [38,52], and assuming that the binding is not significantly modified either for the surfactant structure or for the differences in the ionic strength, takes the binding degree at charge neutralization to a value <1%. This is just the situation found here, where binding isotherms evidence that the amount of free surfactant in solution remains below 10%, even for the highest surfactant concentrations. The low concentration of free surfactant in solution allows us to assume hereinafter that the bulk has a negligible free-surfactant concentration.

Figure 1b shows the dependence of the optical density of the samples on the surfactant concentration for the solutions of PDADMAC and the two surfactants. Similar qualitative concentration dependences of the optical density were found for both polyelectrolyte-surfactant systems. It may safely be expected that all of the studied compositions for PDADMAC-surfactant solutions fall in an equilibrium one-phase region, showing optically transparent solutions. This comes from the fact that the number of surfactant molecules available in solution is not high enough to neutralize the charge of all the monomers in the polyelectrolyte chains, thus leading to the formation of undercompensated cationic complexes in solution. Indeed, considering the high polymer concentration, simple calculations suggest the existence of around 36 monomers for each surfactant molecule for a surfactant concentration of approximately 1 mM. Therefore, assuming the complete binding of surfactant molecules to the polymer chains, around 35 monomers remain positively charged in the complexes, supporting the formation of transparent samples within the entire concentration range. However, contrary to what was expected for solutions with compositions far from the neutralization, the solutions formed by undercompensated complexes show an increase of the turbidity for the highest surfactant concentration. Therefore, for such concentrated solutions, the system should get close to the onset of where the two phase region occurs, even though no signature of charge neutralization was found from electrophoretic mobility measurements. This results from the mixing protocol used for solution preparation, which proceeds during the initial step by mixing a concentrated polymer solution with a concentrated surfactant solution. This precursor solution is them diluted up to the stated bulk composition. It may be expected that this methodology leads, due to the Marangoni stress created, to the formation of persistent kinetically-trapped aggregates that persist even upon dilution, leading to the appearance of a two-phase system far off the real neutralization point of the system [7,54]. These results contrast with those reported in other mixtures studied in the literature. In such systems, the increase on the optical density of the samples results from the formation of charge compensated complexes. The last is associated with the transition from a composition region, in which the charge of the complexes is governed by the excess of charged monomers to another region, in which the excess of bound surfactant to the polymer chain controls the charge of the formed complexes, i.e., a charge inversion transition [20,38,55]. The above results show that the production of kinetically-trapped aggregates during mixing can lead to turbid mixtures far from the real equilibrium phase separation [3]. Preliminary results have shown that the above discussed scenario changes significantly when the interaction of PDADMAC with betaine derived surfactants is considered. In such systems, even the polyelectrolyte-surfactant interactions occur through the negatively charged group in the terminal region of the polar head, the formation of kinetically-trapped aggregates is hindered, probably as a result of the electrostatic repulsion associated with the positively charged groups in the zwitterionic surfactant [56].

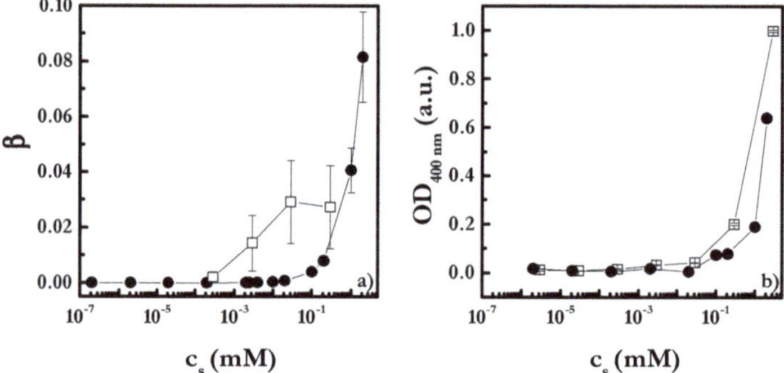

Figure 1. (**a**) Binding isotherms for surfactants on PDADMAC as a function of the initial concentration of surfactant in bulk. (**b**) Surfactant concentration dependences of the optical density of the solution, measured at 400 nm. Note: (□) = PDADMAC-SLMT; (●) = PDADMAC-SLES solutions. Lines are guides for the eyes. The results correspond to PDADMAC-surfactant mixtures containing a fixed PDADMAC concentration of 0.5 wt.%, and left to age for one week prior to measurement.

3.2. Equilibrium Adsorption at the Water–Vapor Interface

The evaluation of the surface pressure of solutions containing surface active compounds helps to understand the mechanisms involved in the equilibration of the water–vapor interface. Figure 2a shows the surface pressure dependences on the surfactant concentrations and on the PDADMAC concentration for the adsorption of the two surfactants and the polymer at the water–vapor interface (note that all solutions were prepared with the same pH and inert salt concentration as the polyelectrolyte-surfactant solutions). The results show that the surface activity of PDADMAC is negligible, at least up to concentrations that are 20-fold the one used in our work. This is in good agreement with the previous study by Noskov et al. [57] and with the negligible surface excess found for PDADMAC using neutron reflectrometry [38].

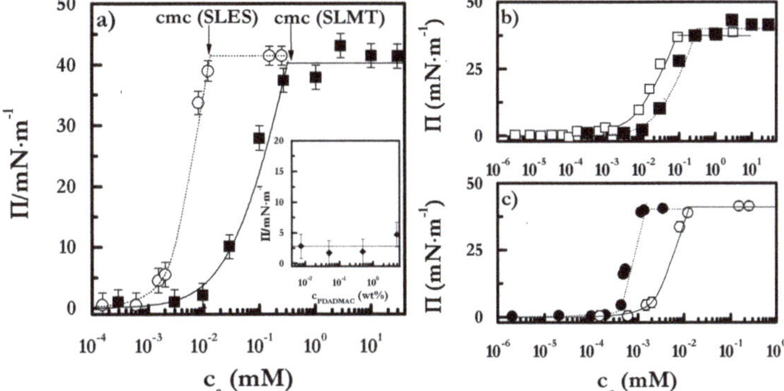

Figure 2. Results obtained using a drop profile analysis tensiometer: (**a**) Surface pressure dependence on surfactant concentration for the adsorption of pure SLES (○) and SLMT (■) at the water–vapor interface; cmc for both surfactants is marked. The inserted panel represents the surface pressure dependence on PDADMAC concentration for the adsorption of pure PDADMAC at the water–vapor interface. (**b**) Surface pressure dependence of SLMT concentration for pure SLMT (■) and PDADMAC–SLMT (□) solutions. (**c**) Surface pressure dependence of SLES concentration for pure SLES (○) and PDADMAC–SLES (●) solutions. The lines are guides for the eyes. The results for PDADMAC-surfactant mixtures correspond to mixtures containing a fixed PDADMAC concentration of 0.5 wt.%, and left to age for one week prior to measurement.

The adsorption behavior of SLMT and SLES is the expected for typical ionic surfactants. The Π increases with the bulk concentration up to the point that the surfactant concentration overcomes the threshold defined by the critical micellar concentration (cmc). Afterwards, Π remains constant with further increases of surfactant concentration. It is worth mentioning that the results obtained using different tensiometers (surface force tensiometer with Pt Wilhelmy as a probe plate and drop profile analysis tensiometer) agree within the combined error bars for the adsorption of both surfactants at the water–vapor interface. The surface pressure isotherms allow one to estimate the cmc of the pure surfactants, which showed values of around 10^{-2} and 10^{-1} mM for SLES and SLMT, respectively.

The comparison of the results obtained for the adsorption of pure surfactants at the water–vapor interface with those obtained for the adsorption of PDADMAC-surfactant solutions shows that for the lowest surfactant concentrations the surface pressure values are similar for pure surfactant and polyelectrolyte-surfactant solutions. This is the result of the low coverage of the interface (see Figure 2b,c). In such conditions, the surface excess is not high enough to produce any significant change in the surface free energy, and hence the Π values remain close to those of the bare water–vapor interface. The increase of the surfactant concentration leads to the increase of Π for both surfactant and polyelectrolyte-surfactant solutions. This increase starts for surfactant concentrations around one order of magnitude lower when polyelectrolyte-surfactant solutions are considered, which is a signature of the existence of a synergetic effect for the increase of the surface pressure as a result of the interaction in the solution of the polyelectrolyte and the surfactant. This is in agreement with previous results reported in the literature for several polyelectrolyte-surfactant systems [3,31,50,58]. The above-mentioned synergetic effects do not influence the adsorption behavior of solutions formed by PDADMAC and zwitterionic surfactants derived from the betaines, as was shown in preliminary results. This could be ascribed to the aforementioned differences in the aggregation process occurring in the bulk [56].

The study of the surface tension isotherms obtained for polymer-surfactant mixtures using different tensiometric techniques can help to understand the complexity of the interfacial behavior appearing when faced with these systems. Figure 3a,b shows that the surface tension isotherms obtained using different tensiometers reveal different features for PDADMAC-SLMT and PDADMAC-SLES solutions. PDADMAC-SLES solutions show similar surface pressure isotherms within the combined error bars, independent of the tensiometer used, and no evidences of the appearance of non-regular trends, either as surface tension peaks [38] or surface tension fluctuations [7], on the dependence of the surface pressure with the surfactant concentration were found. This contrasts with the results obtained for PDADMAC-SLMT solutions, in which the use of a surface force tensiometer with a Pt Wilhelmy plate as probe led to results that were significantly different to those obtained using the other tensiometers. The existence of such differences was previously reported in a study by Noskov et al. [31].

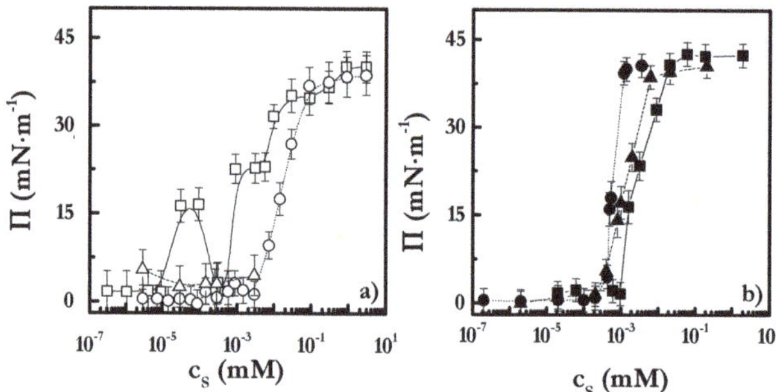

Figure 3. Surface pressure isotherms for solutions of PDADMAC with the two surfactants, obtained using different tensiometers. **(a)** Isotherms for PDADMAC-SLMT solutions. **(b)** Isotherms for PDADMAC-SLES solutions. Note: (□ and ■) Surface force tensiometer with Pt Wilhelmy plate as contact probe; (○ and ●) surface force tensiometer with paper Wilhelmy plate as contact probe; (Δ and ▲) drop profile analysis tensiometer. The lines are guides for the eyes. The results correspond to PDADMAC-surfactant mixtures containing a fixed PDADMAC concentration of 0.5 wt.% left to age for one week prior to measurement.

The differences found in the tensiometric behavior of PDADMAC-SLES and PDADMAC-SLMT solutions are correlated to differences in the equilibration mechanism of the interface. Assuming that the assembly of the polyelectrolyte-surfactant in solutions leads to the formation of kinetically trapped aggregates in both cases, this can evolve following different mechanisms upon adsorption at fluid interfaces. For PDADMAC-SLMT solutions, the appearance of surface tension fluctuations far from the phase separation region may be associated with the fact that upon adsorption at the water–vapor interface of the kinetically trapped aggregates can remain as isolated aggregates embedded at the interface. These do not dissociate spontaneously to form a kinetically trapped film at the interface. As a consequence, the trapped aggregates may adsorb onto the rough surface of the Pt Wilhelmy plate, changing its contact angle, which results in non-reliable surface tension values for the considered aggregates. This scenario is in agreement with the neutron reflectometry results obtained by Llamas et al. [50]. Their results showed a monotonic increase of the surface excess at the interface with the surfactant concentration, confirming that the surface tension fluctuations do not result from fluctuations of the interface composition. The behavior changes significantly when the adsorption of PDADMAC-SLES solutions is considered. In this case, the absence of surface tension fluctuation or significant differences in the results obtained using different tensiometers suggests the existence of dissociation and spreading of the kinetically trapped aggregates upon adsorption at the interface. Thus, the equilibration of the interface after the adsorption of the kinetically trapped aggregates occurs because of its dissociation, which is followed by the spreading of the complexes across the interface as a result of Marangoni flow associated with the lateral heterogeneity of the interface [38,42,50,59].

The differences in the adsorption mechanisms of PDADMAC-SLES and PDADMAC-SLMT complexes at the water–vapor interface may be explained on the bases of the molecular structures of the surfactant and the possibility to establish a cohesion interaction with the surrounding media. SLMT presents a hydrophobic tail formed by an alkyl chain, which tends to minimize the number of contact points with water, which favors the formed aggregates remaining as compact aggregates at the water–vapor interface upon adsorption. On the contrary, the presence of oxyethylene groups in SLES makes the dissociation and spreading of the complexes easier as a result of the possible formation of hydrogen bonds of the surfactant molecules with water. Surprisingly, studies on the adsorption of PDADMAC-SLES and PDADMAC-SLMT mixtures onto solid surfaces have evidenced a scenario compatible with that described for the adsorption at the fluid interfaces, where PDADMAC-SLES films present a topography reminiscent of the formation of extended complexes attached to the interface, whereas PDADMAC-SLMT films present a higher lateral heterogeneity [51,60]. Further confirmation of the discussed mechanisms may be obtained from the analysis of the adsorption kinetics at the water–vapor interface of the polyelectrolyte-surfactant solutions.

3.3. Adsorption Kinetics at the Water–Vapor Interface

The analysis of the adsorption kinetics of polymer-surfactants at the water–vapor interface is a powerful tool for deepening the understanding of the mechanistic aspects of the adsorption of complexes. This is done by studying the time evolution of the surface pressure (dynamic surface pressure) during the adsorption process. The adsorption kinetics have been measured using a drop shape analysis tensiometer. As expected, the adsorption of polymer-surfactant solutions at fluid interfaces is slower than that corresponding to pure surfactant [16,50]. Figure 4 shows the dynamics surface pressure obtained for the adsorption of PDADMAC-SLMT and PDADMAC-SLES solutions at the water–vapor interface.

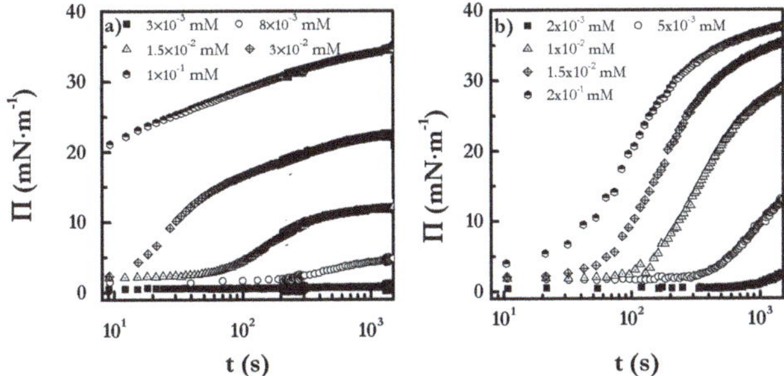

Figure 4. (a) Dynamic surface pressure for PDADMAC-SLMT solutions with different surfactant concentrations. (b) Dynamic surface pressure for PDADMAC-SLES solutions with different surfactant concentrations. The results correspond to PDADMAC-surfactant mixtures containing a fixed PDADMAC concentration of 0.5 wt.%, and left to age for one week prior to measurement.

The analysis of the adsorption kinetics show clearly that the increase of the surfactant concentration leads to the faster increase of the surface pressure, due to the higher hydrophobicity of the formed complexes. A more detailed analysis points out that whereas the adsorption of PDADMAC-SLMT is characterized by the monotonous increase of the surface pressure with time over the whole concentration range, the adsorption of PDADMAC-SLES presents an induction time that is reduced as the SLES concentration increases. Such differences are due to the differences in the processes involved in the equilibration of the interface.

The induction time in the adsorption of PDADMAC-SLES is explained considering that the equilibration of the interface proceeds following a two-step mechanism, as occurs for protein adsorption at fluid interfaces [61]. Firstly, polymer-surfactant complexes attach to the water–vapor interface as kinetically trapped aggregates until the surface excess overcomes a threshold value, after which point the adsorbed complexes undergo a dissociation and spreading process, which is responsible for the surface pressure increase [41,59]. It is worth mentioning that the decrease of the induction time with the increase of surfactant concentration results from the faster saturation of the interface, i.e., the shortening of the time needed to overcome the surface excess threshold, which leads to a prior surface pressure rise. The scenario found for PDADMAC-SLMT solutions is different to that described for PDADMAC-SLES, and the absence of the induction time is a signature of a difference in the equilibration mechanism of the interfacial layer. For PDADMAC-SLMT, the increase of the surface pressure is associated with the adsorption of isolated kinetically trapped aggregates that coalesce as the surfactant concentration increases. In this case, the adsorbed complexes remain compact without any significant dissociation. The above discussed results point out the existence of differences in the mechanisms for the equilibration of the interface of the polycation-anionic surfactant solution as result of the differences in the type of surfactant. The first one involves the dissociation and spreading of the pre-adsorbed kinetically trapped aggregates (PDADMAC-SLES), whereas the second one relies directly on the saturation of the interface with kinetically trapped aggregates. This proves that the adsorption of PDADMAC-SLMT leads to appreciable modifications of the surface pressure for surfactant concentrations one order of magnitude higher than PDADMAC-SLES as a result of the negligible effect of the isolated aggregates over the surface pressure of the bare water–vapor interface until their concentration is high enough. On the contrary, for PDADMAC-SLES, the dissociation and spreading of the aggregates enables the distribution of surface active material along the whole interface, and consequently the surface pressure starts to increase for lower surfactant concentrations as a result of the formation of interfacial layers in which complexes are extended along the interface.

3.4. Interfacial Dilational Rheology

The above discussion was devoted to the study of the adsorption at interfaces with fixed surface areas. However, from a technological point of view, the understanding of the response of the interface against external mechanical perturbations is essential because this provides important insights into the relaxation processes involved in the equilibration of interfacial layers [25,48,62,63]. The dependences of the dilational viscoelastic moduli (ε' represents the dilational elastic modulus and ε'' the viscous modulus) on the surfactant concentration and the deformation frequency provide complementary information for the better understanding of the complexity of the mechanism involved in the equilibration of the interfaces, helping to give a more detailed picture of the physical processes governing the formation of adsorption layers from polymer-surfactant solutions [64]. It must be stressed that for both PDADMAC-SLMT and PDADMAC-SLES solutions, the values of ε'' are negligible in relation to those of ε', with the ratio $\varepsilon''/\varepsilon'$ decreasing as the surfactant concentration increases. Hence, for the sake of simplicity only the behavior of ε' will be discussed. Figure 5 shows the concentration dependences of the elastic modulus for PDADMAC-SLMT and PDADMAC-SLES layers.

The results indicate that the dependence of ε' on the frequency is expected for the formation of layers at fluid interfaces, with ε' increasing with the deformation frequency. Furthermore, the concentration dependence of ε' is similar to that found for layers of surface active materials at fluid interfaces [46], with ε' increasing with the surfactant concentration from the value corresponding to the clean interface, reaching a maximum and then dropping down again to quasi-null values for the highest surfactant concentrations. A careful examination of the values obtained for the elasticity modulus for each system indicate that PDADMAC-SLES layers present values that are more than twice those obtained for PDADMAC-SLMT solutions independent of the considered frequency. This is again indicative of the different features of the interfacial layers. For PDADMAC-SLES layers, the spreading of material along the interface leads to the formation of extended complexes that can build a cross-linked network, increasing the elastic modulus of the interfacial layers. This cross-linking process is not possible when the interfacial layer is formed by compact kinetically trapped aggregates, as in PDADMAC-SLMT layers, leading to lower values of the elastic modulus of the interface.

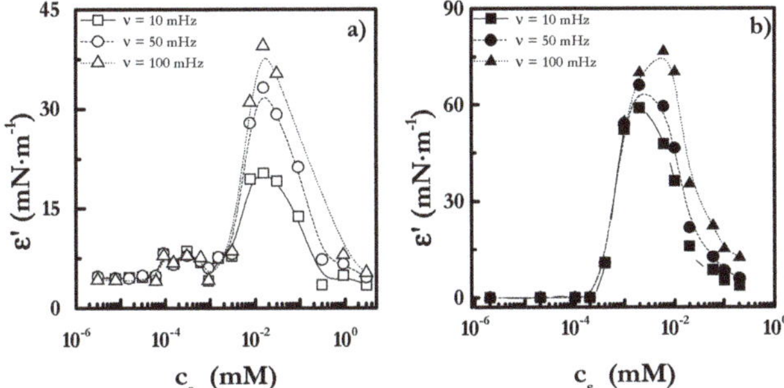

Figure 5. (a) Concentration dependences of the elastic modulus for PDADMAC-SLMT adsorption layers as were obtained from oscillatory barrier experiments performed at different frequencies. (b) Concentration dependences of the elastic modulus for PDADMAC-SLES adsorption layers, obtained from oscillatory barrier experiments performed at different frequencies. Note: ($\square$ and $\blacksquare$) ν = 0.01 Hz; ($\bigcirc$ and $\bullet$) ν = 0.05 Hz; ($\triangle$ and $\blacktriangle$) ν = 0.10 Hz. The lines are guides for the eyes. For the sake of clarity, only results corresponding to some of the explored frequencies (ν) are shown, with the other frequencies presenting similar dependences. The results correspond to PDADMAC-surfactant mixtures containing a fixed PDADMAC concentration of 0.5 wt.%, and left to age for one week prior to measurement.

The frequency dependences of the elasticity modulus can be described in terms of the rheological model proposed by Ravera et al. [64,65] (see Figure 6a for an example). According to this model the frequency dependence of the viscoelastic modulus accounts for the initial adsorption of the polymer-surfactant complexes at the water–vapor interface as a diffusion-controlled process that is coupled to a second step associated with the internal reorganization of the adsorbed layers. Thus, taking into account the aforementioned framework, it is possible to describe the complex viscoelastic modulus with the following expression:

$$\varepsilon^* = \frac{1+\xi+i\xi}{1+2\xi+2\xi^2}\left[\varepsilon_0 + (\varepsilon_1 - \varepsilon_0)\frac{1+i\lambda}{1+\lambda^2}\right] \tag{2}$$

where $\xi = \sqrt{\nu_D/\nu}$, with ν_D and ν being the characteristic frequency of the diffusion exchange and the frequency of deformation, respectively, and $\lambda = \nu_1/\nu$, with ν_1 being the characteristic frequency of the extra relaxation process. Additionally, ε_0 and ε_1 represent the Gibbs elasticity and the high frequency limit elasticity within the frequency range considered, respectively. The validity of the discussed model, beyond confirming the complexity of the mechanisms involved in the equilibration of the interfacial layers formed by polyelectrolyte-surfactant solution, provides a description of the processes involved. It is expected that the equilibration of the interfacial occurs in the first stage by the diffusion-controlled adsorption of the kinetically trapped aggregates, and then such complexes undergo different reorganization processes depending on their nature. The existence of a two-step mechanism is in agreement with the picture proposed by Noskov et al. [45] for the equilibration of adsorption layers of PDADMAC-SDS at the water–vapor interface.

Figure 6b,c show the concentration dependences for the characteristic frequencies of the two dynamic processes appearing for the interfacial layers. As may be expected considering the different nature of the dynamic processes involved in the equilibration of the interfacial layer, ν_1, which is the frequency corresponding to the interfacial relaxation process, presents higher values than those associated with the diffusional transport, ν_D, for both PDADMAC-SLMT and PDADMAC-SLES solutions. This behavior can be explained by assuming that the interfacial relaxation process, involving the reorganization of materials at the interface, occurs only when a certain degree of material is adsorbed at the interface.

The results show that both ν_D and ν_1 increase in concentration for both studied systems. This increase can be explained in the case of ν_D as a result of the enhanced surface activity of the kinetically trapped aggregates, as the surfactant concentration increases due to their higher hydrophobicity. Furthermore, the values of ν_D are in a similar range for PDADMAC-SLMT and PDADMAC-SLES, which is in agreement with the similar origin of the process in both systems and the similarities of the complexes formed according to the above discussion. The slightly smaller values of ν_D found for PDADMAC-SLMT than for PDADMAC-SLES may result from different sizes of the complexes formed in the solution. The dependence of ν_1 is assumed to be because of the increase of surfactant in solution leading to an increase of the surface excess of complexes at the interface, which facilitates their reorganization within the interface. The higher values of ν_1 for PDADMAC-SLMT solution than PDADMAC-SLES solutions, at almost one order of magnitude, are ascribable again to the differences in the structure of the interfacial layers. Thus, the diffusion of extended complexes within the interface can occur across longer distances within the interface than that of compact aggregates, and consequently this process involves longer time scales.

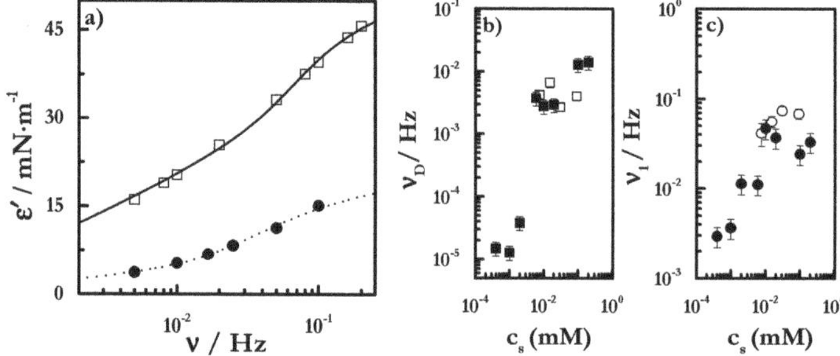

Figure 6. (a) Examples of frequency dependences of the elastic modulus for interfacial layers of PDADMAC-SLMT (□) and PDADMAC-SLES (●) and for solutions with surfactant concentration of 0.1 mM. Symbols represent the experimental data and the lines are the theoretical curves obtained from the analysis of the experimental results in term of the theoretical model described by Equation (2). (b) Concentration dependences of ν_D for PDADMAC-SLMT (□) and PDADMAC-SLES (■). (c) Concentration dependences of ν_1 for PDADMAC-SLMT (○) and PDADMAC-SLES (●). (b,c) Symbols represent the experimental data and the lines are guides for the eyes. The results correspond to PDADMAC-surfactant mixtures containing a fixed PDADMAC concentration of 0.5 wt.%, and left to age for one week prior to measurement.

4. Conclusions

The mechanisms involved in the equilibration of interfacial layers formed by adsorption of PDADMAC and two different anionic surfactants (SLMT and SLES) have been studied by surface tension (equilibrium and dynamics) and interfacial dilational rheology measurements. The combination of the interfacial characterization with studies on the association phenomena occurring in solution has evidenced that even the formation of kinetically trapped aggregates in the bulk occurs following similar patterns in both studied systems. These evolve following mechanisms depending of the specific chemical nature of the surfactant involved.

The equilibration of the interfacial layers formed by polyelectrolyte oppositely charged surfactants can be explained on the basis of a two-step mechanism. The first step is common to the different systems studied and is related with the diffusion-controlled incorporation of kinetically trapped aggregates at the water–vapor interfaces. Such aggregates can remain as compact aggregates at the interface, as in PDADMAC-SLMT solutions, or can undergo dissociation and spreading along the interface due to Marangoni flows, as in PDADMAC-SLES solutions. These different mechanisms result from differences in the hydrophobicity of the formed aggregates and the possibility to establish a cohesion interaction, such as a hydrogen bond, with the interface. On the basis of the obtained results, it can be concluded that there are no general laws governing the equilibration of the interfacial layers formed by the adsorption of polyelectrolyte-surfactant solutions at the fluid interface, with the process being primarily controlled by the specific nature of the chemical compounds involved and the interactions involved in the equilibration of the interface. This study contributes to the understanding of the fundamental basis describing the interfacial behavior of polyelectrolyte-surfactant solutions in conditions similar to that used in industrial application. Thus, the obtained result can help to exploit the interfacial behavior of these systems in technologically relevant conditions.

Author Contributions: Conceptualization, L.F.-P., A.A., and S.L.; Methodology, E.G., L.F.-P., A.A., and S.L.; Software, E.G.; Validation, E.G., F.O. and R.G.R.; Formal Analysis, E.G.; Investigation, E.G., L.F.P., A.A., S.L., F.O., and R.G.R.; Resources, R.G.R. and F.O.; Data Curation, E.G.; Writing—Original Draft Preparation, E.G.; Writing—Review and Editing, E.G., F.O. and R.G.R.; Visualization, E.G.; Supervision, E.G., F.O. and R.G.R.; Project Administration, R.G.R.; Funding Acquisition, R.G.R. and F.O.

Funding: This work was funded by MINECO under grant CTQ-2016-78895-R.

Acknowledgments: The CAI of spectroscopy from Universidad Complutense de Madrid is acknowledged for granting access to its facilities.

Conflicts of Interest: The authors declare no conflict of interest. The funders had no role in the design of the study; in the collection, analyses, or interpretation of data; in the writing of the manuscript, or in the decision to publish the results.

References

1. Llamas, S.; Guzmán, E.; Ortega, F.; Baghdadli, N.; Cazeneuve, C.; Rubio, R.G.; Luengo, G.S. Adsorption of polyelectrolytes and polyelectrolytes-surfactant mixtures at surfaces: A physico-chemical approach to a cosmetic challenge. *Adv. Colloid Interface Sci.* **2015**, *222*, 461–487. [CrossRef] [PubMed]

2. Goddard, E.D.; Ananthapadmanabhan, K.P. *Application of Polymer-Surfactant Systems*; Marcel Dekker, Inc.: New York, NY, USA, 1998.

3. Bain, C.D.; Claesson, P.M.; Langevin, D.; Meszaros, R.; Nylander, T.; Stubenrauch, C.; Titmuss, S.; von Klitzing, R. Complexes of surfactants with oppositely charged polymers at surfaces and in bulk. *Adv. Colloid Interface Sci.* **2010**, *155*, 32–49. [CrossRef] [PubMed]

4. Khan, N.; Brettmann, B. Intermolecular interactions in polyelectrolyte and surfactant complexes in solution. *Polymers* **2019**, *11*, 51. [CrossRef] [PubMed]

5. Gradzielski, M.; Hoffmann, I. Polyelectrolyte-surfactant complexes (PESCs) composed of oppositely charged components. *Curr. Opin. Colloid Interface Sci.* **2018**, *35*, 124–141. [CrossRef]

6. Guzmán, E.; Llamas, S.; Maestro, A.; Fernández-Peña, L.; Akanno, A.; Miller, R.; Ortega, F.; Rubio, R.G. Polymer-surfactant systems in bulk and at fluid interfaces. *Adv. Colloid Interface Sci.* **2016**, *233*, 38–64. [CrossRef] [PubMed]

7. Varga, I.; Campbell, R.A. General physical description of the behavior of oppositely charged polyelectrolyte/surfactant mixtures at the air/water interface. *Langmuir* **2017**, *33*, 5915–5924. [CrossRef] [PubMed]

8. Nylander, T.; Samoshina, Y.; Lindman, B. Formation of polyelectrolyte-surfactant complexes on surfaces. *Adv. Colloid Interface Sci.* **2006**, *123–126*, 105–123. [CrossRef]

9. Ferreira, G.A.; Loh, W. Liquid crystalline nanoparticles formed by oppositely charged surfactant-polyelectrolyte complexes. *Curr. Opin. Colloid Interface Sci.* **2017**, *32*, 11–22. [CrossRef]

10. Piculell, L.; Lindman, B. Association and segregation in aqueos polymer/polymer, polymer/surfactant, and surfactant/surfactant mixtures: Similarities and differences. *Adv. Colloid Interface Sci.* **1992**, *41*, 149–178. [CrossRef]

11. Liu, J.Y.; Wang, J.G.; Li, N.; Zhao, H.; Zhou, H.J.; Sun, P.C.; Chen, T.H. Polyelectrolyte-surfactant complex as a template for the synthesis of zeolites with intracrystalline mesopores. *Langmuir* **2012**, *28*, 8600–8607. [CrossRef]

12. Miyake, M. Recent progress of the characterization of oppositely charged polymer/surfactant complex in dilution deposition system. *Adv. Colloid Interface Sci.* **2017**, *239*, 146–157. [CrossRef] [PubMed]

13. Szczepanowicz, K.; Bazylińska, U.; Pietkiewicz, J.; Szyk-Warszyńska, L.; Wilk, K.A.; Warszyński, P. Biocompatible long-sustained release oil-core polyelectrolyte nanocarriers: From controlling physical state and stability to biological impact. *Adv.Colloid Interface Sci.* **2015**, *222*, 678–691. [CrossRef] [PubMed]

14. Picullel, L. Understanding and exploiting the phase behavior of oppositely charged polymer and surfactant in water. *Langmuir* **2013**, *29*, 10313–10329. [CrossRef] [PubMed]

15. Goddard, E.D. Polymer/surfactant interaction: Interfacial aspects. *J. Colloid Interface Sci.* **2002**, *256*, 228–235. [CrossRef]

16. Llamas, S.; Guzmán, E.; Akanno, A.; Fernández-Peña, L.; Ortega, F.; Campbell, R.A.; Miller, R.; Rubio, R.G. Study of the liquid/vapor interfacial properties of concentrated polyelectrolyte-surfactant mixtures using surface tensiometry and neutron reflectometry: Equilibrium, adsorption kinetics, and dilational rheology. *J. Phys. Chem. C* **2018**, *122*, 4419–4427. [CrossRef]

17. Campbell, R.A.; Arteta, M.Y.; Angus-Smyth, A.; Nylander, T.; Varga, I. Effects of bulk colloidal stability on adsorption layers of poly(diallyldimethylammonium chloride)/sodium dodecyl sulfate at the air/water interface studied by neutron reflectometry. *J. Phys. Chem. B* **2011**, *115*, 15202–15213. [CrossRef] [PubMed]

18. Campbell, R.A.; Arteta, M.Y.; Angus-Smyth, A.; Nylander, T.; Varga, I. Multilayers at interfaces of an oppositely charged polyelectrolyte/surfactant system resulting from the transport of bulk aggregates under gravity. *J. Phys. Chem B* **2012**, *116*, 7981–7990. [CrossRef] [PubMed]

19. Campbell, R.A.; Arteta, M.Y.; Angus-Smyth, A.; Nylander, T.; Noskov, B.A.; Varga, I. Direct impact of non-equilibrium aggregates on the structure and morphology of pdadmac/SDS layers at the air/water interface. *Langmuir* **2014**, *30*, 8664–8774. [CrossRef] [PubMed]

20. Mészáros, R.; Thompson, L.; Bos, M.; Varga, I.; Gilányi, T. Interaction of sodium dodecyl sulfate with polyethyleneiminie: surfactant-induced polymer solution colloid dispersion transition. *Langmuir* **2003**, *19*, 609–615. [CrossRef]

21. Mezei, A.; Pojják, K.; Mészaros, R. Nonequilibrium features of the association between poly(vinylamine) and sodium dodecyl sulfate: The validity of the colloid dispersion concept. *J. Phys. Chem B* **2008**, *112*, 9693–9699. [CrossRef] [PubMed]

22. Pojják, K.; Bertalanits, E.; Mészáros, R. Effect of salt on the equilibrium and nonequilibrium features of polyelectrolyte/surfactant association. *Langmuir* **2011**, *27*, 9139–9147. [CrossRef] [PubMed]

23. Bodnár, K.; Fegyver, E.; Nagy, M.; Mészáros, R. Impact of polyelectrolyte chemistry on the thermodynamic stability of oppositely charged macromolecules/surfactant mixtures. *Langmuir* **2016**, *32*, 1259–1268. [CrossRef] [PubMed]

24. Goddard, E.D.; Hannan, R.B. Cationic polymer/anionic surfactant interactions. *J. Colloid Interface Sci.* **1976**, *55*, 73–79. [CrossRef]

25. Bergeron, V.; Langevin, D.; Asnacios, A. Thin-film forces in foam films containing anionic polyelectrolyte and charged surfactants. *Langmuir* **1996**, *12*, 1550–1556. [CrossRef]

26. Bhattacharyya, A.; Monroy, F.; Langevin, D.; Argillier, J.-F. Surface rheology and foam stability of mixed surfactant-polyelectrolyte solutions. *Langmuir* **2000**, *16*, 8727–8732. [CrossRef]

27. Stubenrauch, C.; Albouy, P.-A.; von Klitzing, R.; Langevin, D. Polymer/surfactant complexes at the water/air interface: A surface tension and x-ray reflectivity study. *Langmuir* **2000**, *16*, 3206–3213. [CrossRef]

28. Braun, L.; Uhlig, M.; von Klitzing, R.; Campbell, R.A. Polymers and surfactants at fluid interfaces studied with specular neutron reflectometry. *Adv. Colloid Interface Sci.* **2017**, *247*, 130–148. [CrossRef]

29. Lu, J.R.; Thomas, R.K.; Penfold, J. Surfactant layers at the air/water interface: Structure and composition. *Adv. Colloid Interface Sci.* **2000**, *84*, 143–304. [CrossRef]

30. Narayanan, T.; Wacklin, H.; Konovalov, O.; Lund, R. Recent applications of synchrotron radiation and neutrons in the study of soft matter. *Crystallography Rev.* **2017**, *23*, 160–226. [CrossRef]

31. Staples, E.; Tucker, I.; Penfold, J.; Warren, N.; Thomas, R.K.; Taylor, D.J.F. Organization of polymer−surfactant mixtures at the air−water interface: sodium dodecyl sulfate and poly(dimethyldiallylammonium chloride). *Langmuir* **2002**, *18*, 5147–5153. [CrossRef]

32. Penfold, J.; Tucker, I.; Thomas, R.K.; Zhang, J. Adsorption of polyelectrolyte/surfactant mixtures at the air−solution interface:poly(ethyleneimine)/sodium dodecyl sulfate. *Langmuir* **2005**, *21*, 10061–10073. [CrossRef] [PubMed]

33. Penfold, J.; Thomas, R.K.; Taylor, D.J.F. Polyelectrolyte/surfactant mixtures at the air–solution interface. *Curr. Opin. Colloid Interface Sci.* **2006**, *11*, 337–344. [CrossRef]

34. Penfold, J.; Tucker, I.; Thomas, R.K.; Taylor, D.J.F.; Zhang, X.L.; Bell, C.; Breward, C.; Howell, P. The interaction between sodium alkyl sulfate surfactants and the oppositely charged polyelectrolyte, polyDMDAAC, at the air-water interface: The role of alkyl chain length and electrolyte and comparison with theoretical predictions. *Langmuir* **2007**, *23*, 3128–3136. [CrossRef] [PubMed]

35. Thomas, R.K.; Penfold, J. Thermodynamics of the air-water interface of mixtures of surfactants with polyelectrolytes, oligoelectrolyte, and multivalent metal electrolytes. *J. Phys. Chem B* **2018**, *122*, 12411–12427. [CrossRef] [PubMed]

36. Bell, C.G.; Breward, C.J.W.; Howell, P.D.; Penfold, J.; Thomas, R.K. A theoretical analysis of the surface tension profiles of strongly interacting polymer–surfactant systems. *J. Colloid Interface Sci.* **2010**, *350*, 486–493. [CrossRef] [PubMed]

37. Bahramian, A.; Thomas, R.K.; Penfold, J. The adsorption behavior of ionic surfactants and their mixtures with nonionic polymers and with polyelectrolytes of opposite charge at the air–water interface. *J. Phys. Chem. B* **2014**, *118*, 2769–2783. [CrossRef] [PubMed]

38. Campbell, R.A.; Angus-Smyth, A.; Yanez-Arteta, M.; Tonigold, K.; Nylander, T.; Varga, I. New perspective on the cliff edge peak in the surface tension of oppositely charged polyelectrolyte/surfactant mixtures. *J. Phys. Chem. Lett.* **2010**, *1*, 3021–3026. [CrossRef]

39. Ábraham, A.; Campbell, R.A.; Varga, I. New method to predict the surface tension of complex synthetic and biological polyelectrolyte/surfactant mixtures. *Langmuir* **2013**, *29*, 11554–11559. [CrossRef] [PubMed]

40. Angus-Smyth, A.; Bain, C.D.; Varga, I.; Campbell, R.A. Effects of bulk aggregation on PEI–SDS monolayers at the dynamic air–liquid interface: Depletion due to precipitation versus enrichment by a convection/spreading mechanism. *Soft Matter* **2013**, *9*, 6103–6117. [CrossRef]

41. Campbell, R.A.; Tummino, A.; Noskov, B.A.; Varga, I. Polyelectrolyte/surfactant films spread from neutral aggregates. *Soft Matter* **2016**, *12*, 5304–5312. [CrossRef] [PubMed]

42. Noskov, B.A.; Loglio, G.; Miller, R. Dilational surface visco-elasticity of polyelectrolyte/surfactant solutions: Formation of heterogeneous adsorption layers. *Adv. Colloid Interface Sci.* **2011**, *168*, 179–197. [CrossRef] [PubMed]

43. Lyadinskaya, V.V.; Bykov, A.G.; Campbell, R.A.; Varga, I.; Lin, S.Y.; Loglio, G.; Miller, R.; Noskov, B.A. Dynamic surface elasticity of mixed poly(diallyldimethylammoniumchloride)/sodium dodecyl sulfate/NaCl solutions. *Colloids Surf. A* **2014**, *460*, 3–10. [CrossRef]

44. Monteux, C.; Fuller, G.G.; Bergeron, V. Shear and dilational surface rheology of oppositely charged polyelectrolyte/surfactant microgels adsorbed at the air-water interface. Influence on foam stability. *J. Phys. Chem. B* **2004**, *108*, 16473–16482. [CrossRef]

45. Noskov, B.A.; Grigoriev, D.O.; Lin, S.Y.; Loglio, G.; Miller, R. Dynamic surface properties of polyelectrolyte/surfactant adsorption films at the air/water interface: Poly(diallyldimethylammonium chloride) and sodium dodecylsulfate. *Langmuir* **2007**, *23*, 9641–9651. [CrossRef] [PubMed]

46. Noskov, B.A. Dilational surface rheology of polymer and polymer/surfactant solutions. *Curr. Opin. Colloids Interface Sci.* **2010**, *15*, 229–236. [CrossRef]

47. Fauser, H.; von Klitzing, R.; Campbell, R.A. Surface adsorption of oppositely charged C14TAB-PAMPS mixtures at the air/water interface and the impact on foam film stability. *J. Phys. Chem. B* **2015**, *119*, 348–358. [CrossRef]

48. Fuller, G.G.; Vermant, J. Complex fluid-fluid interfaces: Rheology and structure. *Annu. Rev. Chem. Biomol. Eng.* **2012**, *3*, 519–543. [CrossRef]

49. Regismond, S.T.A.; Winnik, F.M.; Goddard, E.D. Surface viscoelasticity in mixed polycation anionic surfactant systems studied by a simple test. *Colloids Surf. A* **1996**, *119*, 221–228. [CrossRef]

50. Llamas, S.; Fernández-Peña, L.; Akanno, A.; Guzmán, E.; Ortega, V.; Ortega, F.; Csaky, A.G.; Campbell, R.A.; Rubio, R.G. Towards understanding the behavior of polyelectrolyte—Surfactant mixtures at the water/vapor interface closer to technologically-relevant conditions. *Phys. Chem. Chem. Phys.* **2018**, *20*, 1395–1407. [CrossRef]

51. Llamas, S.; Guzmán, E.; Baghdadli, N.; Ortega, F.; Cazeneuve, C.; Rubio, R.G.; Luengo, G.S. Adsorption of poly(diallyldimethylammonium chloride)-sodium methyl-cocoyl-taurate complexes onto solid surfaces. *Colloids Surf. A* **2016**, *505*, 150–157. [CrossRef]

52. Mendoza, A.J.; Guzmán, E.; Martínez-Pedrero, F.; Ritacco, H.; Rubio, R.G.; Ortega, F.; Starov, V.M.; Miller, R. Particle laden fluid interfaces: Dynamics and interfacial rheology. *Adv.Colloid Interface Sci.* **2014**, *206*, 303–319. [CrossRef] [PubMed]

53. Goddard, E.D.; Gruber, J.V. *Principles of Polymer Science and Technology in Cosmetics and Personal Care*; Marcel Dekker, Inc.: Basel, Switzerland, 1999.

54. Mezei, A.; Mezaros, R. Novel method for the estimation of the binding isotherms of ionic surfactants on oppositely charged polyelectrolytes. *Langmuir* **2006**, *22*, 7148–7151. [CrossRef] [PubMed]

55. Naderi, A.; Claesson, P.M.; Bergström, M.; Dedinaite, A. Trapped non-equilibrium states in aqueous solutions of oppositely charged polyelectrolytes and surfactants: Effects of mixing protocol and salt concentration. *Colloids Surf. A* **2005**, *253*, 83–93. [CrossRef]

56. Akanno, A. *Bulk and Surface Properties of Polyelectrolyte Surfactant Mixtures*; Universidad Complutense de Madrid: Madrid, Spain, 2018.

57. Mezei, A.; Mészáros, R.; Varga, I.; Gilanyi, T. Effect of mixing on the formation of complexes of hyperbranched cationic polyelectrolytes and anionic surfactants. *Langmuir* **2007**, *23*, 4237–4247. [CrossRef] [PubMed]

58. Noskov, B.A.; Bilibin, A.Y.; Lezov, A.V.; Loglio, G.; Filippov, S.K.; Zorin, I.M.; Miller, R. Dynamic surface elasticity of polyelectrolyte solutions. *Colloids Surf. A* **2007**, *298*, 115–122. [CrossRef]

59. Tummino, A.; Toscano, J.; Sebastiani, F.; Noskov, B.A.; Varga, I.; Campbell, R.A. Effects of aggregate charge and subphase ionic strength on the properties of spread polyelectrolyte/surfactant films at the air/water interface under static and dynamic conditions. *Langmuir* **2018**, *34*, 2312–2323. [CrossRef]

60. Llamas, S. *Estudio de Interfases de Interés en Cosmética*; Universidad Complutense de Madrid: Madrid, Spain, 2014.

61. Erickson, J.S.; Sundaram, S.; Stebe, K.J. Evidence that the induction time in the surface pressure evolution of lysozyme solutions is caused by a surface phase transition. *Langmuir* **2000**, *16*, 5072–5078. [CrossRef]

62. Schramm, L.L. *Emulsions, Foams, Suspensions, and Aerosols*; Wiley-VCH Verlag GmbH & Co.: Weinheim, Germany, 2014.

63. Langevin, D. Aqueous foams: A field of investigation at the frontier between chemistry and physics. *ChemPhysChem* **2008**, *9*, 510–522. [CrossRef]

64. Liggieri, L.; Santini, E.; Guzmán, E.; Maestro, A.; Ravera, F. Wide-frequency dilational rheology investigation of mixed silica nanoparticle—CTAB interfacial layers. *Soft Matter* **2011**, *7*, 6699–7709. [CrossRef]

65. Ravera, F.; Ferrari, M.; Santini, E.; Liggieri, L. Influence of surface processes on the dilational visco-elasticity of surfactant solutions. *Adv. Colloid Interface Sci.* **2005**, *117*, 75–100. [CrossRef]

Article

Adsorption Kinetics of a Cationic Surfactant Bearing a Two-Charged Head at the Air-Water Interface

Marcos Fernández Leyes, Santiago Gimenez Reyes, Ezequiel Cuenca, Jhon F. Sánchez Morales and Hernán Ritacco *

Instituto de Física del Sur (IFISUR), Departamento de Física, Universidad Nacional del Sur (UNS), CONICET, Av. L. N. Alem 1253, B8000CPB—Bahía Blanca, Argentina; mfernandezleyes@uns.edu.ar (M.F.L.); sgimenezreyes@gmail.com (S.G.R.); ezequiel.qnk@gmail.com (E.C.); jhon.sanchez@uns.edu.ar (J.F.S.M.)
* Correspondence: hernan.ritacco@uns.edu.ar

Received: 19 December 2019; Accepted: 19 January 2020; Published: 22 January 2020

Abstract: We studied the dynamics of adsorption at the air-water interface of a cationic surfactant bearing two charges, Gemini 12-2-12, at concentrations below and above the critical micelle concentration (cmc). We used maximum bubble pressure and Wilhelmy plate techniques in order to access all time scales in the adsorption process. We found that the adsorption dynamics are controlled by diffusion at the initial stage of the adsorption process (milliseconds) and it is kinetically controlled by an electrostatic barrier (minute) approaching the equilibrium surfactant surface concentration. Between these two extremes, we found several relaxation phenomena, all following exponential decays with characteristic times spanning from one to hundreds of seconds. By means of time-resolved surface potential measurements, we show that these processes involve charge redistribution within the interfacial region. The surface tension data are analyzed and interpreted in the framework of the free energy approach.

Keywords: cationic surfactants; Gemini 12-2-12 surfactant; dynamic surface tension; maximum bubble pressure; surface potential

---

1. Introduction

The dynamics of adsorption-desorption of soluble surfactants at water-air interfaces play a very important role in many phenomena and technological applications such as wetting, detergency, emulsification, and foaming [1–4]. These processes express themselves in the time-dependent surface properties, such as surface tension (dynamic surface tension) and surface dilatational [3] and shear rheology [4]. Although, the phenomenon has been extensively studied, the physics of the adsorption-desorption dynamics of surfactants at fluid-fluid interfaces is far from completely understood [5]. For ionic surfactants, the adsorption dynamics deviates from the diffusion limited model developed by Ward and Tordai [6] in the 1940s. The origin of these deviations can be found in the presence of electrostatic adsorption barriers [7–10]. In a previous paper [10] we studied the adsorption kinetics of a cationic surfactant, dodecyltrimethylammonium bromide (DTAB), using dynamic surface tension measurements. It was found that the adsorption dynamics is controlled by diffusion (DLA) at very short adsorption times but it is kinetically limited (KLA), at times close to the equilibrium, by an electrostatic barrier built-up, as the ionic surfactant molecules adsorb onto the interface. In another very recent article, we explain the appearance of negative surface viscosities for a cationic Gemini surfactant, dimethylene-1,2-bis(dodecyldimethylammonium) bromide (G12-2-12), by the asymmetries found in the adsorption-desorption processes and dynamics [11]. The unphysical result (negative viscosities) is just an artifact due to the procedure used for data analysis, which is invalidated by the existence of those asymmetries.

In this article, we extend the study on G12-2-12 by systematically measuring the dynamic surface tension and time-resolved surface potential, trying to shed light on those processes, their origin, and mechanisms. We used the maximum bubble pressure technique in order to access the very short intermediate adsorption times and Wilhelmy plate technique [12], in order to study the long adsorption time dynamics and to obtain the equilibrium values. We explored the whole range of surfactant concentrations, from very dilute to twice the critical micelle concentration, cmc, of the surfactant. As mentioned, we also performed time-resolved surface potential measurements in the hope to elucidate the role played by charge redistribution within the interface, in the adsorption dynamics and the corresponding evolution of surface tension. We found very complex behavior, with several characteristic times, both for the dynamic surface tension and surface potential. For the more dilute solutions, the initial adsorption step is controlled by diffusion (DLA), and in the last part, close to the equilibrium, the adsorption is kinetically limited (KLA). Between those extremes, the adsorption involves several processes with different characteristic times, where the redistribution of charges at the interface is evident. The existence of these numerous processes is compatible with the observation of several characteristic times in compression surface rheology experiments [11]. The results presented here shed light on the role played by charge organization on the dynamics of adsorption of ionic surfactants.

2. Materials and Methods

2.1. Materials

The Gemini 12-2-12 surfactant (from now on, G12-2-12), was synthetized in our laboratory [13] following a slightly modified method employed by Zana et al. [14]. The success of the synthesis was confirmed by nuclear magnetic resonance (NMR) measurements. After synthesis and purification, the surfactant was kept under vacuum over dried silica to minimize water absorption before use. Milli-Q water (resistance >18.2 MΩ·cm) was used for the preparation of all solutions used in this work.

2.2. Methods

2.2.1. Equilibrium Surface Tension

Surface pressure measurements were performed using the sensor of a KSV LB-5000 Langmuir balance system (KSV-NIMA, Biolin Scientific, Spoo, Finland). The surface pressure is defined as $\Pi = \gamma_0 - \gamma$, being γ_0 and γ the surface tension of pure solvent, and solution, respectively. Disposable 20.6 mm paper (KSV-NIMA) or Pt-Wilhelmy probes were used. The measuring cell (10 cm^3) is made of Teflon and the temperature was controlled by circulating water from a thermostat (Lauda ALPHA RA 8, Lauda-Königshofen, Germany) through a jacket placed at the bottom of the cell. The temperature near the surface is measured with a 0.01 °C precision using a calibrated Pt-100 sensor. All measurements were performed at 22 °C. For the measurements, the Teflon recipient is filled with an accurately measured amount of Mili-Q water, a small accurately measured amount of stock concentrated surfactant solution is then added to the water in order to prepare, in situ, the surfactant solution at the desired concentration. The solution is left for 60 min to stabilize before surface pressure measurements are carried out.

2.2.2. Dynamic Surface Tension

For very short and intermediate adsorption times (from milliseconds to hundreds of seconds), dynamic surface tension, $\gamma(t)$, was measured by means of a homemade apparatus using the maximum bubble pressure technique (MBP). The setup is similar to the one described elsewhere [10,15] and is shown in Figure 1. The technique is based on the measurement of the maximum pressure needed to form a bubble at the tip of a capillary immersed in the surfactant solution. The pressure is related to the surface tension by the Young-Laplace equation [16], $\Delta P = 2\gamma/r$ being r the radius of the capillary. The device consists of a peristaltic pump which pressurizes a 1 dm^3 air reservoir, an electrically controlled needle valve (Aalborg PSV1S-VA with its driver module PSV-D, Organgeburg, NY, USA) that controls the air flow to a micro-pipette tip (Sartorius Optifit 791000, Buenos Aires, Argentina) and acts as the capillary, and is placed 5 mm below the solution's surface. The pressure difference, ΔP, inside the tip was measured with a differential pressure transducer (Cole-Palmer GY-98073-08, Vernon Hills, IL, USA) and transmitted using USB to a PC by a Teensy 3.2 microcontroller board. The Teensy board also senses and controls the reservoir pressure (pressure transducer Ashcroft G2-7-M02-15-G2-30G, Stanford, CT, USA) and the needle valve opening. The tip makes an angle of 30° with respect to the interface and is discarded after each measurement. Each tip is calibrated with pure water prior to measurement by measuring the ΔP as a function of depth, which is precisely controlled by means of the step-motor.

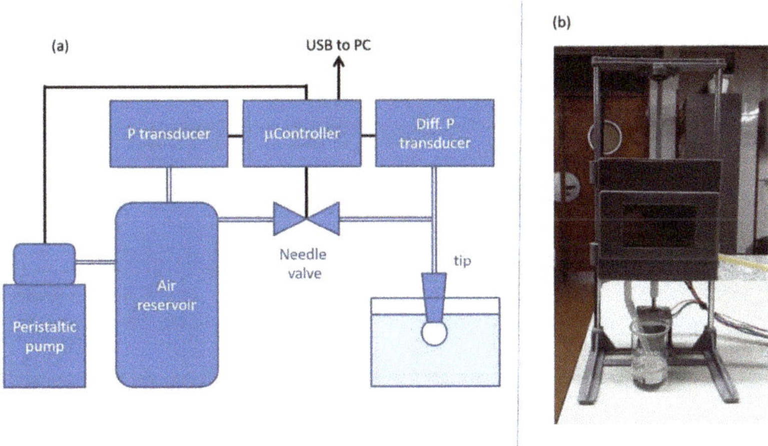

Figure 1. (**a**) Scheme of the Maximum Pressure Apparatus. (**b**) Picture of the device showing the tip and the step-motor.

2.2.3. Time-Resolved Surface Potential Experiments

We measured the surface potential as a function of time by means of a Kelvin probe (KSV-NIMA SPOT). The experiment is schematized in Figure 2 and it is as follows: A concentrated solution of G12-2-12 (surfactant concentration, c_s = 0.1 M), is prepared and let stabilize for 24 h. Separately, a precisely measured volume of pure water (Milli-Q) is placed in a small Teflon cuvette. The surface potential probe is placed at about 2 mm above the water surface. The surface potential of pure water is then measured and the result saved. By means of a Hamilton syringe (50 µL), a certain volume of the concentrated Gemini solution, calculated in order to obtain the desired final concentration, is injected at the bottom of the Teflon cuvette, producing a concentration impulse while the surface potential is continuously monitored. The injection process takes about 5 s. The injection process must be done very carefully, the syringe needle must be perfectly cleaned and any movement that could produce convection must be avoided. For changing the final surfactant concentration, we changed either the initial volume of water or the volume of the surfactant solution injected.

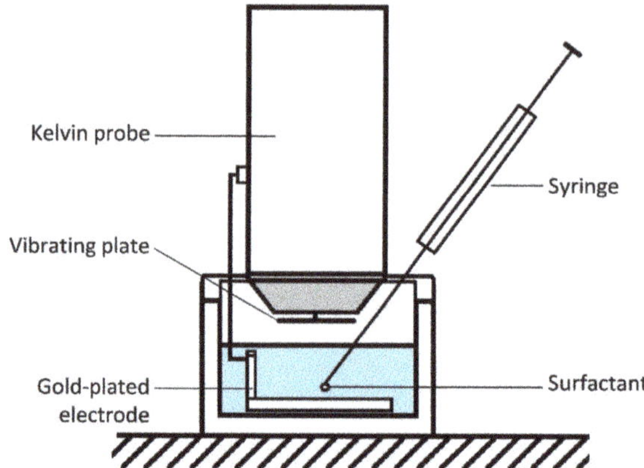

Figure 2. Scheme of the time-resolved surface potential experiment.

3. Results

3.1. Equilibrium Surface Tension Isotherm

In Figure 3a we show the surface tension isotherm for aqueous solutions of the G12-2-12 surfactant. The critical micelle concentration, cmc, obtained from these measurements is 0.9 mM. This value is in good agreement with values previously reported [5,14,17] which are in the range of 0.7 to 1 mM.

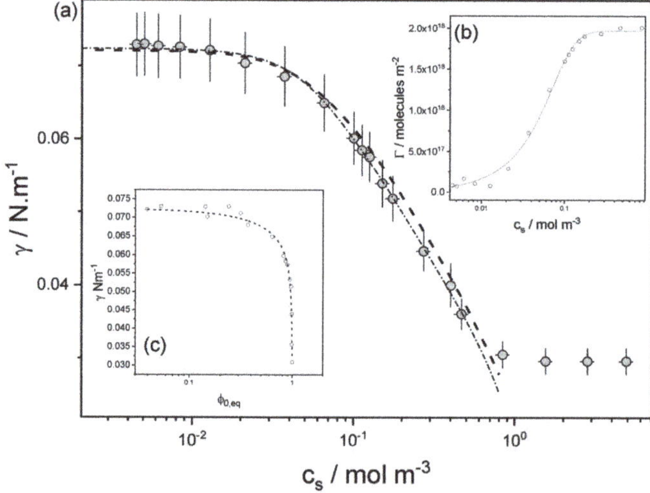

Figure 3. (**a**) Surface tension isotherm for Gemini 12-2-12. The lines are two different fittings with Equation (2), see Discussion section for details. (**b**) Equilibrium surface concentration as a function of bulk concentration calculated from Equation (2) (**c**) Surface tension as a function of surfactant area fraction at equilibrium (see Equation (2)).

3.2. Dynamic Surface Tension

In Figure 4 we present the results of the dynamic surface tension for 6 surfactant concentrations below the cmc. Those concentrations are c_s = 0.1; 0.2; 0.3; 0.4; 0.5, and 0.6 mM. All the curves are well-behaved (smooth) and in all cases, they reach the equilibrium values obtained from Wilhelmy plate technique, considering the errors (in the figures the errors are represented by the dashed regions). In the insets in said figures we show, amplified, the very short adsorption times region (≤ 2 s). The lines shown are fittings using Equation (3) (see discussion below).

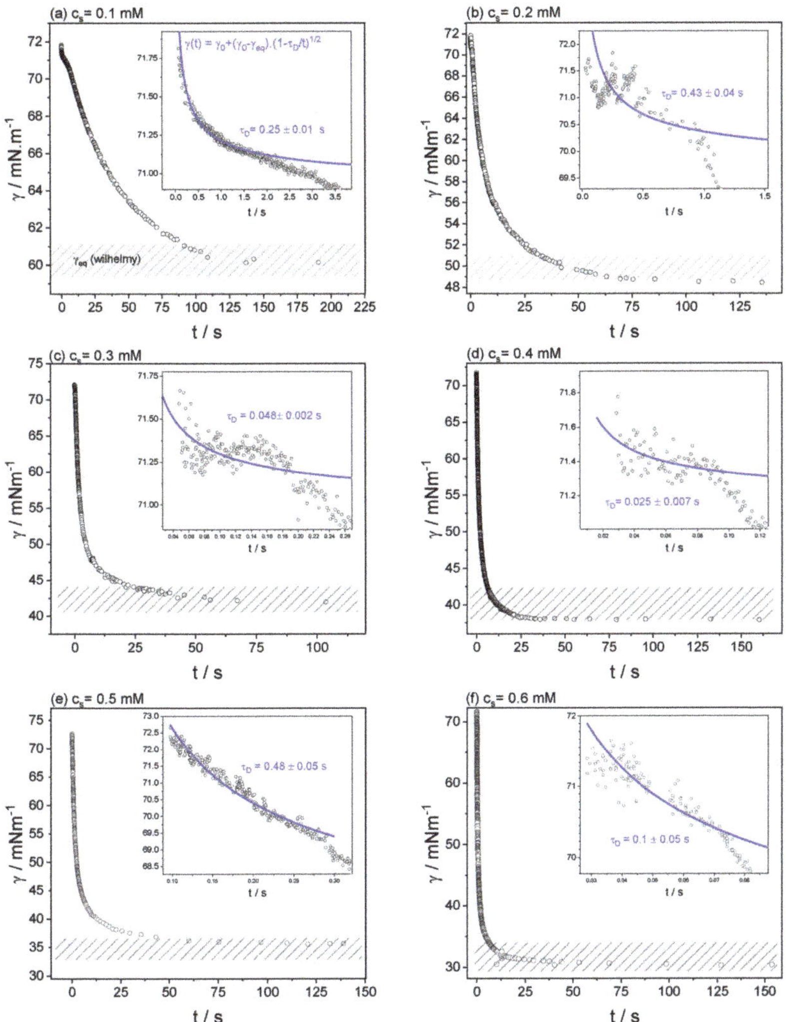

Figure 4. Dynamic surface tension curves for 6 surfactant concentrations well below the cmc (**a**) c_s = 0.1 mM; (**b**) c_s = 0.2 mM; (**c**) c_s = 0.3 mM; (**d**) c_s = 0.4 mM; (**e**) c_s = 0.5 mM; (**f**) c_s = 0.6 mM. The insets are amplifications of the very short adsorption times regions, the curves in them are fittings with Equation (3). The shaded regions represent the equilibrium surface tension (with error) measured by the Wilhelmy plate technique.

In Figure 5 we show the results for the same kind of experiments but for solutions at concentrations close and above the cmc.

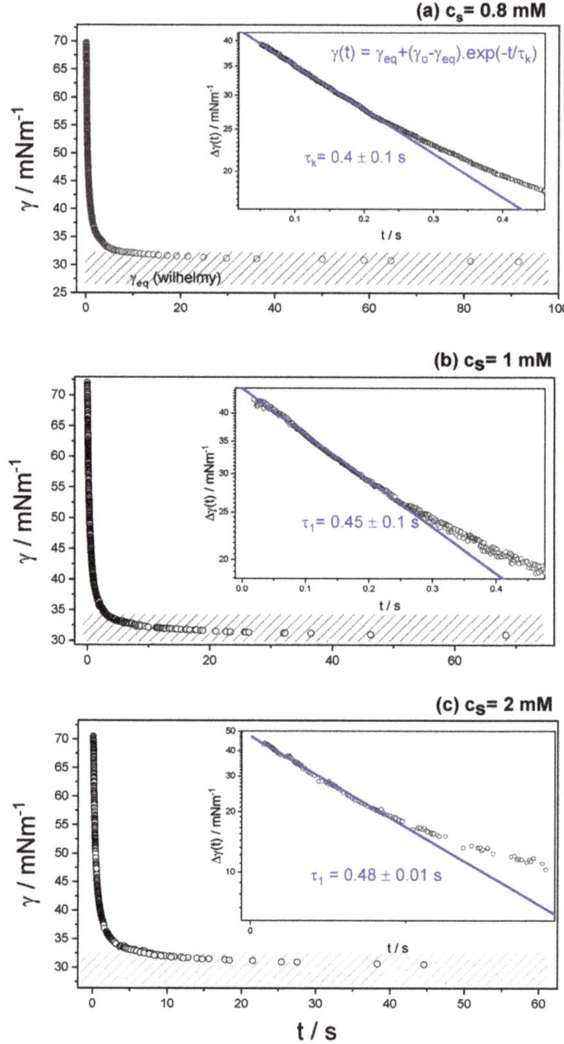

Figure 5. Dynamic surface tension for surfactant concentrations close and above the cmc. (**a**) 0.8 mM; (**b**) 1 mM; (**c**) 2 mM. The insets show the dynamics of adsorption at very short times (<0.5 s). The lines are fittings with Equation (5).

The behavior at short adsorption times (insets) is different from those observed for c_s < cmc. The lines are fittings with exponentials (Equation (5), see discussion).

3.3. Time-Resolved Surface Potential

In order to shed light on the processes involved in the adsorption dynamics, and the role played by the charges at the interface, we performed time-resolved surface potential experiments, as explained in the methods section. In Figure 6 we present the results for c_s = 1 mM, superimposed to the dynamic surface tension curve in a semi-log plot. We show simultaneously three independent measurements of the surface potential as a function of time. The three curves are different but all of them present general common features. Note that, at short times, there is a fast increase of surface potential followed by a decrease and several oscillations, those oscillations were not replicated on the surface tension dynamics. In the figure, we included the relaxation times obtained by fitting certain parts of the curves with exponentials (see discussion).

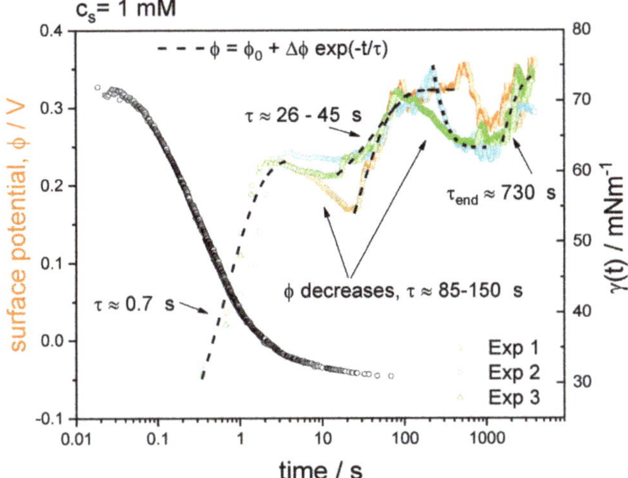

Figure 6. Surface potential curves (symbols in color, left axis) superimposed to the dynamic surface tension, $\gamma(t)$ (circles, black, right axis) for a solution at a surfactant concentration of 1 mM. We show three independent surface potential experiments (Exp 1, 2 and 3) to illustrate the "reproducibility" of such measurements.

In Figure 7 we show the results corresponding to c_s = 0.1 mM, c_s = 0.2 mM and c_s = 0.5 mM. Note that the surface potential curves have the same features mentioned above, a fast increase of the surface potential and oscillations.

For all surfactant concentrations the initial (water) surface potential was about −0.2 volts and reaches the stationary value of 0.3–0.5 volts, depending on surfactant concentration. The insets on those figures are included to show that φ tends to a stationary value, a fact that is difficult to see when time is in logarithmic scale.

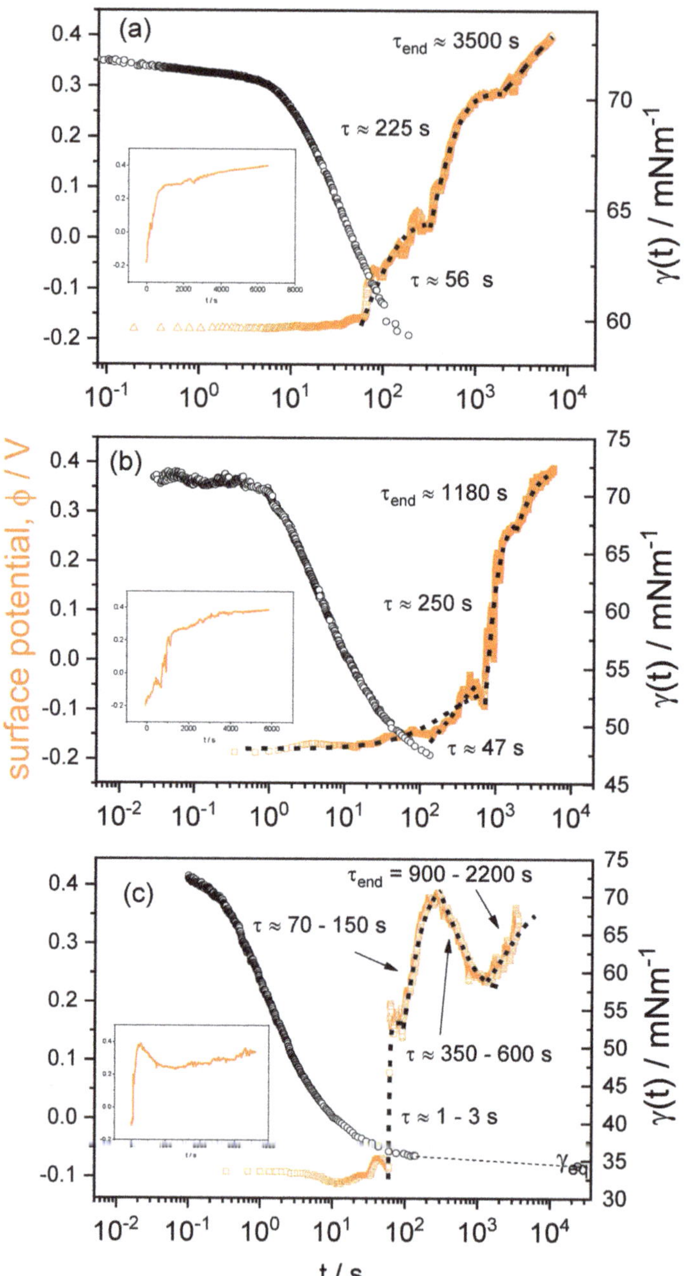

Figure 7. Time dependence of the Surface potential compared with dynamic surface tension (Surface tension in black, right axis) for three surfactant concentrations. (**a**) $c_s = 0.1$ mM, (**b**) $c_s = 0.2$ mM, (**c**) $c_s = 0.5$ mM. The insets show the surface potential but with the time axis in linear scale.

4. Discussion

4.1. Equilibrium Surface Tension

The first step in the discussion of the results shown in previous sections is the equilibrium isotherm of Figure 3a. Our point of departure is the Gibbs adsorption equation [12]:

$$\Gamma = -\frac{1}{nk_\text{B}T}\frac{d\gamma}{d\ln(c_\text{s})} \tag{1}$$

From the dependence of the surface tension, γ, on the surfactant bulk concentration, c_s, the surface excess, Γ, (surface surfactant concentration) can be obtained by means of this equation. Here k_B is the Boltzmann constant and T the absolute temperature. The constant n takes the value 3 for a dimeric surfactant made up of a divalent surfactant ion and two univalent counterions, such as G12-2-12, in the absence of added salt. However, by means of neutron reflectivity measurements [18,19], it was found that, for G12-2-12, $n = 2$. This means that one counterion is condensed onto the Gemini polar head at the interface and thus G12-2-12 behaves like a monovalent ionic surfactant ($n = 2$). In Figure 3b we show the surface excess obtained from the experimental data of Figure 3a and the Gibbs adsorption equation, Equation (1). Note that the surface saturates at about 0.2 mM with a surface concentration of $\Gamma \sim 1.9 \times 10^{18}$ molec.m^{-2}, given an area per surfactant head of 5.3×10^{-19} m^2, which gives 7.25 Å as an estimation for the distance between surfactant heads at the interface. On the other hand, from a free energy approach [10,20,21], the adsorption isotherms and the equation of state for a 1:1 ionic surfactant (note that we consider that one of the surfactant charges is also condensed in bulk because the distance between charges in the surfactant head [22], ~4 Å, is less than the Bjerrum length) can be expressed as,

$$\phi_0 = \frac{\phi_\text{b}}{\phi_\text{b}+\left[b\phi_0+\sqrt{(b\phi_0)^2+1}\,\right]^2 exp(-\alpha-\beta\phi_0)}$$

$$\gamma = \gamma_\text{w} + \frac{k_\text{B}T}{a^2}\left[ln(1-\phi_0) + \tfrac{\beta}{2}\phi_0^2 - \tfrac{2}{b}\left(\sqrt{(b\phi_0)^2+1}-1\right)\right] \tag{2}$$

a being the average size of a surfactant molecule, φ_0 is the surfactant area fraction at equilibrium ($\varphi_0 = \Gamma a^2$), β the Frumkin lateral interaction parameter, α the Langmuir adsorption parameter and b a parameter characterizing the strength of electrostatic interactions [10], $b = [\pi l_\text{B}/(2a\phi_\text{b})]^{1/2}$ (ϕ_b is the surfactant volume fraction, $\phi_\text{b} = a^3 c_\text{s}$). In Figure 3c we show the fitting of the data, $\gamma(\varphi_0)$, with Equation (2). The lines in Figure 3a correspond to the same fitting for the curve $\gamma(c_\text{s})$. We show two fittings, in one case keeping the solvent surface tension constant at 72 mNm^{-1} and for the other leaving this parameter free to adjust the curve. From the former fitting we found: $a = (0.89 \pm 0.2)$ nm, $\beta = (-3.24 \pm 1)\,k_\text{B}T$; $\alpha = (8.9 \pm 2)\,k_\text{B}T$; from the latter, $a = (0.77 \pm 0.2)$ nm, $\beta = (-0.6 \pm 0.2)\,k_\text{B}T$; $\alpha = (10 \pm 2)\,k_\text{B}T$ and $\gamma_0 = 72.8$ mNm^{-1}. The molecular size, a, obtained from the fittings with this model is 7.7–8.9 Å, which is close but larger than that obtained for DTAB solutions [10], for which $a = 7.2$ Å was obtained. This result is somehow expected because G12-2-12 has two DTAB chains linked by the heads with an ethyl group. Note that the Frumkin interaction parameter β is negative, this indicates a repulsive interaction among surfactant molecules at the interface.

4.2. Equilibrium Surface Potential

In Figure 8 we plot the change in surface potential, $\Delta\varphi$, between pure water interface ($\varphi_\text{water} \sim -0.2$ V) and the final stationary value at the surfactant solution-air interface, φ. Note that the surface potential difference seems to go through a maximum at a surfactant concentration between 0.2 and 0.4 mM. Above this concentrations, $\Delta\varphi$ diminishes continuously indicating the condensation of counterions onto the interface. This is consistent with the distance separating surfactant heads at the interface estimated from Gibbs equation, ~7.25 Å, which is very close to the value of the Bjerrum length l_B ($l_\text{B} = \frac{e^2}{\varepsilon k_\text{B}T} \approx 7$Å, e is the electron charge and ε is the water dielectric constant). We recall that, when the distance between charges is less than l_b, condensation of counter-ions is expected, such that the effective distance between charges become equal to the Bjerrum length [23,24].

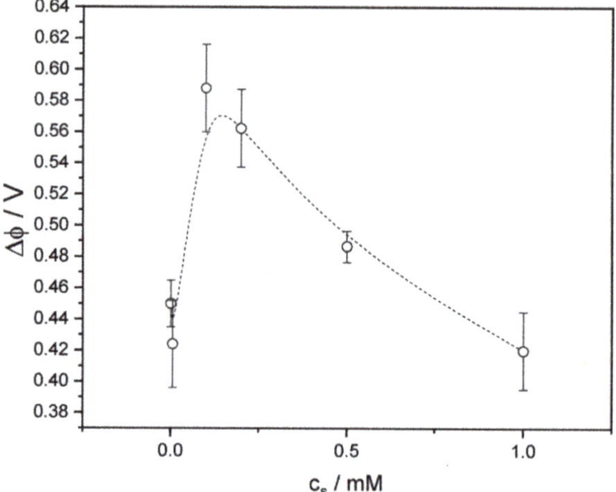

Figure 8. Change in the surface potential between water and surfactant solutions, $\Delta\varphi = \varphi - \varphi_{water}$, as a function of surfactant concentration. This curve includes the values for $c_s = 0.01$ and 0.06 mM which are not included in Figure 7.

4.3. Dynamic Surface Tension

Let us first discuss the short adsorption times. In the insets of Figures 4 and 5 the dynamics in this time range can be observed in an amplified scale. For $c_s = 0.1$ mM and for adsorption times shorter than about 2 s, the dynamic surface tension data is beautifully fitted (see inset in Figure 4a) with the diffusional model [20,21], indicating that the adsorption dynamics is limited by diffusion (DLA),

$$\gamma(t) \cong \gamma_0 + \left(\gamma_0 - \gamma_{eq}\right)\left(1 - \sqrt{\frac{\tau_D}{t}}\right)$$

(3)

where τ_D is the characteristic diffusional time. From the fitting with Equation (3), we obtain $\tau_D = 0.25 \pm 0.01$ s. As the surfactant concentration increases, but always below the cmc, the experimental points become noisy and the fittings less reliable (see insets in Figure 4b–f). The corresponding times, τ_D, obtained from Equation (3) for each concentration are summarized in Table 1.

Table 1. Characteristic times obtained from fittings with Equation (3), and the diffusion coefficients, D, calculated with Equation (4). The surfactant concentrations are all below the cmc.

c_s/mM	τ_D/s	D/m^2 s^{-1}
0.1	0.25 ± 0.01	4.1×10^{-10}
0.2	0.43 ± 0.04	1.1×10^{-10}
0.3	0.048 ± 0.02	5.0×10^{-10}
0.4	0.025 ± 0.01	5.4×10^{-10}
0.5	0.48 ± 0.05	1.8×10^{-11}
0.6	0.1 ± 0.05	6×10^{-11}

From the theory the characteristic time for a DLA process can be estimated [20,21],

$$\tau_D = \frac{\phi_0^4}{\phi_b^2} \frac{a^2}{\pi D} \tag{4}$$

being φ_b and D the bulk surfactant volume fraction and diffusion coefficient respectively. From the bulk concentration, the surfactant volume fraction can be estimated, $\phi_b = a^3 c_s$, for $c_s = 0.1$ mM, $\varphi_b = 3.1 \times 10^{-5}$. Now, making use of the data on Figure 3, we estimate $\varphi_0 \sim 0.83$. With these values, from Equation (4) and $\tau_D = 0.25$ s, we calculate $D = 4.1 \times 10^{-10}$ m^2 s^{-1}. This value coincides perfectly with the literature value [25] of 4×10^{-10} m^2 s^{-1}. The same calculation for larger surfactant concentrations deviates from this value but they fall in the correct order of magnitude except for $c_s = 0.5$ and 0.6 mM. The results are summarized in Table 1.

For surfactant concentrations close and above the cmc, the adsorption dynamics is consistent with an exponential decay of surface tension even at very short adsorption times:

$$\gamma(t) - \gamma_{eq} = \left(\gamma_0 - \gamma_{eq}\right) exp\left(-\frac{t}{\tau}\right). \tag{5}$$

This behavior can be seen in the insets of Figure 5, the characteristic times found by fitting with Equation (5) are labelled as τ_1 and shown in Table 2. In Equation (5), γ_{eq} and γ_0 are the equilibrium surface tension of the solutions, and pure solvent, respectively.

Table 2. Characteristic times, in seconds, found in dynamic surface tension for all surfactant concentrations studied.

c_s/mM	τ_1	τ_2	τ_3	τ_4	τ_k
0.1	–	–	–	6.5 ± 1	37 ± 1
0.2	–	–	—	28.5 ± 1	105 ± 12
0.3	–	–	2.4 ± 0.1	9.5 ± 1	20 ± 3
0.4	–	–	5.1 ± 0.1	9.5 ± 0.5	180 ± 40
0.5	–	10.4 ± 0.5	20 ± 1	40 ± 2	120 ± 15
0.6	–	0.71 ± 0.002	8 ± 1	30 ± 2	67 ± 3
0.8	0.4 ± 0.01	1.34 ± 0.005	3.7 ± 0.1	25 ± 1	44 ± 4
1	0.45 ± 0.001	–	3.6 ± 0.1	15 ± 1	74 ± 4
2	0.48 ± 0.001	–	2.6 ± 0.1	7.8 ± 0.5	33 ± 2

Turning our attention to the last part of the adsorption process, close to equilibrium, the relaxation is consistent with a kinetically limited adsorption (KLA), following an exponential decay [10,20,21] as in Equation (5). We label the characteristic time of this final stage of the adsorption dynamics as τ_k.

In Figures 9 and 10 we plot, in a semi-log scale, $\Delta\gamma = \gamma(t) - \gamma_{eq}$ as a function of time for surfactant concentrations below and close/above the cmc respectively. The results of the fittings with exponentials are shown as lines on the figures and labelled with the corresponding characteristic time. Note that there are several processes (and characteristic times) all consistent with an exponential decay. Those results are summarized in Table 2. Note that the number of processes increases with surfactant concentration until the cmc. The observation of several processes, each with its characteristic time, is consistent with what was found in surface rheology experiment on the same surfactant system [11].

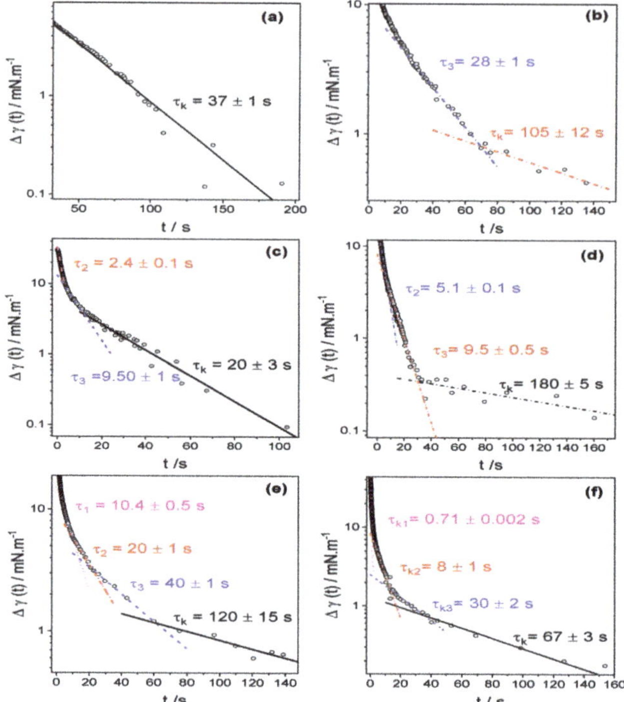

Figure 9. Several relaxation processes following an exponential decay were observed in dynamic surface tension curves at concentrations below the cmc. (**a**) c_s = 0.1 mM, (**b**) 0.2 mM, (**c**) 0.3 mM, (**d**) 0.4 mM, (**e**) 0.5 mM, (**f**) 0.6 mM.

The origin of the intermediate relaxations, between τ_D and τ_k, seems to be related to charge redistribution as observed from surface potential results on Figures 6 and 7. This redistribution of charges may include the formation of aggregates at the interface, condensation of counterions onto those aggregates and further surfactant adsorption, as well as phase transitions [26].

Now, from the values of the characteristics times in the KLA regime, τ_K, we can estimate the average $(\overline{\psi})$, the equilibrium surface (ψ_0) and subsurface (ψ_a) electrostatic potentials, $\overline{\psi} = (\psi_0 + \psi_a)/2$ [10,20,21]:

$$\frac{e\overline{\psi}}{k_B T} = \frac{\left(\alpha + \beta + \ln\left(\pi a^4 D c^2 \tau_k\right)\right)}{2}. \tag{6}$$

From previous calculations we have $a \sim 8 \times 10^{-10}$ m, $\alpha = 8.9\ k_B T$ and $\beta = -3.24\ k_B T$, using for the diffusion coefficient the value of $D = 4 \times 10^{-10}$ m^2 s^{-1} and the τ_K values from the last column on Table 2, we calculate the average surface and subsurface electrostatic potential shown in Table 3. In that table ψ_0 was estimated from the Poisson-Boltzmann theory [21],

$$\frac{e\psi_0}{k_B T} \approx 2\ln(2b\phi_0) \tag{7}$$

where b was defined above, $b = [\pi l_B/(2a\phi_b)]^{1/2}$.

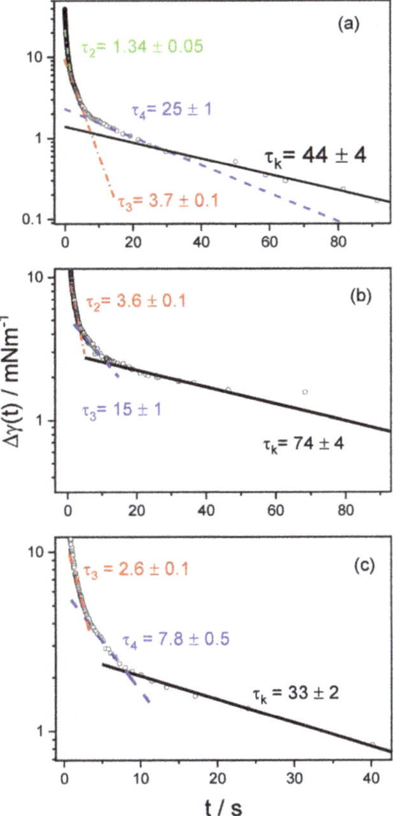

Figure 10. Several relaxation times following an exponential decay at concentrations above the cmc. (**a**) c_s = 0.8 mM, (**b**) 1 mM, (**c**) 2 mM.

Table 3. Average ($\overline{\psi}$), the equilibrium surface (ψ_0) and subsurface (ψ_a) electrostatic potentials from Equations (6) and (7). ψ_a is calculated as $2\overline{\psi} - \psi_0 = \psi_a$.

c_s/mM	$e\overline{\psi}/k_BT$	$e\psi_0/k_BT$	$e\psi_a/k_BT$
0.1	4.96	11.26	−1.34
0.2	6.17	10.93	1.41
0.3	5.75	10.56	0.94
0.4	7.14	10.27	4.01
0.5	7.15	10.05	4.25
0.6	7.05	9.87	4.23
0.8	7.12	9.58	4.66

The obtained theoretical values for the surface electrostatic potential, ψ_0, are larger than the average, $\overline{\psi}$, obtained from dynamic surface tension experiments. Because the subsurface electrostatic potential, ψ_a, should be lower than the surface electrostatic potential, ψ_0, the average should be $\overline{\psi} < \psi_0$, as it is in fact observed on Table 3. Note that the values of ψ_0 and $\overline{\psi}$ approach each other as the surfactant concentration increases, which could be explained by ions self-screening (being G12-2-12 a cationic surfactant, as the surfactant concentration increases, the ionic strength increases, and the Debye length diminishes reducing the distance of the electrostatic interaction, even in the absence of added salt). It is worth mentioning that the obtained values are similar to those found for the cationic surfactant DTAB [10].

We cannot use the previous analysis for surfactant concentrations above the cmc, in this case the adsorption-desorption and aggregation-disaggregation of micelles [27,28] could play a role in the observed dynamics both in $\gamma(t)$ and $\varphi(t)$. The fact that the dynamics of surfactant adsorption at short times follow an exponential relaxation (see insets on Figure 5) could indicate that we are not accessing times short enough to see the diffusion-limited adsorption step.

4.4. Time-Resolved Surface Potential

The main feature we want to stress about the surface potential results is the existence of oscillations in φ during the adsorption process as shown in Figure 6. Despite the variability observed from each independent experiment, the qualitative behavior is always the same and for all surfactant concentrations. The several "relaxations" observed can be roughly fitted with exponentials:

$$\phi = \phi_0 + \Delta\phi \, exp\left(-\frac{t}{\tau}\right) \tag{8}$$

Some of the results of those fittings are shown in the figures. The number of processes observed in surface potential experiments seem to be related to the several processes observed in dynamic surface tension. However, the times and characteristic times are different. The differences found in the time when the surface potential and surface tension change is probably due to the way in which time-resolved surface potential measurements are done. After the concentration impulse, there is a time delay, because of the concentration homogenization in bulk, before adsorption takes place. Despite that, it is clear that the large $\varphi(t)$ oscillations are not replicated as large oscillations in surface tension. This is an indication that the behavior of φ is due to the reorganization of counterions at the interfacial region and not due to the migration of surfactant molecules into and out to the interface, at least in such amounts that could produce appreciable changes in the surface tension. We could rationalize the results as follows: First the G12-2-12 surfactant adsorbs onto the interface producing an increase of the surface potential from that of pure water. This process is fast and consistent with the decrease of the surface tension (see Figures 6 and 7). The drop in surface potential that follows, could be due to the reorganization, redistribution, and condensation of charges, resulting in a temporal decrease of the measured surface potential. After that, and because the condensation of counter ions, a decrease in the charge repulsion among surfactant molecules, at the interface, would allow more surfactant molecules to adsorb, thereby producing a subsequent increase of the surface potential, and a slow decrease of surface tension, until the final equilibrium value is reached. This last process lasts from hundreds to thousands of seconds. The characteristic times obtained from fittings with Equation (8), labelled as τ_{end} in Figures 6 and 7, span from 700 to 3500 s. It is worth mentioning here that these results help to explain the behavior of the surface step-compression rheology observed in these systems [11].

A final comment is needed about the apparent differences observed on the time-dependent surface potential curves of Figure 7. In those figures (see insets), we observe that the behavior for $c_s = 0.5$ mM (and also for $c_s = 1$mM in Figure 6) is different from those at lower surfactant concentrations. First note that, for $c_s = 0.1$ and 0.2 mM, the change in the equilibrium surface potential goes through a maximum (Figure 8). This is probably because the interface is less saturated than for $c_s = 0.5$ (see Figure 3b), therefore, the distances between charged heads are larger and the condensation of counterions onto the interface lower for $c_s = 0.1$ and 0.2 (the distances between charges are larger than the Bjerrum length) than for $c_s = 0.5$ mM. For this last concentration, the distances between charged surfactant heads at the interface are close to the Bjerrum length, thus, what we observe in the time-dependent surface potential after the first surfactant adsorption, could be a fast condensation of counterions onto the interface, followed by a slow surfactant adsorption until equilibrium is reached, a behavior not observed for the two lower concentrations.

5. Summary and Conclusions

We studied the adsorption dynamics of a cationic surfactant bearing two charges, Gemini 12-2-12, onto the air-solution interface. By means of dynamic surface tension and time-resolved surface potential, we have identified several processes with their characteristic times. For the more dilute solutions, well below the cmc, and at very short adsorption times, the adsorption is consistent with a process limited by diffusion (DLA). The final stage of the adsorption dynamics, close to equilibrium, follows the exponential law, derived from the existence of a surface potential, built up by the charged molecules as they adsorb onto the interface. This behavior for ionic surfactants were observed previously [10,20,21,29]. Between the initial diffusion controlled stage and the final kinetically controlled (KLA) adsorption process, several processes have been identified with their characteristic times. The number of processes observed depends on surfactant concentration. In this respect, the observed behavior is consistent with several processes and characteristic times, which are observed in step-compression surface rheology experiments on these systems [11], which help to explain them. Based on the surface potential results, it is clear that those processes are related to a charge reorganization in the interfacial region. The reorganization may include the formation of aggregates at the interface, condensation of counter-ions onto those aggregates, and further surfactant adsorption, as well as phase transitions [26,30]. For the surfactant concentrations above the cmc, some of the processes observed could be related to the presence of micelles and their dynamics (adsorption-desorption, aggregation-disaggregation [27,28]). The redistribution of charges within the interfacial region, as observed in the surface potential experiments, seems to be responsible for the existence of the intermediate relaxations, well-described by exponentials, which are observed on the dynamic surface tension curves. Finally, the existence of phase transitions or surfactant aggregates at the interface, mentioned above, remains to be probed experimentally.

Author Contributions: Conceptualization, H.R.; Methodology, M.F.L., E.C. and S.G.R.; Experimental realization M.F.L., S.G.R., E.C. and J.F.S.M.; Formal analysis, H.R., M.F.L and E.C; Writing—original draft preparation, H.R.; Writing—review and editing, M.F.L., E.C., J.F.S.M., H.R; Supervision, H.R. and M.F.L.; Project administration, H.R. All authors have read and agreed to the published version of the manuscript.

Funding: This research was funded by Universidad Nacional del Sur (UNS, Argentina) under grant PGI-UNS 24/F080; by Agencia Nacional de Promoción Científica y Tecnológica (ANPCyT, Argentina) under grant PICT-2016-0787 and by Consejo Nacional de Investigaciones Científicas y Técnicas (CONICET, Argentina) under grant PIP-GI 2014 Nro 11220130100668CO.

Conflicts of Interest: The authors declare no conflicts of interest.

References

1. Georgieva, D.; Schmitt, V.; Leal-Calderon, F.; Langevin, D. On the Possible Role of Surface Elasticity in Emulsion Stability. *Langmuir* **2009**, *25*, 5565–5573. [CrossRef] [PubMed]
2. Georgieva, D.; Cagna, A.; Langevin, D. Link between Surface Elasticity and Foam Stability. *Soft Matter* **2009**, *5*, 2063. [CrossRef]
3. Ravera, F.; Ferrari, M.; Santini, E.; Liggieri, L. Influence of Surface Processes on the Dilational Visco-Elasticity of Surfactant Solutions. *Adv. Colloid Interface Sci.* **2005**, *117*, 75–100. [CrossRef] [PubMed]
4. Langevin, D. Rheology of Adsorbed Surfactant Monolayers at Fluid Surfaces. *Annu. Rev. Fluid Mech.* **2014**, *46*, 47–65. [CrossRef]
5. He, Y.; Salonen, A.; Lecchi, A.; Veber, M.; Langevin, D. Adsorption of Gemini Surfactants at the Air-Water Surface. *Colloid Polym. Sci.* **2016**, *294*, 483–490. [CrossRef]
6. Ward, A.F.H.; Tordai, L. Time-Dependence of Boundary Tensions of Solutions I. The Role of Diffusion in Time-Effects. *J. Chem. Phys.* **1946**, *14*, 453–461. [CrossRef]
7. Bleys, G.; Joos, P. Adsorption Kinetics of Bolaform Surfactants at the Air/Water Interface. *J. Phys. Chem.* **1985**, *89*, 1027–1032. [CrossRef]

8. Bonfillon, A.; Langevin, D. Viscoelasticity of Monolayers at Oil-Water Interfaces. *Langmuir* **1993**, 2172–2177. [CrossRef]

9. Bonfillon, A.; Langevin, D. Electrostatic Model for the Viscoelasticity of Ionic Surfactant Monolayers. *Langmuir* **1994**, *10*, 2965–2971. [CrossRef]

10. Ritacco, H.; Langevin, D.; Diamant, H.; Andelman, D. Dynamic Surface Tension of Aqueous Solutions of Ionic Surfactants: Role of Electrostatics. *Langmuir* **2011**, *27*, 1009–1014. [CrossRef]

11. Cuenca, V.E.; Fernández Leyes, M.; Falcone, R.D.; Correa, N.M.; Langevin, D.; Ritacco, H. Interfacial Dynamics and Its Relations with "Negative" Surface Viscosities Measured at Water–Air Interfaces Covered with a Cationic Surfactant. *Langmuir* **2019**, *35*, 8333–8343. [CrossRef] [PubMed]

12. Adamson, A.; Gast, A. *Physical Chemistry of Surfaces*, 6th ed.; Jhon Wiley & Sons: New York, NY, USA, 1997.

13. Cuenca, V.E.; Falcone, R.D.; Silber, J.J.; Correa, N.M. How the Type of Cosurfactant Impacts Strongly on the Size and Interfacial Composition in Gemini 12-2-12 RMs Explored by DLS, SLS, and FTIR Techniques. *J. Phys. Chem. B* **2016**, *120*, 467–476. [CrossRef] [PubMed]

14. Zana, R.; Benrraou, M.; Rueff, R. Alkanediyl-.Alpha.,.Omega.-Bis(Dimethylalkylammonium Bromide) Surfactants. 1. Effect of the Spacer Chain Length on the Critical Micelle Concentration and Micelle Ionization Degree. *Langmuir* **1991**, *7*, 1072–1075. [CrossRef]

15. Ritacco, H.; Kurlat, D.; Langevin, D. Properties of Aqueous Solutions of Polyelectrolytes and Surfactants of Opposite Charge: Surface Tension, Surface Rheology, and Electrical Birefringence Studies. *J. Phys. Chem. B* **2003**, *107*, 9146–9158. [CrossRef]

16. Fainerman, V.B.; Miller, R. The Maximum Bubble Pressure Technique. In *Drop and Bubbles in Interfacial Science*; Möbius, M.E., Miller, R., Eds.; Elsevier B.V.: Amsterdam, The Netherlands, 1998; pp. 279–326.

17. Espert, A.; von Klitzing, R.; Poulin, P.; Colin, A.; Zana, R.; Langevin, D. Behavior of Soap Films Stabilized by a Cationic Dimeric Surfactant. *Langmuir* **1998**, *14*, 4251–4260. [CrossRef]

18. Li, Z.X.; Dong, C.C.; Thomas, R.K. Neutron Reflectivity Studies of the Surface Excess of Gemini Surfactants at the Air-Water Interface. *Langmuir* **1999**, *15*, 4392–4396. [CrossRef]

19. Zana, R. Dimeric and Oligomeric Surfactants. Behavior at Interfaces and in Aqueous Solution: A Review. *Adv. Colloid Interface Sci.* **2002**, *97*, 205–253. [CrossRef]

20. Diamant, H.; Ariel, G.; Andelman, D. Kinetics of Surfactant Adsorption: The Free Energy Approach. *Colloids Surf. A Physicochem. Eng. Asp.* **2001**, *183–185*, 259–276. [CrossRef]

21. Diamant, H.; Andelman, D. Kinetics of Surfactant Adsorption at Fluid-Fluid Interfaces. *J. Phys. Chem.* **1996**, *100*, 13732–13742. [CrossRef]

22. Almeida, J.A.S.; Pinto, S.P.R.; Wang, Y.; Marques, E.F.; Pais, A.A.C.C. Structure and Order of DODAB Bilayers Modulated by Dicationic Gemini Surfactants. *Phys. Chem. Chem. Phys.* **2011**, *13*, 13772–13782. [CrossRef]

23. Manning, G.S. Limiting Laws and Counterion Condensation in Polyelectrolyte Solutions II. Self-Diffusion of the Small Ions. *J. Chem. Phys.* **1969**, *51*, 934. [CrossRef]

24. Manning, G.S. Limiting Laws and Counterion Condensation in Polyelectrolyte Solutions I. Colligative Properties. *J. Chem. Phys.* **1969**, *51*, 924. [CrossRef]

25. Nilsson, M.; Cabaleiro-Lago, C.; Valente, A.J.M.; Söderman, O. Interactions between Gemini Surfactants, 12-s-12, and β-Cyclodextrin As Investigated by NMR Diffusometry and Electric Conductometry. *Langmuir* **2006**, *22*, 8663–8669. [CrossRef] [PubMed]

26. Monroy, F.; Giermanska Kahn, J.; Langevin, D. Dilational Viscoelasticity of Surfactant Monolayers. *Colloids Surf. A Physicochem. Eng. Asp.* **1998**, *143*, 251–260. [CrossRef]

27. Bonfillon, A.; Sicoli, F.; Langevin, D. Dynamic Surface Tension of Ionic Surfactant Solutions. *J. Colloid Interface Sci.* **1994**, *168*, 497–504. [CrossRef]

28. Bykov, A.G.G.; Liggieri, L.; Noskov, B.A.A.; Pandolfini, P.; Ravera, F.; Loglio, G. Surface Dilational Rheological Properties in the Nonlinear Domain. *Adv. Colloid Interface Sci.* **2015**, *222*, 110–118. [CrossRef]

29. Miller, R.; Aksenenko, E.V.; Fainerman, V.B. Dynamic Interfacial Tension of Surfactant Solutions. *Adv. Colloid Interface Sci.* **2017**, *247*, 115–129. [CrossRef]

30. Fainerman, V.B.; Mys, V.D.; Makievski, A.V.; Petkov, J.T.; Miller, R. Dynamic Surface Tension of Micellar Solutions in the Millisecond and Submillisecond Time Range. *J. Colloid Interface Sci.* **2006**, *302*, 40–46. [CrossRef]

 *coatings*

Article

Interfacial Properties and Emulsification of Biocompatible Liquid-Liquid Systems

Katarzyna Dziza [1], **Eva Santini** [1], **Libero Liggieri** [1,*], **Ewelina Jarek** [2], **Marcel Krzan** [2], **Thilo Fischer** [3] **and Francesca Ravera** [1]

[1] Institute of Condensed Matter Chemistry and Technologies for Energy, Unit of Genoa, 16149 Genoa, Italy; katarzyna.dziza@ge.icmate.cnr.it (K.D.); eva.santini@ge.icmate.cnr.it (E.S.); francesca.ravera@ge.icmate.cnr.it (F.R.)
[2] Jerzy Haber Institute of Catalysis and Surface Chemistry, Polish Academy of Sciences, 30-239 Krakow, Poland; ncjarek@cyf-kr.edu.pl (E.J.); nckrzan@cyf-kr.edu.pl (M.K.)
[3] 4GENE GmbH, 85354 Freising, Germany; thilo.fischer@4gene.de
* Correspondence: libero.liggieri@ge.icmate.cnr.it

Received: 19 March 2020; Accepted: 10 April 2020; Published: 17 April 2020

Abstract: A comparative study is reported on the interfacial properties of a set of surfactants and is discussed in terms of the effects on the features of the corresponding oil-water emulsions. The surfactants are saponin, Tween 80 and citronellol glucoside (CG), while the oil is Miglyol 812N—A Medium Chain Triglyceride (MCT) oil. Due to their high biocompatibility, all these compounds are variously utilized in food, cosmetic or pharmaceutical products. Among the surfactants, which are all soluble in water, CG presents also an important solubility in oil, as shown by the measured partition coefficient. For these systems, dynamic and equilibrium interfacial tensions and dilational viscoelasticity are measured as a function of the surfactant concentration and analyzed according to available adsorption models. In order to compare these results with the time evolution of the corresponding emulsions, the actual surfactant concentration in the matrix phase of the emulsion is accounted for. This may differ significantly from the nominal concentration of the solutions before dispersing them, because of the huge area of droplets available for surfactant adsorption in the emulsion. Using this approach allows the derivation of the correlations between the observed emulsion behavior and the actual surfactant coverage of the droplet interface.

Keywords: interfacial tensions; dilational rheology; biocompatible emulsions; partition coefficient; Tween 80; saponin; citronellol glucoside; MCT oil; Miglyol 812N

---

1. Introduction

Emulsions in cosmetic, pharmaceutical and food industry rely on biocompatible formulations. Among the various ingredients, surfactants are needed to provide long term stability and suitable features (drop size and distribution, tactile and sensorial attributes, etc.) to the emulsified products. Aiming at these applications and, more generally, to pursue the objectives related to green and sustainable development, surfactants derived from plant products are more and more utilized, in substitution of synthetic and even biocompatible molecules.

The present paper addresses the interfacial properties and emulsification of water-oil system for two surfactants derived from vegetable sources, namely saponin and citronellol glucoside (CG) as compared to the classical biocompatible synthetic surfactant Tween 80.

As the biocompatible oil phase the Medium Chain Triglyceride (MCT) oil is used. MCT is a coconut-oil derivate which, owing to its purity and high and well-defined content of saturated triglycerides, which are metabolized in an easy way by the body, is largely utilized in pharmaceutics and is receiving increasing attention from nutritionists. Applications are therefore also devised by food and cosmetic industry. Emulsions of MCT find, in particular, interesting applications in the development of effective parenteral nutrition formulations, nutraceutics and for the encapsulation of active ingredients and drugs [1–4]. The formulation of fully biocompatible emulsions of MCT represents therefore a valuable target for many applications.

Saponin is a natural surfactant mainly extracted from plants [5,6] characterized, in general, by a hydrophobic aglycone structure with hydrophilic sugar residues. Many species of saponin exist, mostly classified on the basis either of the differences in the aglycone structure or the number of linked sugar chains [7]. The adsorption properties of saponin at water-air interface in relation to its ability to stabilize foams have been widely investigated [8–11]. In particular, it has been evidenced that the appreciable surface activity of saponin at water-air interface together with high value of the dilational viscoelasticity modulus, which has been found to be of the order of hundreds mN/m [8], make this compound a rather effective stabilizer of foams, already in pure water, without any further additives. Recently, saponin is intensively studied also as an emulsifying and emulsion stabilizing agent [12–15].

Citronellol glucoside (CG) is the β-D-glucoside of citronellol, an unsaturated monoterpene widespread in plant essential oils and a common flavor used in cosmetics and body care, as well as an aroma compound of food (Citrus sensation). Its glucoside CG is also occurring in crop plants like grape as a storage form of aroma. Recently, it became available by biotechnological production and can serve a dual role as a flavor precursor split by enzymes (e.g., in oral cavity) and as a non-ionic tenside or emulsifier in consumer products. CG is less complex in comparison to saponin with respect to chemical structure(s) and chemical identity. The aglycon of CG is a monoterpene (C10, branched alkene), while that of saponin is a triterpene (C30, substituted ring structures); the glycosidic parts are monosaccharides in the case of CG, while they are oligosaccharides in case of saponin, containing also an ionic carboxy-function. Furthermore, saponin is a mixture of triterpene glycosides.

Tween surfactants belong to a class of synthetic non-ionic surfactants which are considered non-toxic and to weakly interact with electrolytes and, for that, widely employed in those fields requiring bio-compatibly as emulsion stabilizers, both for domestic use and industrial application [16,17]. Moreover, they are often used as model surfactants in fundamental studies on emulsification and coalescence [18]. In particular, Tween 80 (polyoxyethylene sorbitan monooleate) is a synthetic non-ionic tenside or emulsifier and is composed of a central sorbitol (anhydride) structure derivatized from polyoxyethylene chains and the oleate ester group, the latter representing the lipophilic part of the molecule. It is in wide use in food, cosmetics and pharmacy technology.

Even though surfactants are key ingredients for emulsifiers, there is still a lack of deterministic approaches for the formulation of emulsions with desired characteristics, on the basis of the properties of concerned surfactant adsorption layers. This represents a challenging task, requiring systematic investigations contributing to clarify how the properties measured for single interfaces correlate for example with the stability and the drop size evolution of corresponding emulsions.

Based on these premises, a study on the interfacial properties of the above described surfactants at water-MCT oil interfaces, in relation to their ability to stabilize emulsions, is here presented. The interfacial characterization is carried out by interfacial tension and dilational viscoelasticity measurements, with the aim of assessing the adsorption properties of the different surfactants from the point of view of both thermodynamics, determining the equilibrium relation between the bulk concentration and the adsorption or adsorption isotherm and dynamics, evaluating the dilational rheology response of the adsorbed layers. These properties are then correlated with the behavior of the respective emulsions, monitored during their evolution and analyzed in terms of their structure by optical microscopy observation.

The mitigation of potential adverse impacts of emulsifiers on environment and health is pursued not only by the use of biocompatible surfactants but also by the reduction of their amount in the formulations. At low surfactants concentrations however, the process of emulsification in micro- and sub-micro-sized droplets results in a significant depletion of the surfactant concentration in the liquid phases, caused by the adsorption at the huge amount of liquid-liquid interfacial area produced. The proper correlation between interfacial properties and emulsion features needs then to account for these depletion effects. These aspects, however, have not received so far sufficient attention in the available studies, while they are explicitly accounted for and discussed in the present one.

2. Background on Emulsions and Droplet Interfacial Properties

The relationship between the interfacial properties of adsorbed layers at liquid-liquid interfaces and the process of emulsification and emulsion stability is well-recognized [19,20].

Emulsification is the production of small droplets by fragmentation of the oil phase with a consequent large increase of the surface area. For energetic reasons, it is favored by low surface tensions and fast adsorption processes. In fact, for a given energy provided to the system, smaller droplets are generated with surfactants that are more efficient to adsorb and reduce the interfacial tension. The aging of emulsions is governed by various processes influenced by the bulk and adsorption properties of surfactants, both at equilibrium and in dynamic conditions. These are creaming or sedimentation, which is the gravity-driven phase separation due to the density difference between the two liquids, Ostwald ripening and droplet coalescence [21]. Ostwald ripening consists in a partial dissolution of the dispersed liquid phase induced by the capillary pressure, which results in a net mass transfer from small to big droplets. This process mainly involves small droplets, because of their higher capillary pressure, and, for oils with low solubility in water, as it is case of long-chain alkanes [22,23], presents very long characteristic times. The emulsions reported here are obtained by a low-energy method and are therefore characterized by relatively big droplet sizes, for which Ostwald ripening has a negligible relevance for the emulsion ageing as compared to creaming and coalescence. Coalescence consists on the merging of two droplets into a single larger one, due to the complex hydrodynamic process of thinning and rupture of the liquid film between them. As a consequence of coalescence the droplet size distribution tends to be enlarged and moves towards larger sizes, eventually favoring creaming and destabilization of the emulsion. Among the most important effects relevant for the hindering of droplet coalescence, it is worth to mention the repulsive interaction between adsorbed layers, the interfacial coverage, the steric effects and the high dilational viscoelasticity of the interfacial layers [24,25]. The repulsive interactions are particularly relevant when ionic surfactants are concerned. In that case, it has been shown that a small amount of adsorbed molecules at the drop surface is sufficient to avoid coalescence [25] and improve remarkably the stability of emulsions. The interfacial coverage is the relative area of the interface occupied by the surfactant. High values of this parameter represent another stabilizing factor against coalescence due to the short range interactions of the surfactant molecules adsorbed at the two sides of the film between droplets. Since the total area of the droplet interfaces decreases with droplet coalescence, the surface coverage increases and, when high values of it are achieved through this process, the coalescence eventually stops. Thus surface coverage is implied in determining the droplet size of stable emulsions. Steric effects are especially concerned with composite surface layers, such as surfactant-nanoparticle mixtures or surfactant aggregates but also with large surfactant molecules at high adsorption coverage.

The dilational viscoelasticity, E, is the dynamic response of the interfacial tension, γ, to extensional perturbations of the surface area, A. For small amplitude harmonic perturbations, E is a frequency dependent complex quantity, defined as:

$$E = \frac{\Delta\gamma}{\Delta A/A_0}e^{i\varphi} \tag{1}$$

where $\Delta\gamma$ and ΔA are the amplitudes of the oscillating surface tension and surface area, respectively, A_0 is the reference area and φ is the phase shift between the oscillating surface tension and surface area. According to its definition, high values of the dilational viscoelasticity tend to make the liquid films between drops in emulsions more stable and, in particular, to hinder their local thinning for Marangoni effects.

From these considerations it is clear that a deep characterization of the surfactants used for stabilizing emulsions, at liquid-liquid interfaces, is of fundamental importance to understand the stabilizing mechanisms and the role of the different features of the used surfactants.

3. Materials and Methods

3.1. Materials

For all the experiments, ultrapure water—produced by a MilliRO plus MilliQ (Millipore, Burlington, MA, USA) system and characterized by resistivity larger than 18 MΩ·m—was used. Surface tension measurement provides a value of 72.5 ± 0.2 mN/m, stable for at least two hours at 20 °C meaning negligible amount of surface active impurities.

The Medium Chain Triglyceride (MCT) oil was Miglyol 812N, obtained from IOI OLEO (Hamburg, Germany) and utilized without further purification. Miglyol 812N contains about 95% of triglycerides of the fractionated C8 (50%–65%) and C10 (30%–45%) vegetable fatty acids. Lighter or heavier triglycerides compose the 5% left. The absence of surface active impurities in the used MCT oil was checked by interfacial tension measurements against water, which did not reveal any appreciable kinetics over some hours. At 20 °C the interfacial tension of 25.4 ± 0.2 mN/m was found, which is in agreement with the values available in the literature [17,26–28].

Saponin was reagent grade purchased from VWR (catalogue n. VWRV0163, Radnor, PA, USA) and utilized as supplied. This product has been used also in Reference [11], where its composition has been analyzed and discussed in detail. It is an extract from *Quillaia saponaria* containing a mixture of components with a rather broad distribution of molecular weights (from 1070 to 1700 g/mol). However, in order to analyze the adsorption properties of saponin, in comparison with the other surfactants investigated, we assume, like in many studies of saponin or mixtures of saponins [29–31], the value of 1650 g/mol as the average molecular weight. Moreover, from the results reported in Reference [11] for neutral pH, like in this work, saponin is expected to be predominantly dissociated and to behave as an ionic surfactant.

Tween 80 was purchased from Sigma Aldrich (St. Louis, MO, USA) and used as received. It is a non-ionic surfactant whose molecular weight and cmc are 1310 g/mol and 0.012 mM at 20–25 °C, respectively (as provided by the supplier).

Citronellol-β-D-glucoside, CG, was kindly supplied by the producer (4GENE GmbH, Freising, Germany), which also provided a value of the molecular weight equal to 318.39 g/mol and an estimation of the HLB and cmc, that are 11.3 and 0.01 M, respectively.

The chemical formulas of saponin and CG are given in Figure 1.

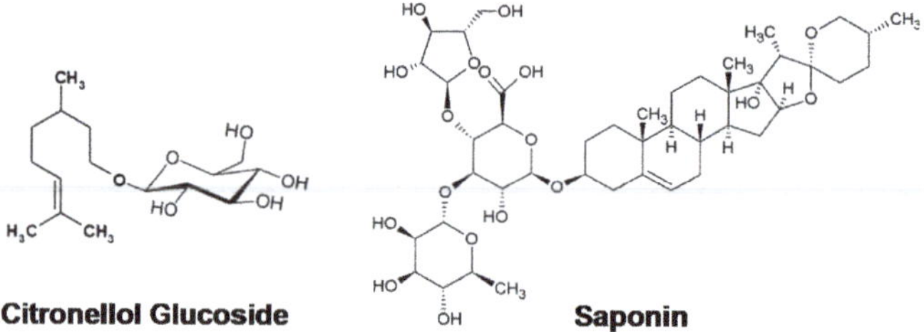

Figure 1. Natural surfactants investigated in the study.

3.2. Methods

For the reported tensiometric experiments, the interfacial tensions were measured with a Pendant Drop Tensiometer (PAT1, Sinterface, Berlin, Germany). As described in large detail in Reference [32], the technique is based on the analysis of the shape of a pendant drop attached to a capillary. The shape of a drop results, in fact, from the competition between interfacial tension and gravity and can be formally described by the Bashfort-Adams equation (B-A). In the utilized apparatus, the drop images are automatically acquired by a 1024 × 1024 pixels camera, equipped with a macro objective and interfaced to a PC. Then, the software of the apparatus fits the B-A equation to the coordinates of the drop profiles extracted from the acquired images, yielding the interfacial tension as one of the best fit parameters. Specifically, the PAT1 apparatus allows for an accurate control of the droplet area by a syringe pump, driven by a feedback loop software based on the drop imaging. Due to this feature, the tensiometer has been utilized to investigate the adsorption processes of the various surfactants by two types of measurements. The dynamic interfacial tension versus time on a freshly formed drop is measured during the ageing of the interface while keeping the drop area constant. The equilibrium interfacial tensions are obtained from these data at long time. The measurement of the dilational viscoelasticity is instead performed according to the oscillating drop method [33]. For this latter, harmonic perturbations of the droplet area are applied once the adsorption equilibrium is achieved. The amplitude of the area oscillation is typically 2% and the frequency range from 0.005 to 0.2 Hz, to warrant the quasi-equilibrium mechanical condition and the linearity of the rheological response, respectively. For each frequency, amplitude and phase of the measured interfacial tension and area signals during the oscillations are calculated by standard Fourier analysis methods and utilized to obtain the dilational viscoelasticity according to Equation (1). Full details on this measurement method and on the appropriate conditions for its applicability are reported elsewhere [34].

Notice that, in order to minimize the possible depletion of the surfactant solution caused by the adsorption itself or by the surfactant transfer into the oil phase (if any partitioning may exist), in all the reported cases the measurement refers to emerging drops of oil (from 30 to 50 mm^3, depending on the surfactant and on its concentration) inside the surfactant aqueous solution.

Emulsions were produced according to a Double Syringe (DS) method. This method, developed and mainly used for foams [11,35], is here adapted for investigating emulsions. In this technique, emulsions are obtained by repetitive exchanging of the liquids between two syringes connected by a short narrow pipe. In this way, the emulsion is formed by the turbulent flow caused by the shearing inside the pipe.

For the tests reported here, two 5 mL standard syringes connected through a rigid 15 mm long pipe and with an inner diameter of 1 mm were used. Each syringe initially contained a volume of 2.5 mL of one liquid phase, aqueous surfactant solution and oil, so that the water-oil volume ratio was V_{oil}:V_{water} = 1:1 for all the emulsions investigated.

To produce the emulsions, twenty sequential cycles were applied consisting of a pushing of the liquid/liquid mixture through the connector from one syringe to the other one. Like in the previous work on foams [11], the number of cycles was optimized to maximize the emulsification obtaining, at the same time, emulsion stability appropriate to perform comparative investigations varying the type of surfactant and the concentration. After formation, the syringes containing the emulsions were stored in a vertical position and monitored for at least two days, measuring the heights of the emulsion and of the separating phases by using a ruler, with a resolution of 0.5 mm.

The size of the droplets in the emulsions, produced by the DS method, were evaluated using the reflected light DVM6-M microscope (Leica, Hamburg, Germany) fitted with the PlanAPO FOV 3.60 objective, warranting a resolution better than 1 micron. For this purpose, small amounts of emulsion (~0.1 mL) taken directly from the syringe immediately after the emulsification and diluted 10 times with the corresponding surfactant solution, were placed between two glass slides separated by a spacer of 0.2 mm thickness to be observed. That allows to distance the emulsion droplets, easing the analysis of the droplet size distribution on the captured microscope images. The droplet radii are obtained by the analysis of these images by purposely available routines of the proprietary microscope software (Leica LAS X, v.3.7.1).

4. Results

4.1. Interfacial Properties

The interfacial tension of surfactant solutions against the MCT oil has been measured at various concentrations in water according to the method described in the previous section. In particular, the measurement of interfacial tension begins just after the formation of a "fresh" droplet interface—that is, an interface with an initially negligible amount of an adsorbed surfactant—and continues during the advancement of the surfactant adsorption process until the achievement of the equilibrium. This kind of measurements, apart from providing information on the adsorption kinetics such as, for example, the adsorption characteristic time, allows determining the equilibrium interfacial tension versus the bulk concentration, which characterizes the thermodynamic adsorption properties of an examined surfactant system. For all the cases investigated, after about two hours from the creation of the interface, the adsorption equilibrium may be assumed to be achieved, as the interfacial tension presents a variation lower than 1 mN/m in one hour. Figure 2 shows the interfacial tension versus time obtained for saponin at various bulk concentrations in water. The corresponding equilibrium interfacial tensions versus concentration are reported in Figure 3.

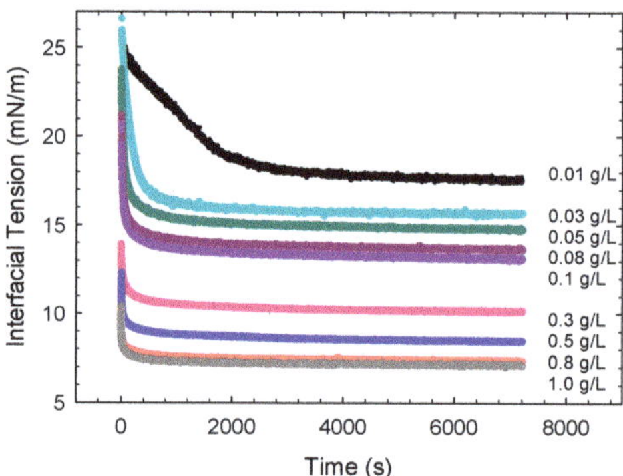

Figure 2. Interfacial tension versus time during the adsorption of saponin at water/Medium Chain Triglyceride (MCT) interface, for different concentrations of saponin in water: from 0.01 to 1 g/L.

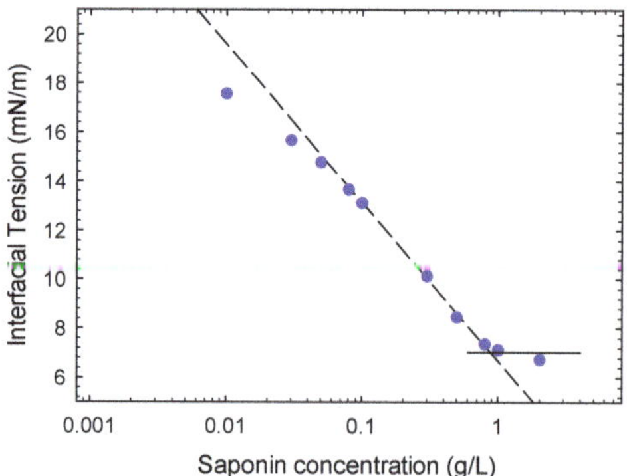

Figure 3. Equilibrium interfacial tension versus saponin concentration (symbols) and linear fitting of the experimental data close to the cmc providing through Equation (1) the maximum adsorption Γ_{max} = 2.6 μM/m^2.

From Figure 3 the achievement of the critical micellar concentration around 0.8 g/L (4.8×10^{-4} M, assuming 1650 g/mol as molecular weight) is evident. This value is in agreement with that found for the same type of saponin in Reference [11], from surface tension measurements at water-air interface.

The slope of the best fit line at concentrations (C) approaching the cmc, provides the maximum value of the adsorption at the water-MCT interface, through the Gibbs adsorption isotherm:

$$\Gamma = -\frac{1}{RT}\frac{\partial \gamma}{\partial \ln C} \tag{2}$$

where T is the temperature and R the gas constant, which provides a value for the maximum adsorption of 2.6 ± 0.1 µmol/m^2. This adsorption value is much lower than that found in Reference [11], for the water-air interfaces, that is 5.1 µmol/m^2. This can be due to the lower affinity of saponin molecules with the water-oil interface or to a much higher occupational area of the adsorbed saponin molecules caused by a different placement of the saponin molecule at the interface between oil and water.

Figure 4 reports the equilibrium interfacial tension versus the molar concentration of surfactants in water for the three systems investigated in this work. The continuous curves reported in the figure for saponin and Tween 80 are the best fit theoretical γ-c isotherms from the reorientation adsorption model [36,37]. According to this model, the surfactant molecules may adsorb assuming different orientation with respect to the interface, corresponding to different occupational area. The average area per mole, ω, may be expressed through two parameters, ω_1 and ω_2, corresponding to the orientations which make maximum and minimum the molecular area, respectively. That is:

$$\omega = \frac{\omega_1\left(\frac{\omega_1}{\omega_2}\right)^{\alpha} e^{-\frac{\Pi(\omega_1-\omega_2)}{RT}} + \omega_2}{1+\left(\frac{\omega_1}{\omega_2}\right)^{\alpha} e^{-\frac{\Pi(\omega_1-\omega_2)}{RT}}} \tag{3}$$

where Π is the surface pressure, $\Pi = \gamma_0 - \gamma$, which is strictly related to the total adsorption Γ, through:

$$\Pi = -\frac{RT}{\omega}\ln(1-\Gamma\omega) \tag{4}$$

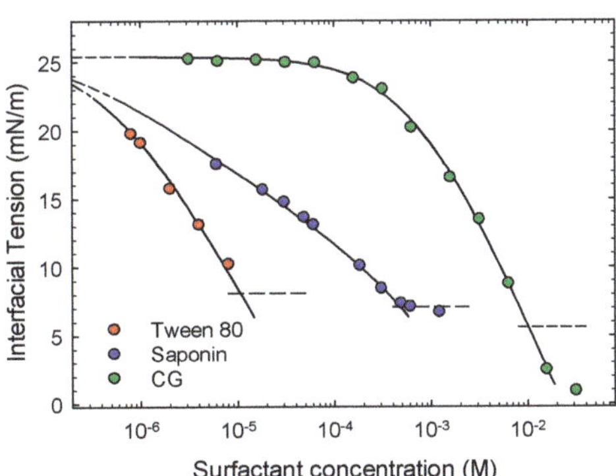

Figure 4. Equilibrium interfacial tension versus molar surfactant concentration in water (for saponin calculated assuming 1650 g/mol as molar weight) and respective best fit isotherm curves (Equations (3)–(5)).

This model provides, moreover, the following equilibrium relation between the interfacial tension and the bulk surfactant concentration, *c*:

$$bc = \frac{1 - e^{-\frac{\Pi\omega}{RT}}}{\left(\frac{\omega_1}{\omega_2}\right)^{\alpha} e^{-\frac{\Pi\omega_1}{RT}} + e^{-\frac{\Pi\omega_2}{RT}}} \tag{5}$$

where b and α are two parameters related to the surface activity of the two orientation states which, together with ω_1 and ω_2, completely describe the equilibrium adsorption properties of the system.

Notice that this model is essentially an extension of the Langmuir model where only one molar surface area is considered in the whole bulk concentration range investigated. The value of the molar area, ω, in this case also defines the saturation adsorption $\Gamma_{sat} = 1/\omega$.

As shown in Figure 4, Equation (5) fits well the equilibrium interfacial tension data obtained for saponin and Tween 80, while the CG behavior is well-described by a Langmuir isotherm. In fact, for all cases the weighted average standard deviation between the calculated and measured values is lower than 0.05, which is considered more than satisfactory. The best fit values are reported in Table 1. The surface activities of the different compounds are expressed through the values of b which change by two orders of magnitude passing from CG to Tween 80. Similar values are found for ω_2 which is related to the saturation adsorption, that is $\Gamma_{sat} = 1/\omega_2$. The tendency of the adsorbed molecules to re-orient found for saponin and Tween 80 is coherent with the more complex structure of these molecules with respect to CG, while the values of ω_1 reflect their difference in molecular weight and size.

Table 1. Adsorption isotherm parameters corresponding to the best fit curves reported in Figure 4 and critical micellar concentration, cmc.

Surfactant	ω_1 (m²/mol)	ω_2 (m²/mol)	b (m³/mol)	α	cmc (M)
Tween 80	5.60×10^5	3.80×10^5	8.69×10^2	0.6	1.2×10^{-5} [a]
Saponin	1.20×10^6	3.60×10^5	15.2	4.4	4.8×10^{-4} [b]
CG	–	3.4×10^5	1.43	–	1.0×10^{-2} [a]

[a] from producer; [b] from interfacial tension measurements.

Notice that, in the case of saponin, the peculiarity of reorienting at the interface explains the different values of maximum adsorption that such surfactant may assume when different interfaces are involved like in the above- reported case of water-air interface.

4.2. Partitioning

When surfactants are investigated at a liquid-liquid interface an important aspect to be taken into account is the possible partitioning between the two liquid phases. In our cases, the transfer of the surfactant across the water-oil interfaces and the possible impoverishment of one phase with respect to the other one, may play an important role in the dynamics of the adsorption as well as in the features of the obtained emulsions. This phenomenon has been widely investigated [38] and methods to evaluate the partition coefficient, that is the ratio between the equilibrium surfactant concentrations in oil and water phases, have been developed [39,40].

The partitioning can be especially relevant for Tween 80 and CG that are non-ionic surfactants, while, we can expect the transfer into the oil phase to be negligible for saponin. In fact, according to the results reported in Reference [11], for neutral pH, like in the present case, saponin is predominantly dissociated and behaves therefore as an ionic surfactant.

An effective way to assess the possible surfactant transfer across the interface is to compare the behavior of the dynamic interfacial tension obtained with a drop of oil immersed in the aqueous surfactant solution with that obtained with the opposite configuration, that is a pendant drop of solution in an initially surfactant-free oil phase. In the second case, due to the much higher volume of the oil phase, the transfer provides the typical minimum trend of the interfacial tension [38].

For saponin and Tween 80 this test indicated negligible solubility of surfactant in MCT, resulting in an almost coincident dynamic interfacial tension variations found in the two different configurations. On the contrary, the dynamic interfacial tension of a drop of GC aqueous solution in MCT shows an evident minimum, as illustrated in Figure 5, indicating the transfer of CG into the oil phase.

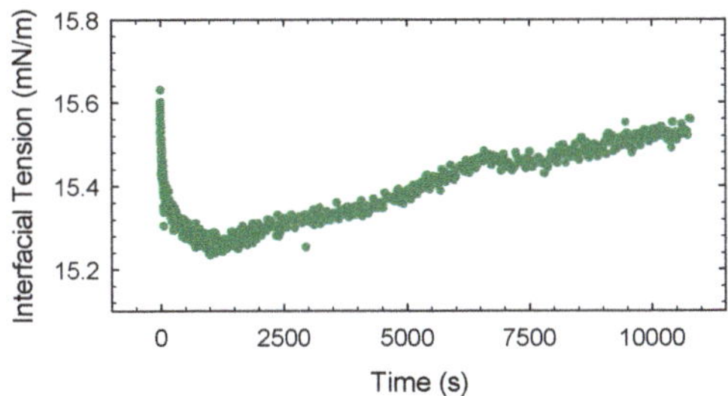

Figure 5. Dynamic interfacial tension versus time obtained with a pendant drop of aqueous solution of citronellol glucoside (CG at $c = 7.6 \times 10^{-4}$ M inside an initially pure MCT phase.

To evaluate the partition coefficient $K_p = c_w/c_{MCT}$, we have adopted the method described in Reference [40]. Accordingly, CG solution in MCT are prepared and the equilibrium interfacial tensions measured using a pendant drop of initially pure water immersed in such solutions. The comparison of these data with the adsorption isotherm obtained in the opposite configuration, that is, equilibrium interfacial tension versus concentration in water, provides for each value of the interfacial tension (see Figure 6) the two concentrations in partition equilibrium. Using the data in Figure 6, the partition coefficient of CG was estimated as $K_p = 0.42 \pm 0.05$, where the value and the uncertainty are obtained by the average of the K_p's determined by the couples of concentrations at the various interfacial tensions.

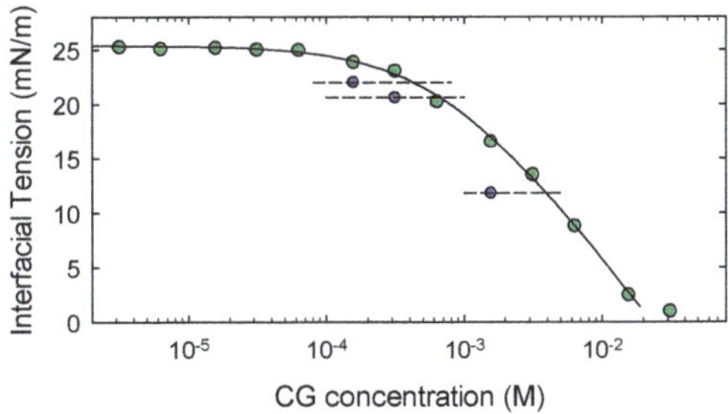

Figure 6. Equilibrium interfacial tension versus molar CG concentration in MCT (blue symbols) and in water (green symbols).

4.3. Dilational Rheology

The dilational viscoelasticity versus frequency of the above-investigated systems has been measured according to the oscillating drop method, as described in Section 3.2, in a range of concentration in water below the cmc where the concentration dependence of the equilibrium interfacial tension is appreciable. Figure 7 reports the results obtained for saponin and Tween 80. For CG, lower values of the dilational viscoelasticity modulus were found, around 4 mN/m and below, without a significant trend with frequency and, for this reason, they are not reported in Figure 7.

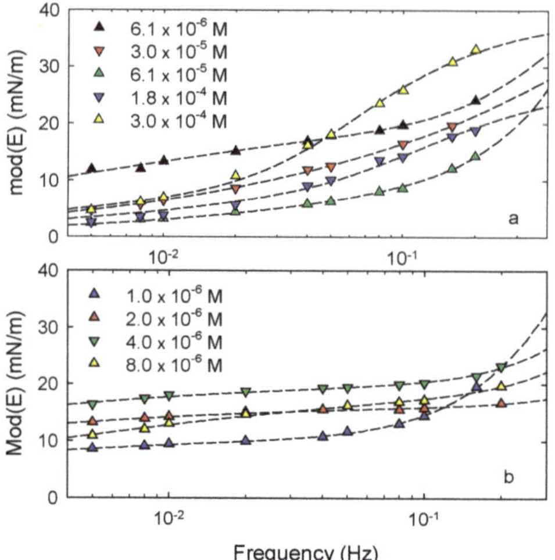

Figure 7. Modulus of the dilational viscoelasticity measured by the Oscillating Drop method for saponin (**a**) and for Tween 80 (**b**). Dashed curves obtained fitting a theoretical expression of mod(E) like in Reference [25] from the model presented in Reference [41].

For saponin and Tween 80 the trend of the dilational viscoelasticity versus frequency is that typical of soluble surfactants at liquid interfaces. The data are, in fact, well fitted by theoretical curves obtained assuming a general model taking into account the bulk diffusion and a re-arrangement process in the adsorbed layer [41]. For these surfactants, on the basis of the results reported in the previous section, this interfacial process is expected to be related to the variation of the average molecular orientation in the adsorbed layer.

In many studies on interfacial rheology of surfactants at liquid interfaces [25,42,43], using this model to interpret the experimental data has been effective to understand the mechanisms governing the adsorption process and to evaluate the associated kinetic parameters. For the systems here investigated, however, the frequency range used is probably not wide enough to provide an accurate description of their dynamic behavior and consequently quantitative information on the processes occurring in the adsorbed layers. Despite this limitation, these results, even if concerned with low frequency perturbations, give important information on the rheological properties of the adsorbed layers and evidence the differences between the investigated surfactants. One difference concerns the values of the dilational viscoelasticity for saponin and Tween 80 which are similar to those of common low-weight surfactants at liquid-liquid interfaces [25,44], while, in comparison, the rheological response of CG is negligible. Moreover, for saponin the increasing rate of the viscoelastic modulus with frequency is higher than that observed for Tween 80. This implies that in the case of saponin much higher values can probably be reached at high frequency.

It is interesting to notice that, as found in previous studies [11,29,30], for saponin at water-air interface the viscoelasticity modulus reaches very high values, compared with common surfactants, already for frequencies below 0.2 Hz. This peculiarity, not observed here for water-MCT interfaces, was associated with the good ability of this natural surfactant to stabilize foams. This means that the better ability of saponin as emulsion stabilizer, compared to the other surfactants, reported in the next section, should be attributed to the not directly observed high frequency elasticity or to other molecular features of saponin such as the higher molar area and the ionic character.

4.4. Emulsions

In order to compare the effectiveness of the different surfactants as emulsion stabilizers and to evidence the effects of their concentrations, all emulsions were produced following exactly the same procedure, that is the double syringe method as described in Section 3.2. These water-in-oil emulsions were then monitored for 2 days by measuring the relative emulsion volume, V_{em}/V_{total}. The results obtained for all surfactants and concentrations are summarized in Figure 8, where the concentrations correspond to those of the aqueous solutions before emulsification.

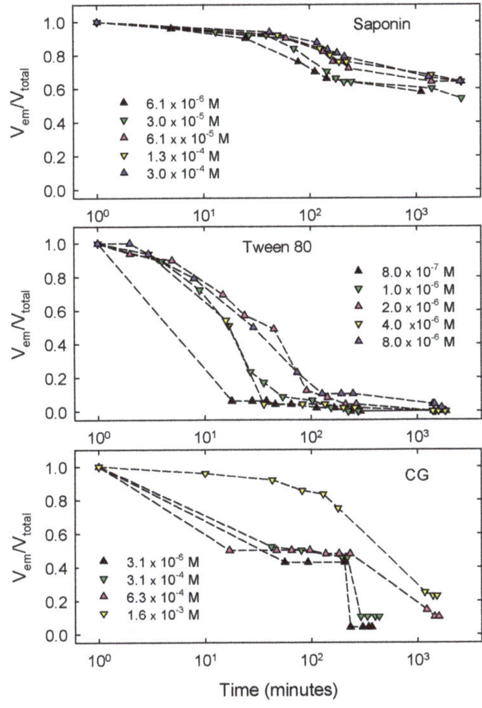

Figure 8. Evolution of the relative oil-in-water emulsion volume for different concentrations of the aqueous surfactant solution before emulsification. Results obtained by the Double Syringe technique using two 5 mL syringes, 20 manual pushing cycles and volume ratio $V_{oil}:V_{water}$ = 1:1.

In all the cases reported a total emulsification of the samples was observed, that is, V_{em}/V_{total} = 1 at the initial time of the emulsion monitoring. Afterwards, emulsions evolve with different trends depending on the surfactants used. As shown in Figure 8, saponin-stabilized emulsions present a slow reduction of the volume during the first two days from the formation, with a slight dependence on the concentration. For the lowest concentration investigated, that is, 6.1×10^{-6} M, a small volume of MCT phase starts to separate after about one day from the formation, reducing the emulsion volume. For the higher concentrations, instead, the MCT phase remains totally emulsified for several days (more than 15), with the relative emulsion volume which after the reduction of the first two days maintains constant at $V_{em}/V_{total} \approx 0.6$. The fact that no separated oil phase appears means that coalescence of MCT droplets is hindered by saponin adsorbed layers and the initial reduction of the emulsion volume is mainly due to an increased droplet compactness related to the aqueous phase drainage.

Tween 80-stabilized emulsions are appreciably less stable than those with saponin. This is not only evidenced by the faster reduction of the emulsion volume (see Figure 8) but also by the fact that, for all the concentrations studied, a separated oil phase appears and grows on the top of the emulsions already after 10 min from the formation. This means that Tween 80 under these conditions is not an efficient stabilizer against coalescence. The emulsions stabilized by CG present a more complex behavior where the time evolution of the emulsion volume significantly depends on the concentration. For this surfactant, like for Tween 80, the reduction of the emulsion volume is associated with the appearance of a separated MCT phase meaning that, also in this case, the droplet coalescence has an important role in the destabilization. Moreover, as evident from Figure 8, only for the emulsions obtained with the highest concentration, that is, 1.6×10^{-3} M, the destabilization is rather slow, during the monitoring time. For the other concentrations investigated, the emulsion volume is reduced by half in about ten minutes maintaining rather constant for several hours before a drastic fall after about one day. In correspondence with this abrupt volume reduction the appearance of a new turbid thin phase is visually observed, just above the emulsion, below the separated MCT phase, with a small volume not easy to evaluate. This particular behavior of the CG-stabilized emulsions as well as the appearance of this new phase during their evolution, is possibly related to the solubility of this surfactant in both water and oil phases. On the basis of the value of the partition coefficient indicated in the previous section, one can suppose that the new phase observed could be a water-in-oil emulsion coexisting with the principal oil-in-water ones. The analysis of these latter ones by optical microscopy gives information that support this hypothesis. In fact, from the image reported in Figure 9 of a typical emulsion obtained with CG, it is evident that a multiple emulsion is produced, where the droplets are, in turn, a disperse water-in-oil phase.

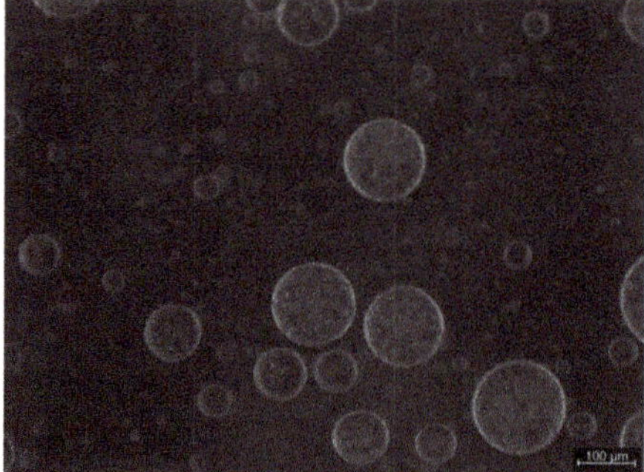

Figure 9. Typical microscope image of an emulsion stabilized by CG at 6.2×10^{-4} M, where the composite feature of the disperse phase is evident.

The emulsions were also analyzed by optical microscopy at the beginning of their evolution, when they are completely emulsified. Figure 10 shows some examples of size distribution obtained by this analysis. In general, the droplets resulted to have a radius distributed in a range from 2 to 15 micron with the average value between 4 and 9 microns, depending on the surfactant type and concentration. Considering the simple emulsification method, the width of these distributions is therefore rather narrow. Moreover, the droplet sizes observed confirm that Ostwald ripening does not play a relevant role over the timescale investigated, so that the change in V_{em} can be attributed to the combined effect of creaming, drainage and droplet coalescence.

In order to summarize the properties of the emulsions obtained with the different surfactants, we report in Figure 11 the relative emulsion volume, V_{em}/V_{total}, reached after one hour from the formation as a function of the surfactant concentration in water before emulsification (panel a) and the corresponding average droplet radius obtained from the size distributions by microscopy analysis (panel b). Apart from the highest concentration of CG, these data evidence a slight decrease of the average droplet size with the concentration and a general tendency to improve the stability with the decrease of the droplet size.

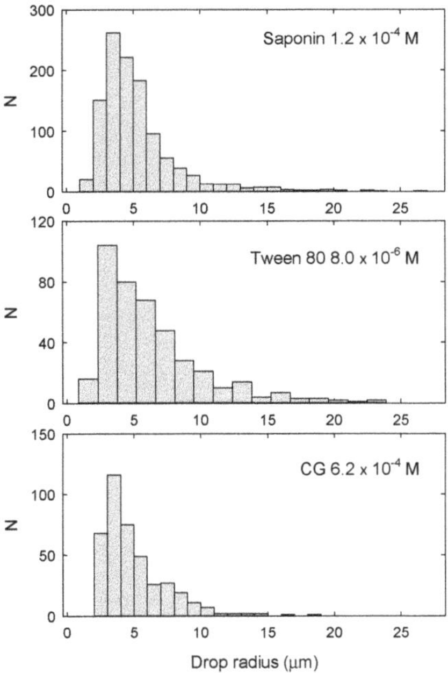

Figure 10. Examples of drop radius distributions from the microscope images acquired just after emulsification for the different surfactants.

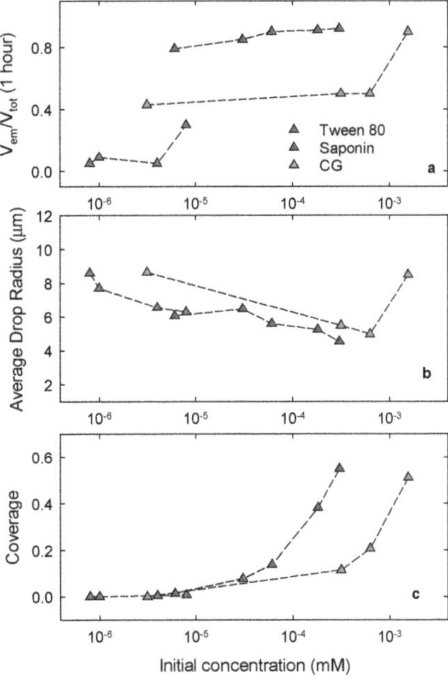

Figure 11. Relative emulsion volume at one hour from the emulsification (**a**), average emulsion drop radius obtained by microscopy images just after the emulsification (**b**) and surface coverage of oil drops in emulsions by the adsorbed surfactant (**c**), against the surfactant concentration in water before emulsification.

5. Discussion

The surfactants here investigated present different features as emulsion stabilizers that can be related to their adsorption properties at the surface of the dispersed MCT droplet after emulsification. During the emulsification process, however, the water-oil interface undergoes a huge area increase resulting in a significant depletion of the aqueous matrix phase induced by the adsorption. This effect is recognized to be very important to investigate the correlation between the interfacial properties and the stability of the corresponding disperse system. In fact, only accounting for this effect it is possible to evaluate the real surfactant adsorption onto the oil droplet surface in emulsions which is related, through the adsorption isotherm, to the real bulk concentration in water.

The calculation of the above depletion can be done by coupling the surfactant mass balance with the equilibrium adsorption isotherm. In our specific case, considering also the surfactant partitioning ($k = c_{oil}/c_{water}$) between the two liquid phases, the mass balance reads:

$$A\Gamma + cV_w\left(1 + k\frac{V_{oil}}{V_w}\right) = c_0 V_w \tag{6}$$

where c is the concentration achieved after emulsification of oil and water volumes, V_{oil} and V_w, c_0 the initial concentration of the aqueous phase and A the total interfacial area that, assuming the complete emulsification of the oil volume in drops with average radius R, can be estimated as $A = 3V_{oil}/R$. Thus, coupling this balance of mass with the Γ-c equilibrium relation, provided by the adsorption isotherm, it is possible to obtain the value of Γ and of the coverage $\omega\Gamma$. The steps of this calculation are reported in more detail in Appendix A. Figure 11c shows the coverage for the three surfactants as a function of the initial concentration in water, calculated using the isotherm parameters found by the analysis of the equilibrium interfacial properties (Table 1) and the average drop radius as obtained by the droplet size distributions (Figure 11b). These data are also reported in Table A1 of the Appendix A.

From these results it is possible to evidence the main differences in the features of the surfactant investigated. It is important to underline again that, in order to obtain reliable comparative results, emulsions have been produced following exactly the same procedure. Emulsions stabilized by saponin are definitely more stable with respect to others and this is expected to be mainly due to the ionic character of this surfactant. In fact, it is well known that ionic surfactants are effective to hinder coalescence in emulsions due to the electrostatic repulsion between the surfactant layers at the two sides of the liquid film between droplets. Nevertheless, the results reported here suggest further reasons that can increase the stabilizing effects, related to the possibility for the large size surfactant molecules to reorient, which provides a large coverage already at low concentration. The evident low ability of Tween 80 as an emulsion stabilizer is surely due to the very low adsorption that can be achieved with this system at the surface of the MCT droplets. In fact, in this case the coverage is very low and the surfactant is non-ionic, hence it is not sufficient to hinder coalescence. Coalescence for this system can possibly occur also during the emulsification process and this could be the reason for the larger size of the drop at the beginning of the emulsion evolution with respect to the other surfactants. The behavior of CG-stabilized emulsions is surely the most complex. The principal reasons for the general unstable behavior of these emulsions are probably the low coverage that, together with the non-ionic character of the surfactant and negligible dilational elasticity of the adsorbed layer, makes this system not appropriate to hinder the droplet coalescence. Moreover, for this system an important role is played by the transfer of surfactants into the MCT phase, that can induce the formation of double emulsions or inverse water-in-oil emulsions which may be a further reason for the observed emulsion instability. However, increasing the CG concentration, such ability to form double emulsions may lead to the formation of more complex disperse systems characterized by a better stability. For example, the more stable emulsion obtained with the highest CG concentration (1.6×10^{-3} M) may be a dispersion of large aggregates of water droplets in MCT. Deeper investigations on this surfactant are surely required to better understand its behavior and its potentiality as a natural emulsion stabilizer.

Using the average drop radius in the estimation of the available interfacial area in the emulsion, for obtaining the data in Figure 11, is clearly a first order approximation. A more accurate estimation of A could be obtained accounting for the droplet size distributions, where available, according to the calculations given in the Appendix A. However, for rather narrow distributions, as in our case, the general conclusions of the above discussion would not change.

As a last point, it can be said that for the systems investigated in this work, the rheological properties of the adsorbed layers at the water-MCT interface seems to have only a marginal relationship with the stabilization of the corresponding emulsions. The negligible values of the dilational viscoelasticity for the CG system is associated with the unstable character of the corresponding emulsions. However, both Tween 80 and saponin present higher values but similar between them, even if their ability as emulsion stabilizers is very different. For these two surfactants a major role is played by the differences in the adsorption properties, ionic character and molar areas. In addition, the slight difference in the viscoelasticity trends, observed at low frequency, let us expect important differences in the rheological response in a higher frequency range.

6. Conclusions

We have investigated the properties of adsorption layers of different water-soluble biocompatible surfactants, namely Tween 80, saponin and citronellol glucoside at the interface between water and a Medium Chain Triglyceride oil, correlating them to the aging of corresponding emulsions.

The measured equilibrium interfacial tensions versus concentration for Tween 80 and saponin are well described within the framework of the reorientation model [37]. For CG the results are sufficiently well-described by the Langmuir model, for which the molar area of the adsorbed molecules is only one. This latter surfactant revealed also an appreciable solubility in MCT oil. The value of the partition coefficient has been then estimated and used in the interpretation of the behavior of the corresponding emulsions.

The results of the comparative emulsion study, performed for surfactant concentrations below the cmc and an observation time of two days, evidence different characteristics of the surfactants as emulsion stabilizers. According to these results, Tween 80 does not present, under the studied conditions, a good stability against droplet coalescence, so that the liquid phases are fully separated within about two hours. A good stability against coalescence is instead obtained with saponin. In fact, the corresponding emulsions, after an initial creaming, remain stable for at least 10 days, that is a time much longer than the observation time. CG emulsions present a relatively better stability only at the highest concentrations. The microscopy studies revealed, however, a much more complicated behavior of the emulsions with this surfactant since the formation of different phases and multiple emulsions was observed.

For the interpretation of these results, particular attention has been paid on the effect of the depletion of the surfactant concentration in the matrix phase of the emulsion caused by the increased available area for adsorption, induced by emulsification. Using the data of droplet sizes in emulsions, obtained by microscopy analysis and the eventual partitioning in the oil phase, such area has been evaluated and resulted to be more than significant. In fact, the observed behavior for the emulsions has been discussed and rationalized on the basis of the surfactant coverages at the droplet interface, corresponding to the real values of the concentrations resulting after emulsification. As expected, the emulsions showing a high stability against coalescence are those characterized by the larger coverages.

Concluding, the results of this work show that investigating both the adsorption properties of surfactants at liquid-liquid interfaces, by complementary techniques and the behavior of the corresponding emulsions, may be effective to determine in a comparative way the efficiency of new natural compounds. This may contribute to the development of innovative formulations of bio-compatible emulsions.

Author Contributions: Conceptualization, E.S., L.L., E.J., M.K. and F.R.; methodology, E.S., L.L., E.J., M.K. and F.R.; investigation, K.D., E.S., E.J. and M.K.; data curation, E.S. and K.D.; formal analysis F.R, M.K., T.F. and L.L.;

writing—original draft preparation, F.R. and L.L.; writing—review and editing, F.R. and L.L.; visualization F.R., E.S., K.D.; supervision F.R. and M.K. All authors have read and agreed to the published version of the manuscript.

Funding: This research was funded by the CNR-PAN Bilateral project "Biocompatible foams and emulsions stabilized by natural surfactants and particles for bio-medical applications" and by the European Space Agency within the MAP projects "Soft Matter Dynamics" and "Emulsion Dynamics and Droplet Interfaces—EDDI".

Conflicts of Interest: T.F. is member of 4GENE, the manufacturer of Citronellol glucoside. He participated to the interpretation and discussion of the data. The other authors declare no conflicts of interest.

Appendix A. Calculations of the Surfactant Concentration Depletion Due to Adsorption on Emulsion Drops

After the formation of a new interface with area A, in a surfactant solution with the initial concentration c_0, an adsorption process starts and the system achieves a new equilibrium state, characterized by an adsorption Γ and a bulk concentration c, satisfying the set of equations provided by the mass balance and by the equilibrium adsorption isotherm:

$$\begin{cases} A\Gamma + cV = c_0 V \\ c = c(\Gamma) \end{cases} \tag{A1}$$

where V is the volume of the surfactant solution. Within the Langmuir model, the adsorption isotherm can be made explicit and the system becomes:

$$\begin{cases} A\Gamma + cV = c_0 V \\ c = \dfrac{\omega\Gamma}{b(1-\omega\Gamma)} \end{cases} \tag{A2}$$

From which, by substitution in the mass balance, the following second-degree equation in Γ can be obtained:

$$Ab\omega\,\Gamma^2 - (bA + \omega V + bc_0\omega V)\Gamma + bc_0 V = 0 \tag{A3}$$

and solved with the condition, $\omega\Gamma < 1$, which is satisfied by the solution with the minus sign.

For more complicated isotherms (Frumkin, 2-states), where an explicit relationship between c and Γ is not provided, numerical schemes must instead be adopted to solve the above set of Equation (A1).

Concerning the complete emulsification of an oil volume V_{oil} in a volume V of water, as a first approximation we can assume the total area of the droplets given by:

$$A = \frac{3}{R} V_{\text{oil}} \tag{A4}$$

where R is the average radius of the oil droplets. Introducing this area, together with the volume fraction $\Phi = V_{\text{oil}}/V$, the mass balance reads:

$$c + \frac{3}{R}\Phi\Gamma = c_0 \tag{A5}$$

In the case the adsorption process co-exists with surfactant partitioning between the volume of water V and the volume of oil V_{oil}, the mass balance is:

$$A\Gamma + cV + kcV_{\text{oil}} = c_0 V \tag{A6}$$

where $k = c_{\text{oil}}/c_{\text{water}}$ is the partition coefficient and c_0 and c refer to the concentrations in water. After defining $\alpha = (1 + kV_{\text{oil}}/V)$, this equation can be recast as:

$$A\Gamma + \alpha cV = c_0 V \tag{A7}$$

Which allows the above equations for the depletion in the absence of partitioning to be easily adapted to account for partitioning.

The values of Γ, calculated using the described approach, allow estimating the realistic surfactant coverage, $\omega\Gamma$, for the droplets in the emulsion, which for the systems investigated here are reported in Table A1.

More accurate values for A can be obtained by considering the drop size distribution. If the number of drops N_i with the radius R_i, normalized on the total number of drops N_T, is distributed accordingly to $N_i/N_T = f_i(R_i)$, the total drop volume is given by:

$$V_{\text{oil}} = \sum_i N_i \frac{4}{3}\pi R_i^3 = N_T \sum_i f_i \frac{4}{3}\pi R_i^3 \tag{A8}$$

Thus

$$N_T = \frac{V_{\text{oil}}}{\sum_i f_i \frac{4}{3}\pi R_i^3} \tag{A9}$$

The total droplet area is thus given by:

$$A = \sum_i N_i 4\pi R_i^2 = N_T \sum_i f_i 4\pi R_i^2 = 3V_{\text{oil}} \frac{\sum_i f_i R_i^2}{\sum_i f_i R_i^3} \tag{A10}$$

So that, the equation equivalent to the mass balance A5 reads:

$$c + 3\Phi \frac{\sum_i f_i R_i^2}{\sum_i f_i R_i^3} \Gamma = c_0 \tag{A11}$$

For the emulsions studied here, the resulting size distributions are however always quite narrow, so that the use of Equation (A11), does not modify substantially the values of the coverage reported in Table A1.

Table A1. Depleted concentrations of the continuous phase (c) and the corresponding surface coverages of the droplets after emulsification. c_0 is the concentration of the aqueous solution before emulsification.

c_0 (M)	Average Drop Radius (μm)	c (M)	Surface Coverage
	Tween 80		
8.0×10^{-7}	8.6	3.6×10^{-10}	1.2×10^{-3}
1.0×10^{-6}	7.7	4.1×10^{-10}	1.4×10^{-3}
4.0×10^{-6}	6.6	1.4×10^{-9}	4.7×10^{-3}
8.0×10^{-6}	6.3	2.7×10^{-9}	9.0×10^{-3}
	Saponin		
6.1×10^{-6}	6.1	4.9×10^{-9}	0.015
3.0×10^{-5}	6.5	2.7×10^{-8}	0.078
6.1×10^{-5}	5.6	5.3×10^{-8}	0.138
1.8×10^{-4}	5.3	2.0×10^{-7}	0.381
3.0×10^{-4}	4.5	4.0×10^{-7}	0.550
	Citronellol Glucoside		
3.1×10^{-6}	8.7	1.1×10^{-6}	1.6×10^{-3}
3.1×10^{-4}	5.5	9.1×10^{-5}	0.115
6.3×10^{-4}	5.0	1.8×10^{-4}	0.207
1.6×10^{-3}	8.5	7.3×10^{-4}	0.511

References

1. Driscoll, D.F.; Nehne, J.; Peterss, H.; Franke, R.; Bistrian, B.R.; Niemann, W. The influence of medium-chain triglycerides on the stability of all-in-one formulations. *Int. J. Pharm.* **2002**, *240*, 1–10. [CrossRef]
2. Chatzidaki, M.D.; Mateos-Diaz, E.; Leal-Calderon, F.; Xenakis, A.; Carrière, F. Water-in-oil microemulsions versus emulsions as carriers of hydroxytyrosol: An in vitro gastrointestinal lipolysis study using the pHstat technique. *Food Funct.* **2016**, *7*, 2258–2269. [CrossRef] [PubMed]
3. Ragelle, H.; Crauste-Manciet, S.; Seguin, J. Nanoemulsion formulation of fisetin improves bioavailability and antitumour activity in mice. *Int. J. Pharm.* **2012**, *427*, 452–459. [CrossRef] [PubMed]
4. Terjung, N.; Löffler, M.; Gibis, M.; Hinrichs, J.; Weiss, J. Influence of droplet size on the efficacy of oil-in-water emulsions loaded with phenolic antimicrobials. *Food Funct.* **2012**, *3*, 290–301. [CrossRef]
5. Oleszek, W.; Hamed, A. Saponin-Based Surfactants. In *Surfactants from Renewable Resources*; Kjellin, M., Johansson, I., Eds.; John Wiley & Sons, Ltd.: Hoboken, NJ, USA, 2010; pp. 239–249. [CrossRef]
6. Güçlü-Üstündağ, Ö.; Mazza, G. Saponins: Properties, applications and processing. *Crit. Rev. Food Sci. Nutr.* **2007**, *47*, 231–258. [CrossRef]
7. Böttcher, S.; Drusch, S. Saponins—Self-assembly and behaviour at aqueous interfaces. *Adv. Colloid Interface Sci.* **2017**, *243*, 105–113. [CrossRef]
8. Jurado Gonzalez, P.; Sörensen, P.M. Characterization of saponin foam from Saponaria officinalis for food applications. *Food Hydrocoll.* **2020**, *101*, 105541. [CrossRef]
9. Giménez-Ribes, G.; Habibi, M.; Sagis, L.M.C. Interfacial rheology and relaxation behavior of adsorption layers of the triterpenoid saponin Escin. *J. Colloid Interface Sci.* **2020**, *563*, 281–290. [CrossRef]
10. Ulaganathan, V.; Del Castillo, L.; Webber, J.L.; Ho, T.T.M.; Ferri, J.K.; Krasowska, M.; Beattie, D.A. The influence of pH on the interfacial behaviour of Quillaja bark saponin at the air-solution interface. *Colloids Surf. B* **2019**, *176*, 412–419. [CrossRef]
11. Santini, E.; Jarek, E.; Ravera, F.; Liggieri, L.; Warszynski, P.; Krzan, M. Surface properties and foamability of saponin and saponin-chitosan systems. *Colloids Surf. B* **2019**, *181*, 198–206. [CrossRef]
12. Zhu, L.; Xu, Q.; Liu, X.; Xu, Y.; Yang, L.; Wang, S.; Li, J.; Ma, T.; Liu, H. Oil-water interfacial behavior of soy β-conglycinin–soya saponin mixtures and their effect on emulsion stability. *Food Hydrocoll.* **2020**, *101*, 105531. [CrossRef]
13. Wei, Y.; Tong, Z.; Dai, L.; Ma, P.; Zhang, M.; Liu, J.; Mao, L.; Yuan, F.; Gao, Y. Novel colloidal particles and natural small molecular surfactants co-stabilized Pickering emulsions with hierarchical interfacial structure: Enhanced stability and controllable lipolysis. *J. Colloid Interface Sci.* **2020**, *563*, 291–307. [CrossRef] [PubMed]
14. Huang, T.; Tu, Z.; Zou, Z.; Shangguan, X.; Wang, H.; Bansal, N. Glycosylated fish gelatin emulsion: Rheological, tribological properties and its application as model coffee creamers. *Food Hydrocoll.* **2020**, *102*, 105552. [CrossRef]
15. Riquelme, N.; Zúñiga, R.N.; Arancibia, C. Physical stability of nanoemulsions with emulsifier mixtures: Replacement of Tween 80 with quillaja saponin. *LWT Food Sci. Technol.* **2019**, *111*, 760–766. [CrossRef]
16. Bak, A.; Podgórska, W. Interfacial and surface tensions of toluene/water and air/water systems with nonionic surfactants Tween 20 and Tween 80. *Colloids Surf. A* **2016**, *504*, 414–425. [CrossRef]
17. Zhu, Z.; Wen, Y.; Yi, J.; Cao, Y.; Liu, F.; McClements, D.J. Comparison of natural and synthetic surfactants at forming and stabilizing nanoemulsions: Tea saponin, Quillaja saponin, and Tween 80. *J. Colloid Interface Sci.* **2019**, *536*, 80–87. [CrossRef]
18. Kumar, N.; Mandal, A. Thermodynamic and physicochemical properties evaluation for formation and characterization of oil-in-water nanoemulsion. *J. Mol. Liq.* **2018**, *266*, 147–159. [CrossRef]
19. Leal-Calderon, F.; Schmitt, V.; Bibette, J. *Emulsion Science, Basic Principles*; Springer: Berlin/Heidelberg, Germany, 2007; pp. 143–172. ISBN 978-0-387-39682-8.
20. Georgieva, D.; Schmitt, V.; Leal-Calderon, F.; Langevin, D. On the possible role of surface elasticity in emulsion stability. *Langmuir* **2009**, *25*, 5565–5573. [CrossRef]
21. Schmitt, V.; Cattelet, C.; Leal-Calderon, F. Coarsening of Alkane-in-water emulsions stabilized by nonionic poly(oxyethylene) surfactants: The role of molecular permeation and coalescence. *Langmuir* **2004**, *20*, 46–52. [CrossRef]
22. Taisne, L.; Walstra, P.; Cabane, B. Transfer of oil between emulsion droplets. *J. Colloid Interface Sci.* **1996**, *184*, 378–390. [CrossRef]

23. Kabalnov, A.S.; Makarov, K.N.; Pertzov, A.V.; Shchukin, E.D. Ostwald ripening in emulsions: 2. Ostwald ripening in hydrocarbon emulsions: Experimental verification of equation for absolute rates. *J. Colloid Interface Sci.* **1990**, *138*, 98–104. [CrossRef]

24. Boos, J.; Preisig, N.; Stubenrauch, C. Dilational surface rheology studies of n-dodecyl-β-ᴅ-maltoside, hexaoxyethylene dodecyl ether, and their 1:1 mixture. *Adv. Colloid Interface Sci.* **2013**, *197–198*, 108–117. [CrossRef] [PubMed]

25. Llamas, S.; Santini, E.; Liggieri, L.; Salerni, F.; Orsi, D.; Cristofolini, L.; Ravera, F. Adsorption of sodium dodecyl sulfate at water-dodecane interface in relation to the oil in water emulsion properties. *Langmuir* **2018**, *34*, 5978–5989. [CrossRef] [PubMed]

26. Chung, C.; Sher, A.; Rousset, P.; Decker, E.A.; McClements, D.J. Formulation of food emulsions using natural emulsifiers: Utilization of quillaja saponin and soy lecithin to fabricate liquid coffee whiteners. *J. Food Eng.* **2017**, *209*, 1–11. [CrossRef]

27. Taarji, N.; da Rabelo Silva, C.A.; Khalid, N.; Gadhi, C.; Hafidi, A.; Kobayashi, I.; Neves, M.A.; Isoda, H.; Nakajima, M. Formulation and stabilization of oil-in-water nanoemulsions using a saponins-rich extract from argan oil press-cake. *Food Chem.* **2018**, *246*, 457–463. [CrossRef]

28. Bai, L.; McClements, D.J. Formation and stabilization of nanoemulsions using biosurfactants: Rhamnolipids. *J. Colloid Interface Sci.* **2016**, *479*, 71–79. [CrossRef]

29. Wojciechowski, K. Surface activity of saponin from Quillaja bark at the air/water and oil/water interfaces. *Colloids Surf. B* **2013**, *108*, 95–102. [CrossRef]

30. Stanimirova, R.; Marinova, K.; Tcholakova, S.; Denkov, N.D.; Stoyanov, S.; Pelan, E. Surface rheology of saponin adsorption layers. *Langmuir.* **2011**, *27*, 12486–12498. [CrossRef]

31. Mitra, S.; Dungan, S.R. Micellar properties of quillaja saponin. 1. Effects of temperature, salt, and pH on solution properties. *J. Agric. Food Chem.* **1997**, *45*, 1587–1595. [CrossRef]

32. Kairaliyeva, T.; Aksenenko, E.V.; Mucic, N.; Makievski, A.V.; Fainerman, V.B.; Miller, R. Surface tension and adsorption studies by drop profile analysis tensiometry. *J. Surfactants Deterg.* **2017**, *20*, 1225–1241. [CrossRef]

33. Ravera, F.; Loglio, G.; Kovalchuk, V.I. Interfacial dilational rheology by oscillating bubble/drop methods. *Curr. Opin. Colloid Interface Sci.* **2010**, *15*, 217–228. [CrossRef]

34. Loglio, G.; Pandolfini, P.; Liggieri, L.; Makievski, A.V.; Ravera, F. *Bubble and Drops Interfaces*; Miller, R., Liggieri, L., Eds.; Brill: Leiden, The Netherlands, 2011; pp. 7–38.

35. Gaillard, T.; Rochéa, M.; Honorez, C.; Jumeau, M.; Balan, A.; Jedrzejczyk, C.; Drenckhan, W. Controlled foam generation using cyclic diphasic flows through a constriction. *Int. J. Multiph. Flow* **2017**, *96*, 173–187. [CrossRef]

36. Fainerman, V.B.; Miller, R.; Wustneck, R.; Makievski, A.V. Adsorption isotherm and surface tension equation for a surfactant with changing partial molar area. 1. Ideal surface layer. *J. Phys. Chem.* **1996**, *100*, 7669–7675. [CrossRef]

37. Fainerman, V.B.; Zholob, S.A.; Lucassen-Reynders, E.H.; Miller, R. Comparison of various models describing the adsorption of surfactant molecules capable of interfacial reorientation. *J. Colloid Interface Sci.* **2003**, *261*, 180–183. [CrossRef]

38. Ravera, F.; Ferrari, M.; Liggieri, L. Adsorption and partitioning of surfactants in liquid–liquid systems. *Adv. Colloid Interface Sci.* **2000**, *88*, 129–177. [CrossRef]

39. Ravera, F.; Liggieri, L.; Ferrari, M.; Miller, R.; Passerone, A. Measurement of the partition coefficient of surfactants in water/oil systems. *Langmuir* **1997**, *13*, 4817–4820. [CrossRef]

40. Fainerman, V.B.; Sharipova, A.A.; Aidarova, S.B.; Kovalchuk, V.I.; Aksenenko, E.V.; Makievski, A.V.; Miller, R. Direct determination of the distribution coefficient of tridecyl dimethyl phosphine oxide between water and hexane. *Colloids Interfaces* **2018**, *2*, 28. [CrossRef]

41. Ravera, F.; Ferrari, M.; Liggieri, L. Modelling of dilational visco-elasticity of adsorbed layers with multiple kinetic processes. *Colloids Surf. A* **2006**, *282*, 210–216. [CrossRef]

42. Ravera, F.; Ferrari, M.; Liggieri, L.; Loglio, G.; Santini, E.; Zanobini, A. Liquid-liquid interfacial properties of mixed nanoparticle-surfactant systems. *Colloids Surf. A* **2008**, *323*, 99–108. [CrossRef]

43. Santini, E.; Liggieri, L.; Sacca, L.; Clausse, D.; Ravera, F. Interfacial rheology of Span 80 adsorbed layers at paraffin oil–water interface and correlation with the corresponding emulsion properties. *Colloids Surf. A* **2007**, *309*, 270–279. [CrossRef]

44. Whitby, C.P.; Fornasiero, D.; Ralston, J.; Liggieri, L.; Ravera, F. Properties of fatty amine-silica nanoparticle interfacial layers at the hexane-water interface. *J. Phys. Chem. C* **2012**, *116*, 3050–3058. [CrossRef]

Article

Interaction of Particles with Langmuir Monolayers of 1,2-Dipalmitoyl-Sn-Glycero-3-Phosphocholine: A Matter of Chemistry?

Eduardo Guzmán [1,2,*], Eva Santini [3], Michele Ferrari [3], Libero Liggieri [3] and Francesca Ravera [3,*]

[1] Departamento de Química Física, Universidad Complutense de Madrid, 28040 Madrid, Spain
[2] Instituto Pluridisciplinar, Universidad Complutense de Madrid, 28040 Madrid, Spain
[3] Istituto di Chimica della Materia Condensata e di Tecnologia per l'Energia (ICMATE)-U.O.S. Genova, Consiglio Nazionale delle Ricerche (CNR), 16149 Genova, Italy; eva.santini@ge.icmate.cnr.it (E.S.); michele.ferrari@ge.icmate.cnr.it (M.F.); libero.liggieri@ge.icmate.cnr.it (L.L.)
* Correspondence: eduardogs@quim.ucm.es (E.G.); francesca.ravera@ge.icmate.cnr.it (F.R.); Tel.: +34-91-394-4107 (E.G.)

Received: 22 March 2020; Accepted: 7 May 2020; Published: 10 May 2020

Abstract: Lipid layers are considered among the first protective barriers of the human body against pollutants, e.g., skin, lung surfactant, or tear film. This makes it necessary to explore the physico-chemical bases underlying the interaction of pollutants and lipid layers. This work evaluates using a pool of surface-sensitive techniques, the impact of carbon black and fumed silica particles on the behavior of Langmuir monolayers of 1,2-dipalmitoyl-sn-glycero-3-phosphocholine (DPPC). The results show that the incorporation of particles into the lipid monolayers affects the surface pressure–area isotherm of the DPPC, modifying both the phase behavior and the collapse conditions. This is explained considering that particles occupy a part of the area available for lipid organization, which affects the lateral organization of the lipid molecules, and consequently the cohesion interactions within the monolayer. Furthermore, particles incorporation worsens the mechanical performance of lipid layers, which may impact negatively in different processes presenting biological relevance. The modification induced by the particles has been found to be dependent on their specific chemical nature. This work tries to shed light on some of the most fundamental physico-chemical bases governing the interaction of pollutants with lipid layers, which plays an essential role on the design of strategies for preventing the potential health hazards associated with pollution.

Keywords: lipids; pollutants; Langmuir monolayers; particles; rheology

---

1. Introduction

The continuous ejection of pollutants into the atmosphere as a result of the industrial activity and combustion processes has raised many questions related to the potential impact of pollution on human health [1–4]. This is even more important analyzing the World Health Organization (WHO) statistics which ascribe one third of the deaths caused by strokes, lung cancer, or cardiac diseases to the air pollution [5]. Therefore, the severity of this problem makes it necessary to deepen the study of the impact of pollutants on biological systems [6].

Lipid layers, e.g., skin, tear film, or lung surfactant, provide one of the first protective barriers of the human body against environmental pollution. Therefore, it should be expected that lipid layers should be considered among the most important biological structures where pollution may impact negatively [7–9]. This creates the careful examination of the impact of different chemical species, such as nanoparticles, on the physico-chemical properties of lipid layers that may be used as a tool for a preliminary understanding of the most fundamental bases governing the alteration of their physiological response as a result of the incorporation of pollutants [9]. However, the direct in vivo evaluation of the impact of pollutants in the behavior of lipid films is difficult in most of the cases, which makes the use of model systems necessary [10]. The use of such models provides the bases for establishing preliminary assays for evaluating the alterations induced by pollutant species on the physico-chemical properties of lipid layers with potential biological relevance.

Langmuir monolayers of lipids at the water/vapor interface are probably among the most widespread models used as tools on the evaluation of the effects of different chemicals on the behavior of biologically relevant systems [11–16]. This is because Langmuir monolayers allow for performing physico-chemical studies on ordered lipid films, which are reminiscent of different biological layers, e.g., a single cellular membrane leaflet or the lung surfactant film [11,17,18]. However, the use of Langmuir monolayers only allows for mimicking some specific aspects of the physico-chemical behavior of biological relevant systems that are relatively complex. Thus, the use of Langmuir monolayers helps with the study of minimal systems, i.e., the model including a limited number of chemical species, which allows for exploring the potential role of each single species in the interaction with pollutants, and in the modification of the physico-chemical properties of the whole system. The most common lipid used as a model for studying the interaction of biological relevant layers and pollutants is the 1,2-Dipalmitoyl-sn-glycerol-3-phosphocholine (DPPC) [17,19–26]. This is because this lipid is one of the main components of many biological membranes and fluids, e.g., DPPC accounts for 40 wt % of the total weight of lung surfactant in mammals [27]. Despite the simplicity of the models based only on DPPC, they are useful tools for a preliminary evaluation of the worsening of the physico-chemical properties of lipid layers as a result of the incorporation of solid particles. However, the extrapolation of the results obtained from such studies to real biophysical situations studies may require some cautions and additional considerations [28,29]. This is especially important because the specific characteristic of the method used for evaluating the incorporation of particles into the lipid layer impacts the modifications of the interfacial behavior of DPPC layers strongly due to the incorporation of colloidal particle [9,30]. The effect of the methodology used for the incorporation of particles has been recently explored in relation to the interaction of ceria particles with DPPC at the water/vapor interface, using for such evaluation up to three different methodologies for preparing the monolayers containing both DPPC and particles: (i) particles deposited from dispersions in chloroform onto preformed DPPC monolayers, (ii) mixed monolayers prepared from simultaneous co-spreading of the particles and the DPPC at the interface, and (iii) aerosolized particles deposited onto the preformed DPPC monolayer [30]. The obtained results in such study provided evidence that the use of different delivery methods leads to different modifications of the surface tension behavior. Furthermore, special caution must be taken with the use of techniques based on the dispersion of the particles from dispersions because they can impact the agglomeration of the particles and the distribution of such agglomerated within the lipid layer, which may lead to a situation different to that occurring upon the in vivo interaction of pollutants with lipid layers.

The investigation of the impact of particles on lipid layers takes particular importance because several studies have provided evidence that the interaction of colloidal particles with surfactant layers modifies both the lateral organization and the mechanical of such surfactant layers [31–33]. Therefore, a strong modification of the physico-chemical behavior of biologically-relevant lipid layers as results of the particles incorporation may be expected [34,35]. Our previous study showed the different impact of particles with different hydrophobicity on the interfacial properties of DPPC monolayers, which is ascribed to the nature of the interactions involved in the incorporation of the particles [36]. Furthermore, previous studies have shown that hydrophobic particles present a strong impact on the lateral packing and mechanical properties of DPPC layers [21,23,37,38]. Despite this impact, there is not any systematic study comparing the impact of carbonaceous and silica particles on the interfacial properties of biologically-relevant lipid layers, with such comparative evaluation being of particular interest due to the widespread of nanomaterials based in carbon and silica in different technologies and industries. This work tries to shed light on the potential effect of the above-mentioned particles on the behavior of DPPC Langmuir monolayers to perform a preliminary evaluation of the potential risks and hazards associated with their incorporation into biologically-relevant systems. This has been performed analyzing the modifications in the 2D lateral packing of the lipid molecules at the interface, and the cohesion interactions within the interfacial layers. It is expected that the picture obtained from this work may be useful as foundations for a broader study aimed to elucidate impact of pollutants on the physiological response of biological relevant layers.

2. Materials and Methods

2.1. Materials

1,2-Dipalmitoyl-sn-glycerol-3-phosphocholine (DPPC), with a molecular weight of 734.1 g/mol, was purchased from Avanti Polar Lipids, Inc. (Alabaster, AL, USA) at 99.9% purity, and used as received. Hydrophobic fumed silica particles *Aerosil R972* (SiO_2) purchased from Evonik-Degussa (Essen, Germany) and carbon black particles CB N110 (CB) supplied by Phillips Petroleum Co. (Bartlesville, OK, USA) were chosen as pollutant models. Table 1 summarizes some physico-chemical characteristic of the particles [21,23,39]. It is worth mentioning that, whereas SiO_2 particles lead to the formation of chain-like aggregates of primary particles, the primary particles of CB aggregates to form spherical-like agglomerates [21,39,40]. It is worth mentioning that, even though the densities and the average diameter (*d*) of the primary particles are similar for SiO_2 and CB, the different geometry of their aggregates helps may provide an explanation for the difference of the values of the BET surface area.

Table 1. Physico-chemical characteristics of the used particles.

Particles	BET Surface Area [1] (m²/g)	Density [1] (g/cm³)	*d* (nm)
SiO_2	110 ± 20	2.2	16 ± 4 [1]
CB	51 ± 20	1.8	22 ± 7 [2]

[1] BET surface areas and densities values for both types of particles, as well as the diameter of the primary (*d*) particles for SiO_2, were obtained from the suppliers manufacturers; [2] the diameter of the primary particles (*d*) of the primary particles of CB was obtained from Transmission Electron Microscopy image analysis in our previous publication [39].

Chloroform (CHROMASOLV™, for High Performance Liquid Chromatography, stabilized with ethanol) purchased from Sigma-Aldrich (Saint Louis, MO, USA) was used for preparing the spreading solutions of DPPC and the particles' dispersions.

Ultrapure deionized water for cleaning and experiments was obtained by a multi-cartridge purification system Elix + Milli-Q (Millipore, Burlington, MA, USA). This water presents a resistivity higher than 18 MΩ·cm, a total organic content lower than 6 ppm and a surface around 72 mN/m at 22 °C without evidence of surface tension kinetics were found for the used water over several hours. The pH of the water was around 6.5, and no salts were used for fixing the ionic strength.

2.2. Monolayers Preparation

The lipid monolayers were prepared at the water/vapor interface by dropping controlled volumes of DPPC from its solution in chloroform (concentration about 1 mg/mL or 1.36 mM) using a high-precision Hamilton syringe (Hamilton Company, Reno, NV, USA). This methodology ensures the control of the interfacial density of DPPC, Γ, upon solvent evaporation. The initial interfacial density of DPPC spread at the water/vapor interface Γ_0 was fixed in all the experiments in a value of 1.7 µmol/m², corresponding to an area per molecule of about 98 Å².

The preparation of the mixed monolayers was done following a two-step approach: (i) a DPPC monolayer was obtained from the spreading of the lipid from its solution in chloroform (concentration 1 g/L) at the bare water/vapor interface, and (ii) a given amount of the particle dispersion (concentration

1 g/L) was spread onto the preformed DPPC monolayer, again using chloroform as the spreading solvent (Notice that particles dispersions were sonicated during 15 min using a laboratory ultrasound bath; this allows for minimizing particles aggregation before their spreading). This procedure allows for obtaining monolayers with specific DPPC: particles weight ratio at the interface. Once monolayers including particles and DPPC are obtained, it is necessary to wait during 1 h before starting the experiments to ensure the complete evaporation of the solvent and, in the case of mixed monolayer, the achievement of the equilibrium of the composite system which is driven by the nanoparticle–lipid interactions [21,23]. It is worth noting that the monolayers studied here cannot be a strictly considered mixed monolayer because they were not obtained after the spreading of a mixed dispersion containing the DPPC and the particles (first, the DPPC is spread at the pure water/vapor interface, and then the addition of the particles is made onto the preformed DPPC monolayer). However, for the sake of simplicity, the term mixed monolayers will be used for monolayers involving the DPPC and the particles.

The temperature was fixed 22.0 ± 0.1 °C in all the experiments. Even though this temperature is far from the physiological one (37 °C), the main conclusions obtained in our study are extrapolated, at least from a semi-quantitative perspective, to the physiological conditions. This is because the phase transition of DPPC appears above the physiological temperature, thus only a shift of the phase behavior with the temperature is expected, without any significant impact on the main physico-chemical insights extracted from the experimental results [41].

It is worth mentioning that the properties of the mixed monolayers obtained upon chloroform evaporation will be affected by the conditions in which DPPC and particles are mixed [30]. This work has used a methodology for the preparation of the mixed layers where the interaction between particles and lipid layers occurs only at the water/vapor interface. This may be considered similar to that which happens during the interaction between environmental pollutants and lipid layers. However, this study does not consider two aspects that may have impact when in vivo conditions are concerned: (i) the presence of chloroform during particles addition may affect both the lateral packing of the DPPC molecules and the DPPC–particle interactions [30], and (ii) the interaction of particles and the lipid layers may be affected for specific mass transport boundary conditions which cannot be included in our studies using Langmuir monolayers [42,43].

2.3. Methods

The Langmuir monolayers were studied using a Langmuir trough KSV Nima model KN2002 (Biolin Scientific, Espoo, Finland), equipped with two Delrin barriers allowing for symmetric compression/expansion of the free liquid surface. The total surface area of the Teflon trough is 243 cm^2. The surface tension, γ, was measured using a force balance fitted with a paper Wilhelmy plate (Whatman CHR1 chromatography paper, effective perimeter 20.6 mm, supplied by Sigma Aldrich, St. Louis, MO, USA), ensuring a zero contact angle. The surface pressure, Π, is obtained as the difference between the surface tension of the pure water/vapor interface γ_w and γ, i.e., $\Pi = \gamma_w - \gamma$.

The quasi-equilibrium isotherms for the monolayers were obtained measuring the surface pressure as the interfacial area available for the monolayer, A, is reduced at a fixed compression velocity of 2 cm^2/min, which is equivalent to a compression rate $(\Delta A/A^0)/\Delta t$ of about 10^{-5} s^{-1}, with $\Delta A/A^0$ being the amplitude of the deformation, represented as the ratio between the change of area ΔA (amplitude of deformation) and the reference interfacial area A^0 (generally the area in which the compression is started), and Δt the time needed for the deformation. This compression rate allows for avoiding an undesired non-equilibrium effects during the determination of the isotherms [44].

The Langmuir trough also enables the study of the effects of the incorporation of nanoparticles on the dilational rheology of the DPPC monolayers using the oscillatory barrier method. A detailed description of the foundations of this method can be found elsewhere [45–47]. The oscillatory barrier method allows for evaluating the modulus of the complex dilational viscoelasticity $E = \Delta\gamma/(\Delta A/A)$, i.e., the variation, of the surface tension γ as a result of a harmonic change at a controlled frequency ν (in a range of frequencies from 10^{-3} to 0.15 Hz) of the interfacial area which is written as follows:

$$A(t) = A^0 + \Delta A \sin(2\pi\nu t). \tag{1}$$

The harmonic change of the interfacial area (strain) results in a stress response $\Delta\Pi = \Pi^0 - \Pi(t)$, which is defined as the change of surface pressure between the reference state Π^0 and the instantaneous value of the surface pressure $\Pi(t)$. When the deformation presents a small amplitude, i.e., deformation within the linear regime, the stress response also follows a sinusoidal profile with the same frequency than the deformation:

$$\Pi(t) = \Delta\Pi \sin(2\pi\nu t + \phi), \tag{2}$$

where ϕ is a phase shift accounting for a possible delay of stress response (surface pressure change) in relation to the strain (area deformation). Considering the above-mentioned linear response, the stress can be considered proportional to the deformation $u(t) = \Delta A/A^0$ (elastic term) and to the rate of deformation $du(t)/dt$ (viscous term), which allows one to write the stress as:

$$\Pi(t) = \varepsilon u(t) + \eta(du(t)/dt), \tag{3}$$

with ε and η being the dilational elasticity and viscosity, respectively. Considering the definition given by Equation (3) and assuming a generic harmonic perturbation, it is possible to obtain a definition for the complex dilational viscoelasticity:

$$\varepsilon^* = \varepsilon + 2\pi\nu\eta i \tag{4}$$

where $i = (-1)^{1/2}$. The analysis of the curves corresponding to the strain and stress in terms of Equations (1) and (2) provide information about their amplitudes and the phase shift, enabling for the calculation of the dilational viscoelasticity. In the reported experiments, the amplitude of the dilational deformation $u(t) = 0.01$ was adopted, which allows the response to remain within the linear regime. It is worth noting that the conditions considered in our work for the evaluation of the mechanical response of lipid layers, and how particles' incorporation impact such response, are far from those corresponding to the characteristic values of the dynamics processes involved in biologically relevant systems, e.g., respiratory cycle, where higher values for the frequency of the deformation and its amplitude are expected. In the particular case of the respiratory cycle, the frequency and the deformation amplitude assume values around 0.3 Hz and 0.30–0.40, respectively [48]. However, the evaluation of the dilational response within the linear regime provides helpful information for analyzing the impact of incorporation of particles on the relaxation mechanisms leading to equilibration of the lipid layers, which serve as a preliminary assay towards the understanding of more complex dynamics situations than those appearing in biologically relevant systems.

A Brewster Angle Microscope Mulstiskop from Optrel (Sinzing, Germany) fitted with a He-Ne laser (λ = 614 nm) and coupled to the Langmuir through was used to obtain information about of the lateral organization of lipids and particles at the interface on the basis of Brewster Angle Microscopy (BAM) images of the interfacial textures.

3. Results

3.1. Study of Particle Monolayers at the Bare Water/Vapor Interface

The hydrophobic nature of the particles suggests that they may be spread at the bare interface leading to the formation of a particle-laden interface. Therefore, the evaluation of the incorporation of the particles at the bare water/vapor interface, in terms of their surface pressure (Π)-area (A) isotherms, can be useful as a preliminary step for understanding their incorporation into DPPC monolayers. Figure 1 shows the Π-A isotherms for SiO_2 and CB particles at the bare water/vapor interface.

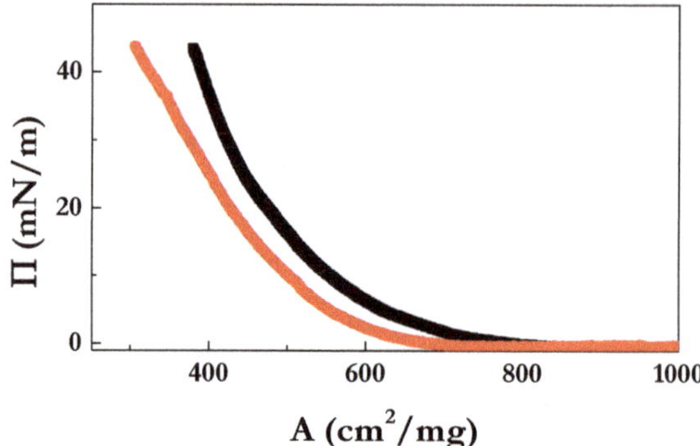

Figure 1. Π–A isotherms for CB (■) and SiO_2 (•) particles at the bare water/vapor interface.

The isotherms point out that neither CB particles nor SiO_2 ones affect the surface tension of the bare water/vapor significantly until the interfacial density of particle is large enough to form a highly packed film, which leads to the increase of the surface pressure of the monolayer.

3.2. Evaluation of Particles Incorporation into DPPC Monolayers

The evaluation of the incorporation of particles into DPPC monolayers, and their potential impact on the lateral organization of the lipid layers, was done on the basis of the modifications associated with the presence of particles in the Π-A isotherms of DPPC monolayers. This is important because the isotherms provide information related to the phase behavior of the mixed monolayers, and how the lateral packing of DPPC monolayers and their mechanical characteristics are modified due to the particles' incorporation, with such aspects being relevant on the biological function of the lipid layers. Figure 2 reports Π versus the area per DPPC molecule, A, normalized to its initial value, A_0 (Note that this initial value A_0 was fixed as a reference in all the experiments in 98 Å^2/molecule), after the lipid spreading, corresponding to DPPC monolayers containing particles in a broad range of particles' weight fractions, x_p. Notice that the weight fractions of the particles spread onto the DPPC monolayers correspond to estimated doses in the 6–115 mg/mL range (assuming a realistic thickness for the interface of 10 nm), which are compatible with the values reported for the dose of deposited particles in the lung surfactant layer upon inhalation [49–51].

The isotherm for DPPC spread at the bare water/vapor interface presents the typical features reported for monolayers of this lipid [44,52–54]. At the highest value of the reduced area (gas and liquid expanded –LE– phases), a mild increase of the surface pressure with the increase of the interfacial density, i.e., with the A/A_0 ratio decreases, was found. This proceeds up to a threshold value of the interfacial density, which defines the onset of the coexistence region between LE and liquid compressed (LC) phases (LE–LC coexistence is reached). This coexistence region is characterized by an almost vanishing change of the surface pressure (pseudo-plateau) as the interfacial density increases, which is associated with the disappearance of the LE phase as the re-orientation of the DPPC molecules occurs driving the system to a more ordered phase (LC). Once the LE-LC coexistence is overcome ($A/A_0 \sim 0.45$), a sharp increase of the surface pressure with the increase of the packing density within the LE expanded and solid phases was found until the rupture of the monolayer occurs at the collapse surface pressure, Π_c.

The incorporation of particles into the DPPC monolayers does not modify substantially the shape of the isotherm in relation to that found for the pure lipid at the water/vapor interface, with this being almost independent from the chemical nature of the particles and the x_p value. However, the incorporation of particles into the DPPC monolayers leads to the emergence of two effects: (i) the shifting of the Π—A/A_0 isotherms to more expanded states, i.e., to higher values of the reduced area, and (ii) the modification of the collapse pressure of the monolayer, i.e., the maximum surface pressure that a monolayer can reach before its rupture. The former aspect is evidenced from the earlier lifting-off of the LE phase in the mixed monolayers than in pure DPPC, which can be understood considering that particles take up a part of the area available for the reorganization of the lipid molecules [55]. This leads to a situation in which the monolayers upon particles incorporation behave in such a way which is reminiscent of a monolayer with a higher effective interfacial density of DPPC. Therefore, it is possible to assume that particles' incorporation induces excluded area effects in the DPPC monolayer, which are strongly dependent on both the chemical nature of the particles and the x_p value. Such dependences, and in particular that on the chemical nature of the particles, give an indication that the excluded area effects alone cannot account for the changes of the behavior of DPPC monolayers due to the incorporation of particles. This makes it necessary to analyze the role of the interactions between the different components forming the mixed layer (particle–particle, lipid–lipid and particle–lipid) to obtain a complete picture of the influence of particles in the behavior of DPPC monolayers. Such interactions affect the lipid lateral packing, and the aggregation and distribution of the particles at the interface, leading to a different behavior than that expected for systems where only the role of the area exclusion is considered.

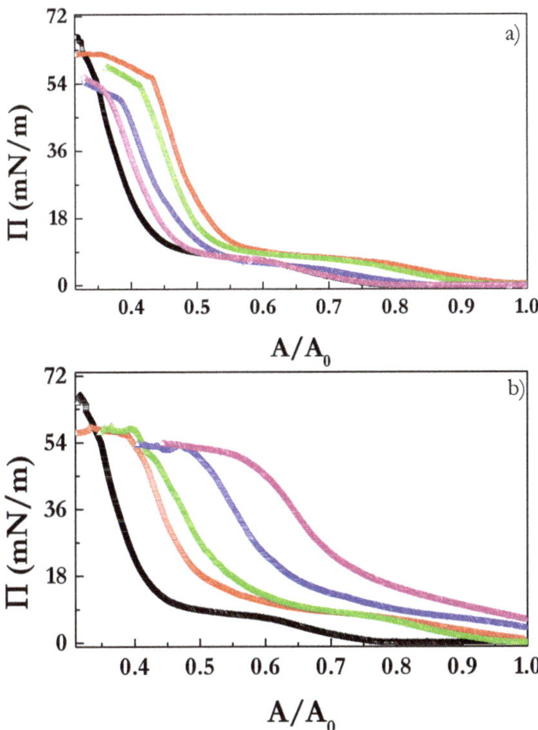

Figure 2. Π–A/A_0 isotherms for DPPC Langmuir monolayers upon the incorporation of different weight fraction of particles at the interface (x_p): CB (**a**) and SiO (**b**). Each curve corresponds to DPPC monolayers with a different weight fraction of particles at the interface (x_p): (Δ) 0.00, (Δ) 0.10, (Δ) 0.33, (Δ) 0.75, and (Δ) 0.90.

The results show different dependences on the x_p value for DPPC monolayers upon the incorporation of particles with different chemical nature. The increase of the amount of SiO_2 particles incorporated into the DPPC monolayer shifts the isotherm to higher values of A/A_0, which is explained as result of enhanced importance of the excluded area effects with the increase of x_p.

However, the situation is different when CB particles are concerned. The incorporation of CB particles into DPPC monolayer leads to two different regimes of behavior as a function of x_p–on the impact of CB on the behavior of DPPC monolayers—were found: (i) for the smallest values of CB weight fraction, a strong shift of the isotherm to higher values of A/A_0 than those corresponding to pure DPPC monolayer was found as result of the area exclusion effect, and (ii) for the highest CB amounts at the interface, even though particles' incorporation leads to excluded area effects, its importance is decreased as x_p increases.

The above-mentioned differences suggest the existence of different distributions for SiO_2 and CB particles upon incorporation into DPPC monolayers. Thus, whereas the SiO_2 particles may be incorporated into the DPPC monolayer as pseudo-2D aggregates which tend to occupy the maximum area available at the interface, the incorporation of CB particles leads to the formation of 3D particle-stacking with the increase of x_p. Thus, the incorporation of CB particles in concentrations above a threshold value of x_p leads to their stacking onto the preformed mixed monolayer, which may result in the formation of out-of-plane structures, such as wrinkles, folds, or buckles. This leads to a situation in which the effective concentration of particles at the interface is lower than that expected from the complete spreading of the particles within the area available, and, as matter of fact, to a reduction of the importance of the excluded area effects with the increase of x_p [56–59]. The differences in the behavior of the DPPC monolayers upon the incorporation of SiO_2 and CB particles are explained considering the different chemical nature of the particles. Thus, even though both types of particles are hydrophobic, the presence of dissociated silanol groups onto the surface of the SiO_2 particles may introduce a repulsive electrostatic interaction between the particles, which facilities its dispersion within the DPPC films. However, when CB particles are considered, a strong hydrophobic attraction should be expected, which favors their agglomeration with the increase of x_p [60].

The above discussion shows the strong impact of the chemical nature of the particles in their incorporation into DPPC monolayers, and its role in the excluded area effects. However, a complete picture of the impact of the particles on the DPPC monolayer behavior also needs a closer look at the the role of the interactions. For this purpose, a simple geometrical consideration may be useful. Assuming the incorporation of spherical particles which can cover a maximum area of the water/vapor interface defined as $N\pi r^2$, with r being the radius of a single particle and N the number of particles incorporated into the monolayer, it would be expected that the fraction of interfacial area occupied by particles oscillate between a value lower than 1% for the lowest value of x_p and a value around 10% for the highest one. However, the results show a higher impact than what was expected on the basis of the above simple geometrical considerations, with area expansions in the ranges 20–50% and 5–25% for SiO_2 and CB particles, respectively. Thus, it is possible to assume that that the impact of particles in the lateral packing of DPPC monolayers results from a complex interplay between the excluded area effects, steric hindrance, and different types of interactions. The role of the interactions is clear from the analysis of the Π_c dependence on x_p shown in Figure 3. The results show that the incorporation of particles decreases the maximum surface pressure that the DPPC monolayers can reach before its rupture. This decrease of Π_c indicates an irreversible incorporation of particles into the DPPC monolayers, which results in a reduction of the ability of DPPC to form highly condensed phases. This is in contrast with what is found when the incorporation of hydrophilic particles in DPPC monolayers is concerned; in those cases, an effective refinement of the interfacial composition is generally found for the highest compression degree, with a partial squeezing-out of the particles from the interface [24,61].

The decrease of Π_c with the increase of x_p results from the impact of particles on the lateral cohesion interactions of the molecules within the interface. Thus, the incorporation of particles, independently of their chemical nature, reduces the strength of the cohesion interactions between the lipid molecules as a result of the emergence of heterogeneities on the lateral organization within the film. The higher decrease of Π_c, found when SiO_2 particles are incorporated into the DPPC monolayer, in relation to those cases in which the incorporation of CB is considered, is explained considering the differences of the steric hindrance associated with the presence of each type of particles at the interface [62]. Thus, whereas SiO_2 particles tend to occupy the maximum area available at the interface, 3D stacking of particles are expected for CB particles, which results in a lower occupancy of the interface by CB particles. This leads to a situation in which SiO_2 particles modify strongly the lateral organization of the lipid layer.

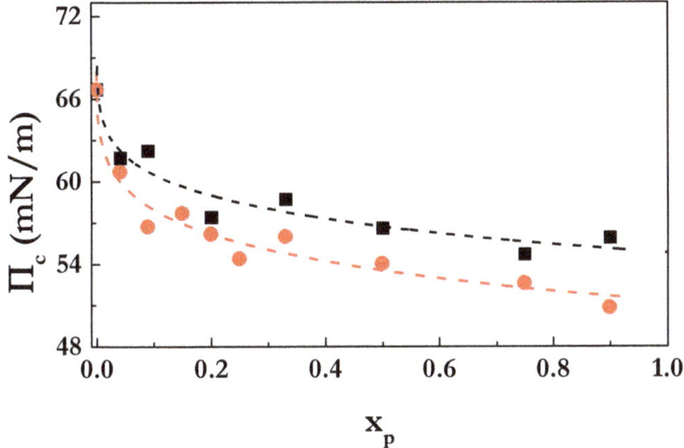

Figure 3. Dependence on x_p of the collapse pressure, Π_c, of DPPC monolayers upon the incorporation of CB (■) and SiO_2 (●) particles. Note that the lines are guides for the eyes.

Additional information related to the incorporation of particles into the DPPC films are obtained from the changes of the quasi-static dilational elasticity ε_0 obtained from the isotherm as

$$\varepsilon_0 = -A\,(\partial\Pi/\partial A)_T. \tag{5}$$

Figure 4 shows the surface pressure dependence of ε_0 for DPPC monolayers after the incorporation of different weight fractions of particles (data for the incorporation of CB and SiO_2 are shown in panels a and b, respectively). The results show three different features when the elasticity of monolayers of pure DPPC are considered: (i) an increase of the elasticity up to a first maximum associated with the formation of the disordered LE phase, which presents a weak lateral packing (for the lowest values of Π), (ii) a drop of the elasticity, with the increase of Π, down to reach a quasi-null value for the LE–LC coexistence, and (iii) an increase of the elasticity within the LC phase up to reach its maximum value associated with an enhanced lateral packing of the monolayer, and then a drop of the elasticity as the monolayer approaches the collapse.

The incorporation of particles modifies dramatically the elasticity of the DPPC monolayers, with the average elasticity of the monolayer decreasing upon the incorporation of particles. This reduction of the monolayer rigidity is associated with a worsening of the lateral packing of the lipid molecules at the interface as a result of the weakening of the lateral cohesion interactions within the monolayer. A more detailed analysis of the impact of the particles' incorporation into the organization of the lipids within the interface is obtained from the changes of the quasi-static dilational elasticity corresponding to the maximum values of the elasticity for the LE and LC phases and to the LE–LC coexistence with x_p (see Figure 5).

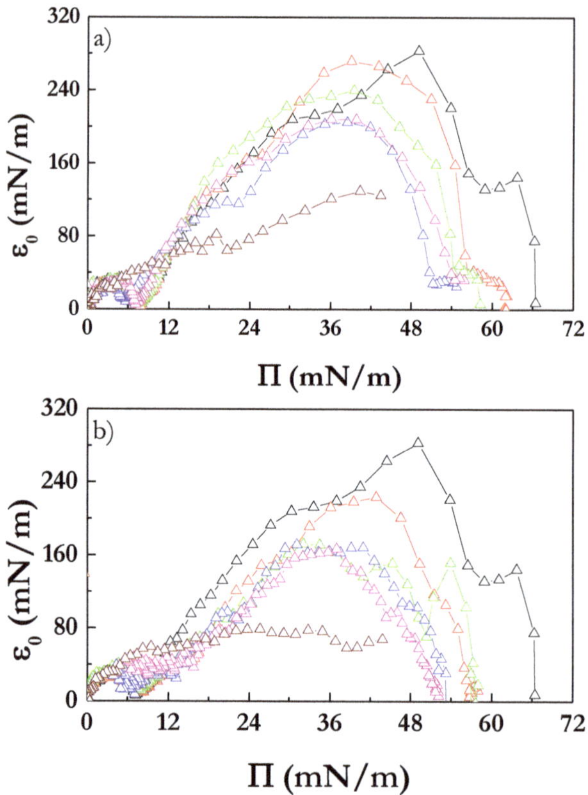

Figure 4. Quasi-static dilational elasticity ε_0 dependences on the surface pressure Π for DPPC Langmuir monolayer upon the incorporation of different weight fraction of particles at the interface (x_p): CB (**a**) and SiO$_2$ (**b**). Each curve corresponds to DPPC monolayers with a different weight fraction of particles at the interface (x_p): ($\triangle$) 0.00, ($\triangle$) 0.10, ($\triangle$) 0.33, ($\triangle$) 0.75, ($\triangle$) 0.90 and ($\triangle$) 1.00. Notice that the lines are guides for the eyes.

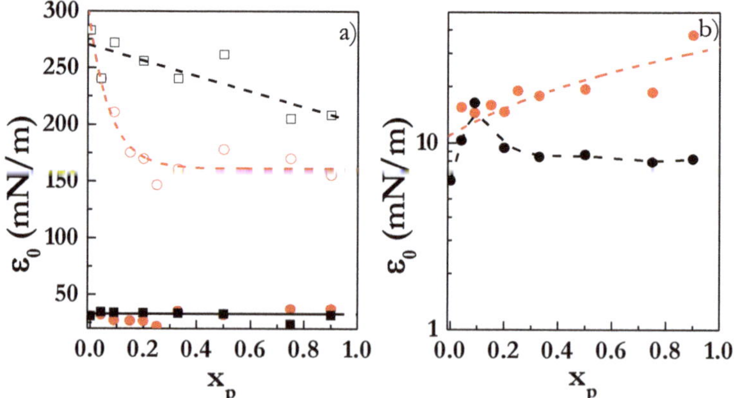

Figure 5. (**a**) Dependences on x_p of the maximum values of ε_0 for the LE (CB (■) and SiO$_2$ (•) particles) and LC (CB (□) and SiO$_2$ (○) particles) phases; (**b**) dependence on x_p of the maximum values of ε_0 for the phase coexistence region (CB (■) and SiO$_2$ (•) particles). Notice that the lines are guides for the eyes.

The absence of any noticeable change of the elasticity corresponding to the LE phases provide evidence that the lateral organization of the DPPC monolayer within this phase is not significantly modified, neither upon the incorporation of SiO_2 particles nor after the incorporation of CB ones. This may be understood considering that the LE phase is an intrinsically disordered phase in which the role of the lateral van der Waals interactions between the lipids molecules is almost negligible, thus it may be expected that a slight modification of such interactions due to the inclusion of the particles does not modify significantly the lateral packing of the lipids within the LE phase. However, a closer look at the elasticity dependences for the LE phases provides evidence of a slight increase of ε_0 with x_p as result of the incorporation of SiO_2 particles, whereas the incorporation of CB results in an initial increase of ε_0 with x_p up to a maximum for a x_p value about 0.10, which is followed by a decrease of ε_0 down to a value slightly higher to that corresponding to pure DPPC. Thus, even though the impact of the particles is very limited in the LE phase, a certain degree of disorder is expected considering the experimental dependences, which is correlated to the differences in the particles' organization as function of their chemical nature. On the other side, the elasticities for the LE-LC coexistence and LC phases are strongly modified in relation to those corresponding to the DPPC monolayers. The incorporation of both types of particles reduces the maximum lateral packing of the monolayer, i.e., the quasi-static dilational elasticity for the LC phase decreases, independently on the nature of the particles. However, the impact of the incorporation of SiO_2 particles is again stronger than that found when the incorporation of CB ones is considered. The effect of particles in the ε_0 value of the LE–LC coexistence results in being more intriguing, whereas the incorporation of CB particles into DPPC monolayers does not modify significantly the phase coexistence, and a strong increase of the elasticity of such region is found with the increase of x_p for SiO_2 particles. This allows one to assume that the impact of particles in the lateral packing of DPPC is driven by a complex balance involving different contributions, including the interactions involved in the mixed monolayers (hydrophobic vs. electrostatic), and the chemical nature and wettability of the particles (hydrophobicity vs. hydrophilicity of the particles). This leads to a hindering of the phase coexistence when SiO_2 particles are incorporated as evidenced in the BAM images shown in Figure 6.

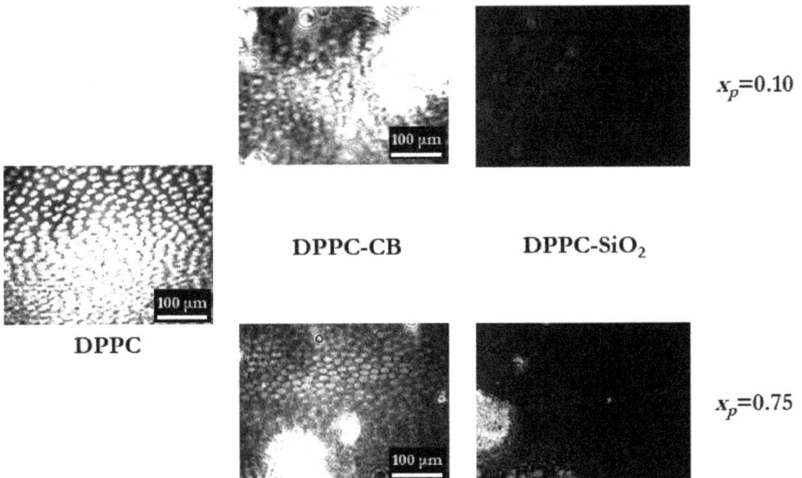

Figure 6. BAM images of DPPC monolayers upon the incorporation of CB and SiO_2 particles at two different values of x_p for surface pressure about 7.5 mN/m, corresponding to the LE–LC phase coexistence.

The BAM images show that, whereas the DPPC monolayers upon the incorporation of CB particles presents ellipsoidal-like domains which are similar to those found for pure DPPC monolayers, the incorporation of SiO_2 particles leads to the disappearance of such domains, i.e., the incorporation of SiO_2 drives to a hindering of the LC domains formation, which is compatible with the increase of the quasi-static dilational elasticity corresponding to such region. Furthermore, the BAM images also make clear the different distribution of the particles as a function of their chemical nature. Thus, a decrease of the size of the bright spots associated with particle agglomerates was found when CB particles are considered. However, no bright spots are found when the incorporation of SiO_2 is analyzed for the lowest value of x_p, which provides evidence that the distribution of the particles within the interface is better.

3.3. Interactions of Particles with DPPC at the Interface

Additional insights on the impact of the particles incorporation into DPPC monolayers can be obtained using the concepts of the thermodynamics of interfacial mixtures [63,64]. This approach helps to understand how the interactions between DPPC and particles modify the behavior of the mixed monolayer in relation to what happens in those cases without DPPC–particles interactions, i.e., ideal mixture conditions. The interfacial area of an ideally mixed monolayer at a fixed value of the surface pressure is defined as follows:

$$A^{id} = x_{DPPC}A_{DPPC} + x_pA_p \tag{6}$$

where A_{DPPC} and A_p correspond to the areas per mass unit of DPPC and particles at the considered values of surface pressure for a monolayer of the pure compounds, respectively, and x_{DPPC} and x_p are referred to the weight fractions of DPPC and particles at the interface in the mixed monolayer, respectively. The differences associated with the mixing process at a fixed value of the surface pressure can be evaluated in terms of the excess area A^E defined as follows:

$$A^E = A_{12} - A^{id} \tag{7}$$

where A_{12} is referred to the area per mass unit corresponding to the mixed monolayer at a fixed value of the surface pressure. The A^E provides information related to the mutual miscibility between the compounds forming the monolayer, which is governed by the cohesion forces existing between them. Figure 7 shows the dependences of the A^E on the weight fraction of particles at the interface for different values of the surface pressure.

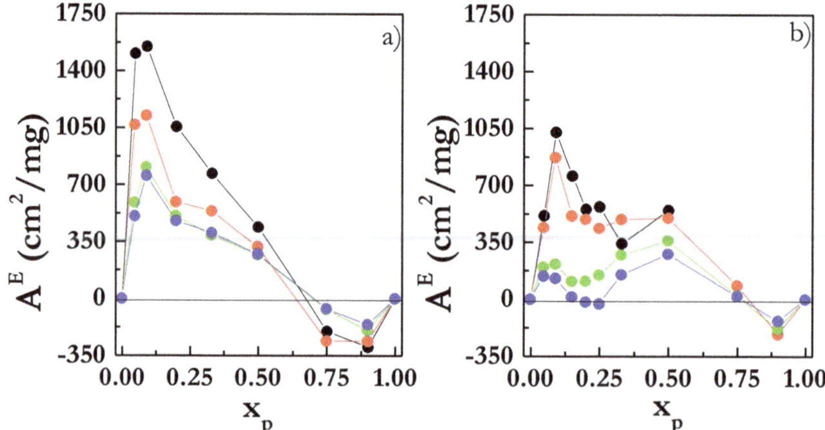

Figure 7. Dependences of the A^E on the particle weight fraction, x_p, for different values of the surface pressure Π for DPPC monolayers upon incorporation of CB (**a**) and SiO_2 (**b**): (—•—) 3 mN/m, (—•—) 7.5 mN/m, (—•—) 20 mN/m and (—•—) 40 mN/m. Notice that the lines are guides for the eyes.

The dependences of A^E on the surface pressure and the values of x_p are common to DPPC monolayers upon the incorporation of SiO_2 and CB particles. The results provide evidence of an enhanced interaction between the particles and the DPPC molecules at the interface with the increase of the packing of the film, i.e., A^E decreases with the increase of the compression degree of the monolayers. This may be explained considering the existence of a forced cohesion of lipids and particles at the interface as a result of the reduction of the available interfacial area. On the other side, the increase of the weight fraction of particles at the interface leads to a decrease of A^E, which may be ascribed to the enhancing of the DPPC–particles cohesion interaction. However, it is worth mentioning that the average cohesion between the DPPC molecules is reduced with respect to what happens in monolayers of pure DPPC at the water/vapor interface. The experimental results show that the interactions between DPPC and particles are repulsive (A^E values > 0) until x_p has reached high values when the excess area becomes negative. The repulsive interactions between particles and DPPC may be explained assuming the existence of a hindered lateral packing of the lipid due to the particles' incorporation. The enhanced miscibility appearing at the highest values of x_p is explained considering the emergence of many-bodies interactions favoring the lateral packing of the monolayer [65].

The differences of the A^E values found between monolayers after the incorporation of SiO_2 and CB particles may be again ascribed to the different organization of the particles within the DPPC monolayers as a function of their chemical nature. Therefore, the lowest values of A^E found upon the incorporation of SiO_2 are explained by the better distribution of these particles within the lipid layer, and the possible role of the electrostatic interactions between the silanol groups on the surface of the particles and the ammonium terminal group of the DPPC.

3.4. Effect of Particles' Incorporation in the Dilational Response of DPPC Langmuir Monolayers

The above discussion was focused on the impact of the particles' incorporation on the equilibrium properties of DPPC monolayers. However, biological systems are highly dynamic system, and hence the study of the effects associated with particles incorporation into DPPC monolayers in the response against mechanical deformations is a useful tool for a preliminary evaluation of the impact of the particles on the functionality of lipids layers. For this purpose, the influence of particles in the response of the DPPC to dilational deformations has been studied using the oscillatory barrier method at a fixed deformation amplitude within the linear response regime (1% of the initial area). These studies inform the modification of the relaxation mechanisms involved in the re-equilibration of the lipid layer as a result of the incorporation of particles [45,59]. For this purpose, the analysis of the frequency (ν) dependences of the viscoelastic modulus ($|E|$) for the pure DPPC monolayers and upon the incorporation of the particles is performed. Figure 8 shows, for the sake of an example, some of the frequency dependences obtained for the interfacial dilational viscoelasticity at different surface pressures for the mixed monolayers containing different weight fractions of particles.

Most of the viscoelastic modulus–deformation frequency curves obtained show the existence of inflexion points which are associated with the characteristic frequency of the reorganization of the molecules and particles within the interface The incorporation of particles, independently of their chemical nature, modifies the relaxation mechanism of the lipid monolayer, i.e., the characteristic relaxation frequencies, as is evidenced from the experimental curves. The characteristic relaxation frequencies can be estimated fitting the experimental data to the following theoretical expression which enables the description of interfacial relaxation occurring in insoluble adsorption layers [66]

$$|E| = [(E_1{}^2 + \lambda^2 E_0{}^2)/(1 + \lambda^2)]^{1/2}, \tag{8}$$

where $\lambda = \nu_R/\nu$, with ν_R being the characteristic relaxation frequency, and E_0 and E_1 are the lower and upper limits of the elasticity within the considered frequency range. Note that, when insoluble monolayers are concerned, E_0 coincides with the quasi-static dilational elasticity obtained from the isotherm. The theoretical curves obtained using the model defined by Equation (8) are shown in

Figure 8 together with the experimental data, and the values of the characteristic frequencies ν_R of the dilational response of mixed monolayers are shown in Figure 9.

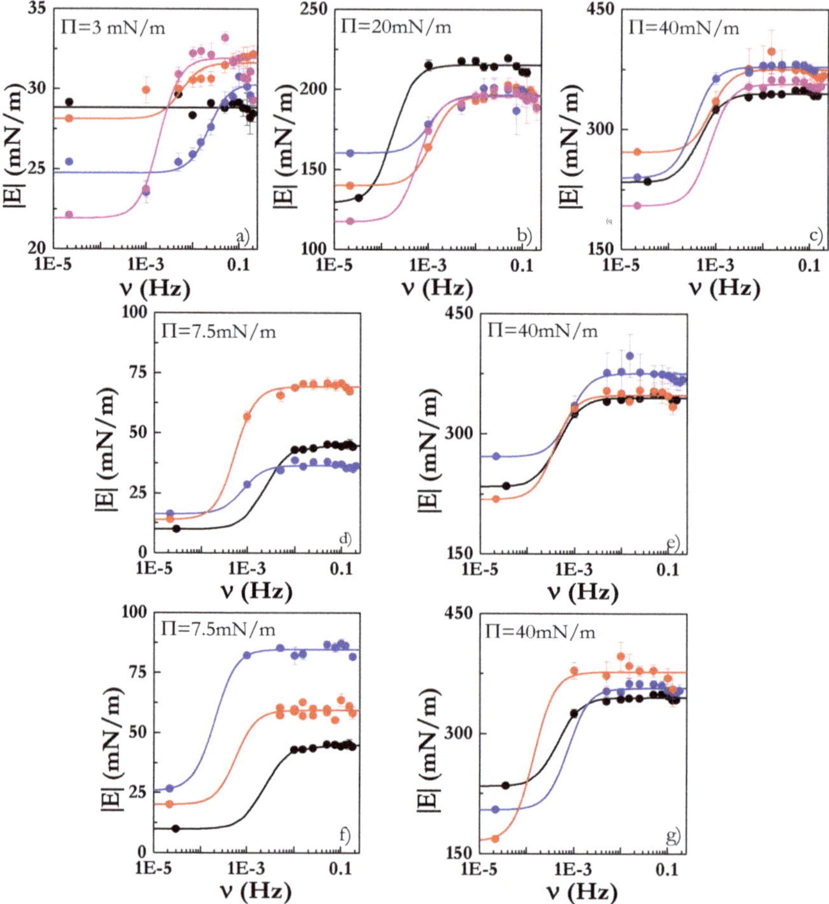

Figure 8. Experimental (symbols) and calculated using Equation (8) (lines) dependences of the interfacial dilational viscoelasticity modulus on frequency. (**a–c**) dependences of $|E|$ on ν for DPPC after the incorporation of different x_p values at fixed surface pressure values (data for CB): (●, —) 0, (●,—) 0.10, (●,—) 0.33 and (●,—) 0.75. (**d,e**) dependences of $|E|$ on ν for DPPC after the incorporation of particles with different chemical nature at fixed surface pressure values and x_p = 0.10: (●, —) pure DPPC, (●,—) DPPC upon incorporation of SiO_2 particles and (●,—) DPPC upon incorporation of CB particles. (**g,f**) Dependences of $|E|$ on ν for DPPC after the incorporation of particles with different chemical nature at fixed surface pressure values and x_p =0.75: (●, —) pure DPPC, (●,—) DPPC upon incorporation of SiO_2 particles and (●,—) DPPC upon incorporation of CB particles. (Note that the value of the $|E|$ for the lowest frequency value was assumed as the quasi-static dilational elasticity obtained from the isotherm, with the value of the frequency being the compression ratio which for the here experiments is about 10^{-5} Hz).

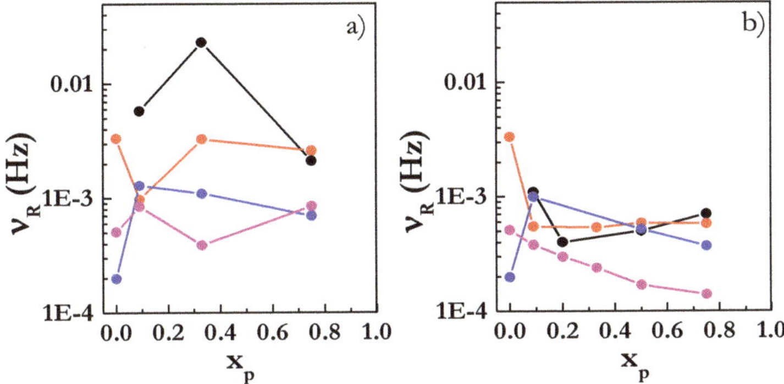

Figure 9. ν_R dependences, obtained using Equation (4), on the particle weight fraction, x_p, for different values of the surface pressure Π for DPPC monolayers upon incorporation of CB (**a**) and SiO$_2$ (**b**): (—•—) 3 mN/m, (—•—) 7.5 mN/m, (—•—) 20 mN/m and (—•—) 40 mN/m. Notice that the lines are guides for the eyes.

The analysis of the viscoelastic modulus-deformation frequency curves obtained for monolayers of pure DPPC shows the existence of an inflexion point, with the exception of that obtained for the lowest value of Π. The absence of such relaxation process for the lowest value of the surface pressure is explained assuming the low interfacial density, which allows the free reorganization of the DPPC molecules within the interface. This takes the relaxation process to time-scales below (higher frequencies) those tested by oscillatory barrier experiments.

The results show that the incorporation of particles modifies the relaxation mechanism of DPPC molecules at the water/vapor interface from the lowest values of surface pressure. SiO$_2$ and CB particles' incorporation leads to the emergence of a relaxation process at Π values about 3 mN/m (LE phase for pure DPPC monolayers), with the time-scale for such relaxation process being faster when the incorporation of CB particles is concerned (0.001–0.01 Hz for CB particles incorporation vs. 0.0001–0.001 Hz for SiO$_2$ ones). This results from the most important role of the steric hindrance associated with the incorporation of SiO$_2$ particles, which makes it the lateral reorganization of the lipid molecules within the monolayer more difficult than when the incorporation of CB particles is considered. The origin of the emergence of a relaxation process may be explained considering the increase of the interfacial density associated with the presence of the particles, which reduces the time-scales involved in the reorganization of the molecules at the interface. The decrease of the time-scale involved in the reorganization process is stronger as the interfacial density of the particles increases. Therefore, it is possible to assume a slowing down of the velocity of this relaxation as a result of an increased role of the steric hindrance.

The approaching of the LE–LC coexistence phase provides evidence again of the differences on the effect of the incorporation of SiO$_2$ and CB particles. The incorporation of CB particles does not modify significantly the relaxation mechanism, with a relaxation process presenting $\nu_R \sim 10^{-3}$–10^{-4} Hz appearing independently of the considered monolayer, i.e., for pure DPPC monolayers and upon the incorporation of CB particles. This is explained considering that, within the phase coexistence region, the nucleation of the LC phase associated with the disappearance of the LE one is found in both cases. Thus, the relaxation process should be ascribed to the exchange of the lipid molecules between the LC and LE phase. However, the introduction of SiO$_2$ particles leads to a slowdown of the relaxation process for almost one order of magnitude [31]. This may be explained considering that the incorporation of SiO$_2$ particles hinders partially the phase coexistence as result of the stronger steric hindrance associated with the particles incorporation which modifies the lateral reorganization of the lipid molecules at the interface.

Once the phase coexistence is overcome, the incorporation of CB does not lead to any significant change of the relaxation mechanism, with the relaxation frequency remaining in values about 10^{-4} Hz, independently of the considered state and the x_p value. This may be understood assuming that CB particles are simple obstacles that limit the average cohesion of the DPPC monolayers, shifting the phase behavior, without any significant change on the lipid lateral packing. However, the situation appears to be different when the incorporation of SiO_2 particles is concerned, and the relaxation mechanism was found to be dependent on the x_p value. The incorporation of SiO_2 particles increases the value of ν_R with the x_p value up to reach a maximum value, and then a decrease of ν_R with the increase of the x_p was observed with the enhancing of the lateral packing of the monolayer. This is explained assuming the complexity of the interactions balance involved in DPPC monolayers upon the incorporation of SiO_2 particles, which leads to the existence of coupled dynamics on the rheological response of the mixed monolayers against dilational deformations. It is worth mentioning that the increase of the interfacial density of the particles may induce a similar lateral packing of the monolayer for lower values of the surface pressure, and this may explain the characteristic features found for the x_p dependence of the relaxation frequencies. Notice that, for the highest values of the lateral packing, the characteristic relaxation frequency appears to be similar for pure DPPC monolayers and upon incorporation of SiO_2 that corresponds to the mixed layer appearing in larger time-scales. This subtle difference can be again ascribed to the role of the steric hindrance interactions, which makes it possible that the relaxation process may include complex rearrangements involving both the particles and the lipid molecules.

Figure 10 shows the dependences of E_1 on the x_p value. The E_1 values obtained prove clearly an increase with the x_p value when the incorporation of CB particles is considered; this may be understood considering that particles occupy partially the area available for the lipid organization, i.e., behave as obstacles, driving to a prior packing of the DPPC at the interface (packing occurs at higher values of the reduced area). About the dependences E_1, the impact of SiO_2 approximates the CB particle one. This aspect can be explained assuming the importance of the occupancy of the interfacial area by the incorporated particles in their impact, independently of their chemical nature.

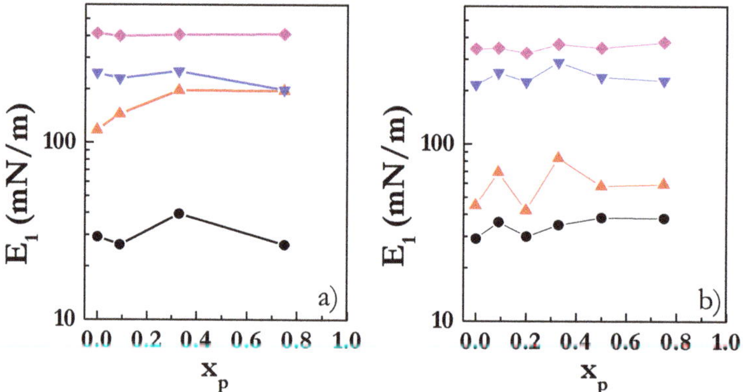

Figure 10. E_1 dependences, obtained using Equation (4), on the particle weight fraction, x_p, for different values of the surface pressure Π for DPPC monolayers upon incorporation of CB (**a**) and SiO_2 (**b**): (—•—) 3 mN/m, (—•—) 7.5 mN/m, (—•—) 20 mN/m and (—•—) 40 mN/m. Notice that the lines are guides for the eyes.

4. Conclusions

This study has shown the influence of the chemical nature of particles interacting with lipid layers. The results provided evidence that CB and SiO_2 particles alter the lateral packing of the DPPC monolayers, modifying the cohesion between the lipid molecules within the interface and worsening the mechanical performance of the lipid film. However, the mechanisms driving such modifications are strongly dependent on the specific chemistry of the considered particles, and, in particular, on the interactions involved in their incorporation. This recalls a framework needing the consideration of the complex balance between the contribution of the hydrophobic and electrostatic interactions and the steric hindrance in the behavior of the lipid layer. Therefore, even though particles' most fundamental physical properties can appear to be similar, their specific chemistry, and possibly their geometry and aggregation at the interface, determine their impact on their incorporation into lipid layers, which may even result in the modification of the lipid ability to form ordered phases. This distortion of the lateral order of the lipid monolayers is also associated with the modification of the relaxation mechanism driving the re-equilibration of the interfacial layers after periodic dilational deformations, i.e., the reorganization of molecules within the interface. Despite the simplicity of the considered model, the results have shown that the study of the changes of the equilibrium properties and the rheological response of lipid layers due to their interaction can be used as a powerful for evaluating the impact of pollutants in the functioning of biological layers.

Author Contributions: Conceptualization, E.G., L.L., and F.R.; methodology, E.G., E.S., and M.F.; software, E.G.; validation, L.L. and F.R.; formal analysis, E.G.; investigation, E.G., E.S., M.F., L.L., and F.R.; resources, L.L. and F.R.; data curation, E.G.; writing—original draft preparation, E.G.; writing—review and editing, E.G., E.S., and M.F., L.L., and F.R.; visualization, E.G.; supervision, L.L. and F.R.; project administration, L.L. and F.R.; funding acquisition, E.G., L.L., and F.R. All authors have read and agreed to the published version of the manuscript.

Funding: This work was financially supported by IIT-Istituto Italiano di Tecnologia within the Project SEED 2009 "Nanoparticle Impact of Pulmonary Surfactant Interfacial Properties-NIPS", by MINECO under grant CTQ2016-78895-R and by Banco Santander-Universidad Complutense grant PR87/19-22513.

References

1. Ghorani-Azam, A.; Riahi-Zanjani, B.; Balali-Mood, M. Effects of air pollution on human health and practical measures for prevention in Iran. *J. Res. Med. Sci.* **2016**, *21*, 65. [CrossRef] [PubMed]

2. Curtis, L.; Rea, W.; Smith-Willis, P.; Fenyves, E.; Pan, Y. Adverse health effects of outdoor air pollutants. *Environ. Int.* **2006**, *32*, 815–830. [CrossRef] [PubMed]

3. Park, M.; Joo, H.S.; Lee, K.; Jang, M.; Kim, S.D.; Kim, I.; Borlaza, L.J.S.; Lim, H.; Shin, H.; Chung, K.H.; et al. Differential toxicities of fine particulate matters from various sources. *Sci. Rep.* **2018**, *8*, 17007. [CrossRef] [PubMed]

4. Pavese, G.; Alados-Arboledas, L.; Cao, J.; Satheesh, S.K. Carbonaceous particles in the atmosphere: Experimental and modelling issues. *Adv. Meteorol.* **2014**, *2014*, 529850. [CrossRef]

5. How Air Pollution Is Destroying Our Health. Available online: https://www.who.int/airpollution/news-and-events/how-air-pollution-is-destroying-our-health (accessed on 2 March 2020).

6. Iannelli, R.; Bianchi, V.; Macci, C.; Peruzzi, E.; Chiellini, C.; Petroni, G.; Masciandaro, G. Assessment of pollution impact on biological activity and structure of seabed bacterial communities in the Port of Livorno (Italy). *Sci. Total Environ.* **2012**, *426*, 56–64. [CrossRef]

7. Van Smeden, J.; Bouwstra, J.A. Stratum Corneum Lipids: Their Role for the Skin Barrier Function in Healthy Subjects and Atopic Dermatitis Patients. *Curr. Probl. Dermatol.* **2016**, *49*, 9–26. [CrossRef]

8. Donaldson, K.; Seaton, A. A short history of the toxicology of inhaled particles. *Part. Fibre Toxicol.* **2012**, *9*, 13. [CrossRef]

9. Kondej, D.; Sosnowski, T.R. Effect of clay nanoparticles on model lung surfactant: A potential marker of hazard from nanoaerosol inhalation. *Environ. Sci. Pollut. Res.* **2016**, *23*, 4660–4669. [CrossRef]

10. Lecas, S.; Boursier, E.; Fitoussi, R.; Vié, K.; Momas, I.; Seta, N.; Achard, S. In vitro model adapted to the study of skin ageing induced by air pollution. *Toxicol. Lett.* **2016**, *259*, 60–68. [CrossRef]

11. Stefaniu, C.; Brezesinski, G.; Möhwald, H. Langmuir monolayers as models to study processes at membrane surfaces. *Adv. Colloid Interface Sci.* **2014**, *208*, 197–213. [CrossRef]

12. Podgórski, A.; Sosnowski, T.R.; Gradoń, L. Deactivation of the Pulmonary Surfactant Dynamics by Toxic Aerosols and Gases. *J. Aerosol Med.* **2001**, *14*, 455–466. [CrossRef] [PubMed]

13. Farnoud, A.M.; Fiegel, J. Low concentrations of negatively charged sub-micron particles alter the microstructure of DPPC at the air–water interface. *Colloids Surf. A* **2012**, *415*, 320–327. [CrossRef]

14. Sosnowski, T.R.; Kolinski, M.; Gradon, L. Alteration of Surface Properties of Dipalmitoyl Phosphatidylcholine by Benzo[a]pyrene: A Model of Pulmonary Effects of Diesel Exhaust Inhalation. *J. Biomed. Nanotech.* **2012**, *8*, 818–825. [CrossRef] [PubMed]

15. Dwivedi, M.V.; Harishchandra, R.K.; Koshkina, O.; Maskos, M.; Galla, H.-J. Size Influences the Effect of Hydrophobic Nanoparticles on Lung Surfactant Model Systems. *Biophys. J.* **2014**, *106*, 289–298. [CrossRef]

16. Sosnowski, T.R.; Kubski, P.; Wojciechowski, K. New experimental model of pulmonary surfactant for biophysical studies. *Colloids Surf. A* **2017**, *519*, 27–33. [CrossRef]

17. Guzmán, E.; Santini, E. Lung surfactant-particles at fluid interfaces for toxicity assessments. *Curr. Opin. Colloid Interface Sci.* **2019**, *39*, 24–39. [CrossRef]

18. Garcia-Mouton, C.; Hidalgo, A.; Cruz, A.; Pérez-Gil, J. The Lord of the Lungs: The essential role of pulmonary surfactant upon inhalation of nanoparticles. *Eur. J. Pharm. Biopharm.* **2019**, *144*, 230–243. [CrossRef]

19. Sosnowski, T.R. Particles on the lung surface—Physicochemical and hydrodynamic effects. *Curr. Opin. Colloid Interface Sci.* **2018**, *36*, 1–9. [CrossRef]

20. Guzman, E.; Orsi, D.; Cristofolini, L.; Liggieri, L.; Ravera, F. Two-Dimensional DPPC Based Emulsion-like Structures Stabilized by Silica Nanoparticles. *Langmuir* **2014**, *30*, 11504–11512. [CrossRef]

21. Guzman, E.; Santini, E.; Ferrari, M.; Liggieri, L.; Ravera, F. Interfacial Properties of Mixed DPPC-Hydrophobic Fumed Silica Nanoparticle Layers. *J. Phys. Chem. C* **2015**, *119*, 21024–21034. [CrossRef]

22. Orsi, D.; Guzman, E.; Liggieri, L.; Ravera, F.; Ruta, B.; Chushkin, Y.; Rimoldi, T.; Cristofolini, L. 2D dynamical arrest transition in a mixed nanoparticle-phospholipid layer studied in real and momentum spaces. *Sci. Rep.* **2015**, *5*, 14. [CrossRef] [PubMed]

23. Guzman, E.; Santini, E.; Zabiegaj, D.; Ferrari, M.; Liggieri, L.; Ravera, F. Interaction of Carbon Black Particles and Dipalmitoylphosphatidylcholine at the Water/Air Interface: Thermodynamics and Rheology. *J. Phys. Chem. C* **2015**, *119*, 26937–26947. [CrossRef]

24. Guzman, E.; Santini, E.; Ferrari, M.; Liggieri, L.; Ravera, F. Effect of the Incorporation of Nanosized Titanium Dioxide on the Interfacial Properties of 1,2-Dipalmitoyl-sn-glycerol-3-phosphocholine Langmuir Monolayers. *Langmuir* **2017**, 10715–10725. [CrossRef] [PubMed]

25. Guzman, E.; Ferrari, M.; Santini, E.; Liggieri, L.; Ravera, F. Effect of silica nanoparticles on the interfacial properties of a canonical lipid mixture. *Colloids Surf. B* **2015**, *136*, 971–980. [CrossRef]

26. Bykov, A.G.; Guzman, E.; Rubio, R.G.; Krycki, M.M.; Milyaeva, O.Y.; Noskov, B.A. Influence of temperature on dynamic surface properties of spread DPPC monolayers in a broad range of surface pressures. *Chem. Phys. Lipids* **2019**, *225*, 6. [CrossRef]

27. Lopez-Rodriguez, E.; Pérez-Gil, J. Structure-function relationships in pulmonary surfactant membranes: From biophysics to therapy. *Biochim. Biophys. Acta Biomembr.* **2014**, *1838*, 1568–1585. [CrossRef]

28. Yang, Y.; Wu, Y.; Ren, Q.; Zhang, L.G.; Liu, S.; Zuo, Y.Y. Biophysical Assessment of Pulmonary Surfactant Predicts the Lung Toxicity of Nanomaterials. *Small Methods* **2018**, *2*, 1700367. [CrossRef]

29. Valle, R.P.; Wu, T.; Zuo, Y.Y. Biophysical Influence of Airborne Carbon Nanomaterials on Natural Pulmonary Surfactant. *ACS Nano* **2015**, *9*, 5413–5421. [CrossRef]

30. Miguel Diez, M.; Buckley, A.; Tetley, T.D.; Smith, R. The method of depositing CeO2 nanoparticles onto a DPPC monolayer affects surface tension behaviour. *NanoImpact* **2019**, *16*, 100186. [CrossRef]

31. Maestro, A.; Guzmán, E. Colloids at Fluid Interfaces. *Processes* **2019**, *7*, 942. [CrossRef]

32. Maestro, A.; Santini, E.; Guzmán, E. Physico-chemical foundations of particle-laden fluid interfaces. *Eur. Phys. J. E* **2018**, *41*, 97. [CrossRef] [PubMed]

33. Maestro, A.; Santini, E.; Zabiegaj, D.; Llamas, S.; Ravera, F.; Liggieri, L.; Ortega, F.; Rubio, R.G.; Guzman, E. Particle and Particle-Surfactant Mixtures at Fluid Interfaces: Assembly, Morphology, and Rheological Description. *Adv. Cond. Matter Phys.* **2015**, *2015*, 917516. [CrossRef]

34. Dasgupta, S.; Auth, T.; Gompper, G. Nano- and microparticles at fluid and biological interfaces. *J. Phys. Condens. Matter* **2017**, *29*, 373003. [CrossRef] [PubMed]

35. Beddoes, C.M.; Case, C.P.; Briscoe, W.H. Understanding nanoparticle cellular entry: A physicochemical perspective. *Adv. Colloid Interface Sci.* **2015**, *218*, 48–68. [CrossRef] [PubMed]

36. Guzmán, E.; Liggieri, L.; Santini, E.; Ferrari, M.; Ravera, F. Effect of Hydrophilic and Hydrophobic Nanoparticles on the Surface Pressure Response of DPPC Monolayers. *J. Phys. Chem. C* **2011**, *115*, 21715–21722. [CrossRef]

37. Borozenko, O.; Faral, M.; Behyan, S.; Khan, A.; Coulombe, J.; DeWolf, C.; Badia, A. Silica Nanoparticle-Induced Structural Reorganizations in Pulmonary Surfactant Films: What Monolayer Compression Isotherms Do Not Say. *ACS App. Nano Mat.* **2018**, *1*, 5268–5278. [CrossRef]

38. Hu, J.; Li, X.; Li, M.; Shang, Y.; He, Y.; Liu, H. Real-time monitoring of the effect of carbon nanoparticles on the surface behavior of DPPC/DPPG Langmuir monolayer. *Colloids Surf. B* **2020**, *190*, 110922. [CrossRef]

39. Santini, E.; Ravera, F.; Ferrari, M.; Alfè, M.; Ciajolo, A.; Liggieri, L. Interfacial properties of carbon particulate-laden liquid interfaces and stability of related foams and emulsions. *Colloids Surf. A* **2010**, *365*, 189–198. [CrossRef]

40. Spahr, M.E.; Rothon, R. Carbon Black as a Polymer Filler. In *Polymers and Polymeric Composites: A Reference Series*; Palsule, S., Ed.; Springer: Berlin/Heidelberg, Germany, 2016; pp. 1–31.

41. Wu, H.-L.; Tong, Y.; Peng, Q.; Li, N.; Ye, S. Phase transition behaviors of the supported DPPC bilayer investigated by sum frequency generation (SFG) vibrational spectroscopy and atomic force microscopy (AFM). *Phys. Chem. Chem. Phys.* **2016**, *18*, 1411–1421. [CrossRef]

42. Gradoń, L.; Podgórski, A. Hydrodynamical model of pulmonary clearance. *Chem. Eng. Sci.* **1989**, *44*, 741–749. [CrossRef]

43. Gradon, L.; PodgoRski, A.; Sosnowski, T.R. Experimental and Theoretical Investigations of Transport Properties of DPPC Monolayer. *J. Aerosol Med.* **1996**, *9*, 357–367. [CrossRef]

44. Hifeda, Y.F.; Rayfield, G.W. Evidence for first-order phase transitions in lipid and fatty acid monolayers. *Langmuir* **1992**, *8*, 197–200. [CrossRef]

45. Guzmán, E.; Liggieri, L.; Santini, E.; Ferrari, M.; Ravera, F. Influence of silica nanoparticles on dilational rheology of DPPC–palmitic acid Langmuir monolayers. *Soft Matter* **2012**, *8*, 3938–3948. [CrossRef]

46. Liggieri, L.; Santini, E.; Guzmán, E.; Maestro, A.; Ravera, F. Wide-frequency dilational rheology investigation of mixed silica nanoparticle–CTAB interfacial layers. *Soft Matter* **2011**, *7*, 7699–7709. [CrossRef]

47. Mendoza, A.J.; Guzmán, E.; Martínez-Pedrero, F.; Ritacco, H.; Rubio, R.G.; Ortega, F.; Starov, V.M.; Miller, R. Particle laden fluid interfaces: Dynamics and interfacial rheology. *Adv. Colloid Interface Sci.* **2014**, *206*, 303–319. [CrossRef] [PubMed]

48. Schürch, S. Surface tension at low lung volumes: Dependence on time and alveolar size. *Respir. Physiol.* **1982**, *48*, 339–355. [CrossRef]

49. Günther, A.; Schmidt, R.; Harodt, J.; Schmehl, T.; Walmrath, D.; Ruppert, C.; Grimminger, F.; Seeger, W. Bronchoscopic administration of bovine natural surfactant in ARDS and septic shock: Impact on biophysical and biochemical surfactant properties. *Eur. Resp. J.* **2002**, *19*, 797–804. [CrossRef]

50. López-Rodríguez, E.; Ospina, O.L.; Echaide, M.; Taeusch, H.W.; Pérez-Gil, J. Exposure to polymers reverses inhibition of pulmonary surfactant by serum, meconium, or cholesterol in the captive bubble surfactometer. *Biophys. J.* **2012**, *103*, 1451–1459. [CrossRef]

51. Beck-Broichsitter, M.; Ruppert, C.; Schmehl, T.; Günther, A.; Seeger, W. Biophysical inhibition of pulmonary surfactant function by polymeric nanoparticles: Role of surfactant protein B and C. *Acta Biomat.* **2014**, *10*, 4678–4684. [CrossRef]

52. Phillips, M.C.; Chapman, D. Monolayer characteristics of saturated 1,2-diacyl phosphatidylcholines (lecithins) and phosphatidylethanolamines at the air-water interface. *Biochim. Biophys. Acta Biomembr.* **1968**, *163*, 301–313. [CrossRef]

53. Klopfer, K.J.; Vanderlick, T.K. Isotherms of Dipalmitoylphosphatidylcholine (DPPC) Monolayers: Features Revealed and Features Obscured. *J. Colloid Interface Sci.* **1996**, *182*, 220–229. [CrossRef]

54. Arriaga, L.R.; López-Montero, I.; Ignés-Mullol, J.; Monroy, F. Domain-Growth Kinetic Origin of Nonhorizontal Phase Coexistence Plateaux in Langmuir Monolayers: Compression Rigidity of a Raft-Like Lipid Distribution. *J. Phys. Chem. B* **2010**, *114*, 4509–4520. [CrossRef] [PubMed]

55. Grigoriev, D.O.; Krägel, J.; Dutschk, V.; Miller, R.; Möhwald, H. Contact angle determination of micro- and nanoparticles at fluid/fluid interfaces: The excluded area concept. *Phys. Chem. Chem. Phys.* **2007**, *9*, 6447–6454. [CrossRef] [PubMed]

56. Wang, Z.; Yang, S. Effects of Fullerenes on Phospholipid Membranes: A Langmuir Monolayer Study. *ChemPhysChem* **2009**, *10*, 2284–2289. [CrossRef]

57. Wang, Z.; Li, X.; Yang, S. Studies of Dipalmitoylphosphatidylcholine (DPPC) Monolayers Embedded with Endohedral Metallofullerene (Dy@C82). *Langmuir* **2009**, *25*, 12968–12973. [CrossRef] [PubMed]

58. Sheridan, A.J.; Slater, J.M.; Arnold, T.; Campbell, R.A.; Thompson, K.C. Changes to DPPC Domain Structure in the Presence of Carbon Nanoparticles. *Langmuir* **2017**, *33*, 10374–10384. [CrossRef] [PubMed]

59. Muñoz-López, R.; Guzmán, E.; Velázquez, M.M.; Fernández-Peña, L.; Merchán, M.D.; Maestro, A.; Ortega, F.; G Rubio, R. Influence of Carbon Nanosheets on the Behavior of 1,2-Dipalmitoyl-sn-glycerol-3-phosphocholine Langmuir Monolayers. *Processes* **2020**, *8*, 94. [CrossRef]

60. Medalia, A.I. Reinforcement of Elastomers. In *Encyclopedia of Materials: Science and Technology*; Buschow, K.H.J., Cahn, R.W., Flemings, M.C., Ilschner, B., Kramer, E.J., Mahajan, S., Veyssière, P., Eds.; Elsevier: Oxford, UK, 2001; pp. 2475–2480.

61. Guzmán, E.; Liggieri, L.; Santini, E.; Ferrari, M.; Ravera, F. Influence of silica nanoparticles on phase behavior and structural properties of DPPC—Palmitic acid Langmuir monolayers. *Colloids Surf. A* **2012**, *413*, 280–287. [CrossRef]

62. Farnoud, A.M.; Fiegel, J. Calf Lung Surfactant Recovers Surface Functionality After Exposure to Aerosols Containing Polymeric Particles. *J. Aerosol Med. Pulm. Drug Deliv.* **2016**, *29*, 10–23. [CrossRef]

63. Gaines, G.L. *Insoluble Monolayers at Liquid–Gas Interfaces*; Interscience Publishers: New York, NY, USA, 1966.

64. Dynarowicz-Łątka, P.; Kita, K. Molecular interaction in mixed monolayers at the air/water interface. *Adv. Colloid Interface Sci.* **1999**, *79*, 1–17. [CrossRef]

65. Qiao, R.; Roberts, A.P.; Mount, A.S.; Klaine, S.J.; Ke, P.C. Translocation of C60 and Its Derivatives Across a Lipid Bilayer. *Nano Lett.* **2007**, *7*, 614–619. [CrossRef] [PubMed]

66. Ravera, F.; Ferrari, M.; Santini, E.; Liggieri, L. Influence of surface processes on the dilational visco-elasticity of surfactant solutions. *Adv. Colloid Interface Sci.* **2005**, *117*, 75–100. [CrossRef] [PubMed]

Article

Structure of DPPC Monolayers at the Air/Buffer Interface: A Neutron Reflectometry and Ellipsometry Study

Javier Carrascosa-Tejedor [1,2,†], Andreas Santamaria [1,3,†], Daniel Pereira [1] and Armando Maestro [1,*]

[1] Large Scale Structures Group, Institut Laue-Langevin, 71 Avenue des Martyrs, CEDEX 9, 38042 Grenoble, France; carrascosa-tejedor@ill.fr (J.C.-T.); santamariaa@ill.fr (A.S.); pereirad@ill.fr (D.P.)

[2] Division of Pharmacy and Optometry, University of Manchester, Manchester M13 9PT, UK

[3] Departamento de Química - Física I, Facultad de Ciencias Químicas, Universidad Complutense de Madrid, Ciudad Universitaria s/n, 28040 Madrid, Spain

* Correspondence: maestro@ill.fr

† These authors contributed equally to this work.

Received: 1 April 2020; Accepted: 21 May 2020; Published: 26 May 2020

Abstract: Langmuir monolayers of 1,2-dipalmitoyl-sn-glycerol-3-phosphocholine, known as DPPC, at the air/water interface are extensively used as model systems of biomembranes and pulmonary surfactant. The properties of these monolayers have been mainly investigated by surface pressure–area isotherms coupled with different complementary techniques such as Brewster angle microscopy, for example. Several attempts using neutron reflectometry (NR) or ellipsometry have also appeared in the literature. Here, we report structural information obtained by using NR and ellipsometry on DPPC monolayers in the liquid condensed phase. On one side, NR can resolve the thickness of the aliphatic tails and the degree of hydration of the polar headgroups. On the other side, ellipsometry gives information on the refractive index and, therefore, on the physical state of the monolayer. The thickness and surface excess obtained by multiple-angle-of-incidence ellipsometry (MAIE) is compared with the results from NR measurements yielding a good agreement. Besides, a novel approach is reported to calculate the optical anisotropy of the DPPC monolayer that depends on the orientation of the aliphatic chains. The results from both NR and ellipsometry are also discussed in the context of the existing results for DPPC monolayers at the air/water interface. The differences observed are rationalized by the presence of buffer molecules interacting with phospholipids.

Keywords: neutron reflectometry; ellipsometry; DPPC; lipid monolayers; air/water interface

---

1. Introduction

Synthetic in vitro lipid mono- and bilayers, as well as uni- and multi-lamellar vesicles, can be considered as simple biomembrane models. For more than 20 years, 1,2-dipalmitoyl-sn-glycerol-3-phosphocholine (DPPC) has been widely exploited to mimic plasma membranes or lung surfactant, mostly in the form of monolayers at the air/water interface. Indeed, phospholipids with 16- and 18-carbon fatty acids chains are the most abundant in plasma membranes [1,2]. DPPC is, therefore, a well-studied lipid. Besides, DPPC has being used alone or in combination with other lipids, to study the interaction of biomembrane with proteins [3,4], anticancer [5–8] and antifungal [9] compounds, and small molecules of biological relevance, such as cholesterol [10], hormones [11], and antibiotics [12]. Its handling simplicity, relatively low price, and stability at room temperature as well as when exposed to air, make DPPC a versatile model system for biomedical research.

In general, phospholipid monolayers at the air/water interface can be characterized by different techniques such as microscopy (e.g., atomic force microscopy, AFM, and Brewster angle microscopy,

BAM), scattering (e.g., neutron reflectometry, NR, and X-ray reflectometry, XRR), and ellipsometric and spectroscopic (e.g., polarization modulation infrared reflection–absorption spectroscopy) techniques, together with surface pressure-area (Π-A) isotherms. Currently, there are many outstanding examples of the usefulness of this combinatorial approach, exploited to study DPPC monolayers structure and properties [13,14] as well as their interaction with different cations [15], nanoparticles [16], graphite-based compounds [17], and molecules of biological relevance, such as proteins [4,18], small peptides [19], and antibiotics [12].

Different techniques have been used to study the structure and optical properties of lipid monolayers. Indeed, the refractive index of the film (n_F) in combination with the reflectivity allows obtaining the thickness (d_F) of the monolayers. Kienle et al. successfully used multiple beam interferometry to determine simultaneously n_F and d_F of DPPC and DPPE (1,2-dipalmitoyl-sn-glycero-3-phosphoethanolamine) supported monolayers at high surface pressures [20]. The results obtained by this technique were in good agreement with those obtained by XRR and AFM. Another approach to determine n_F of monolayers at the air/water interface is looking at the minimum of the reflectivity as a function of the refractive index of the subphase. Pusterla et al. used this approach varying the concentration of glycerol or sucrose in the subphase, with known refractive indices [21]; the refractive index of the monolayer is equal to the one of the subphase when the reflectivity is minimum. Besides, knowing the refractive index and the reflectivity they calculated d_F. The determination of the increment of the refractive index with concentration (dn/dc) constitutes another alternative to obtain n_F. The subsequent application of the lipid density to the dn/dc provides n_F [22]. However, n_F can slightly vary with the surface pressure, especially from one phase to another, so it is necessary to take into account the physical state of the monolayer to know its n_F [23]. NR has also been widely exploited to perform studies at the air/water interface to investigate the structure and properties of lipid monolayers [24–26]. This technique enables complete structural characterization of the monolayers, giving information about the chemical composition along the axis normal to the interface, the thickness of both polar headgroups layer and hydrophobic tails layer when a two-layer model is used to interpret the data. Through NR it is also possible to determine the hydration degree of the lipids, as well as the surface excess and the area per molecule at a certain value of surface pressure [27]. Hence, NR is a very powerful tool to study lipid monolayers at the air/water interface.

Traditionally, ellipsometry has been one of the most exploited techniques to study surfactant and lipid monolayers at the air/water interface [28,29]. It gives access to the determination of the thickness and the refractive index of films, and it is useful to investigate the interaction of monolayers with different molecules such as proteins [30,31] or nanoparticles [32,33] through time/spatial resolved experiments. Nevertheless, the simultaneous determination of the n_F and the d_F for Langmuir films at the air/water interface with $d_F \ll \lambda$ by the measurement of the phase shift at a fixed angle of incidence, can give inaccurate results due to the strong coupling of the parameters [29,34]. The combination of ellipsometry with other techniques that give access to the determination of one of the parameters allows the accurate determination of the other one by ellipsometry [35,36]. Benjamins et al. developed a method for the study of films at liquid interfaces by ellipsometry without assumptions of the thickness or the refractive index [37]. They demonstrated that the combination of measurements performed for the same system using D_2O and H_2O as the subphase, i.e., different refractive indices of the subphase, give enough additional information to determine the amount of material adsorbed.

In this work, we use the combination of two reflection techniques, NR and ellipsometry, and surface pressure measurements to determine the interfacial structure and the optical properties, including the refractive index anisotropy, of a condensed DPPC monolayer at the air/buffer interface (see Figure 1). Besides, we show how HKM buffer molecules (see composition below), widely exploited in biological assays, are responsible for the differences observed in the structure and density of DPPC monolayers.

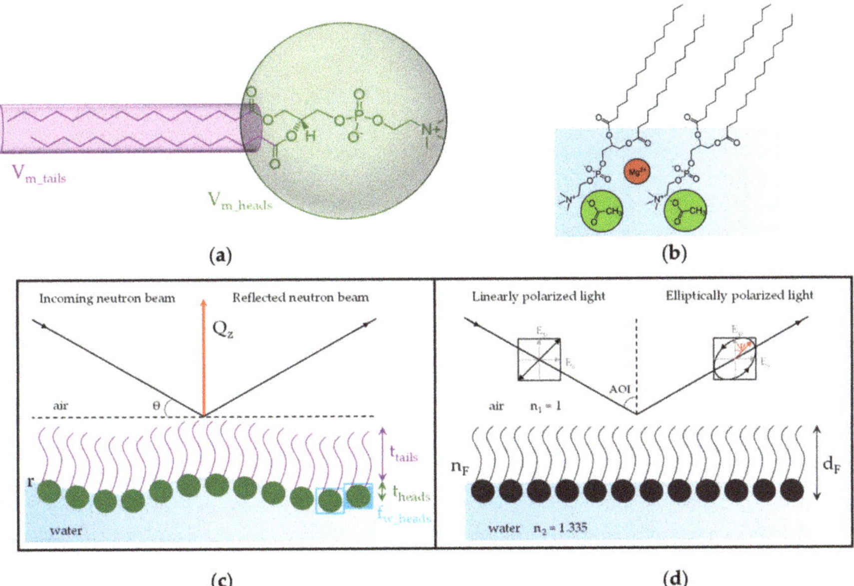

Figure 1. (**a**) Structure formula of DPPC. The polar headgroup is depicted in green and the hydrophobic tails in violet following a two-layer model proposed in the main text. A green sphere and a violet cylinder are drawn to better show V_{m_heads} and V_{m_tails}, respectively. (**b**) The scheme representing the main electrostatic interactions between the phospholipids molecules with the ions composing the buffer: the magnesium cation is depicted as a red sphere, while the acetate anion is depicted as a green sphere. We deduce that these interactions have an influence on DPPC monolayer structure. Bottom panel compares neutron reflectometry (**c**) and ellipsometry (**d**) principles of measurement. In the case of NR, a two-layer model can be exploited in order to get both the thickness of the headgroups (t_{heads}) and the one of the tails (t_{tails}); indeed, the two-layer model is shown, depicting the heads in green and the tails in violet. Moreover, NR also gives information about the fraction of water per polar headgroup (f_{w_heads}) and the roughness of the interfaces (r), whose value depends on the water capillary waves. The angle of incidence of the neutron beam (θ) and the scattering vector, or momentum transfer, (Q_Z) are shown. On the other hand, ellipsometry does not disentangle the contribution of the headgroup and the aliphatic chains to the thickness as NR. In addition, the roughness is considered to be equal to zero (as an ideal interface). The incoming and the reflected light beams, whose electric field is divided in parallel (E_p) and perpendicular (E_s) components, are shown. Besides, the angle of incidence (AOI) and the ellipsometric angle Ψ are shown.

This work reports complementary measurements from two different methods. On one hand, NR, which has been demonstrated to be a suitable technique for the investigation of the structure at the sub-nm scale of thin films, allowed us to determine the thickness of the aliphatic chains and the level of hydration of the polar headgroups of the DPPC monolayer. On the other hand, we propose a novel ellipsometric method to determine both the surface excess and the optical anisotropy shown by the DPPC monolayer in the condensed phase that depends on the orientation of the aliphatic chains.

2. Materials and Methods

2.1. Materials

Hydrogenous DPPC (h-DPPC) and chain-deuterated DPPC (d62-DPPC) were purchased as powder from Avanti Polar Lipids (>99%, Alabaster, AL, USA). Solutions of 1 mg·mL^{-1} of h-DPPC, d62-DPPC, and h-DPPC/d62-DPPC (95:5 mol %) mixture, from now on contrast matched DPPC (cm-DPPC), were prepared in chloroform stabilized with ethanol (99.8%; Sigma Aldrich, St. Louis, MO, USA). Ultra-pure water was generated by passing deionized water through a Milli-Q unit (total organic content = 4 ppb; resistivity = 18 mΩ·cm, Millipore, Burlington, MA, USA). D_2O (99.9%) was purchased from Sigma Aldrich and used as received.

The experiments were performed in HKM buffer pH = 7.2, whose composition is the following: 25 mM HEPES (4-(2-hydroxyethyl)piperazine-1-ethanesulfonic acid and N-(2-hydroxyethyl)piperazine-N′-(2-ethanesulfonic acid)), 125 mM potassium acetate, 5 mM magnesium acetate, and 1 mM DTT (threo-1,4-dimercapto-2,3-butanediol, DL-dithiothreitol, Cleland's reagent, and DTT). HEPES (in solution, 1 M in H_2O, and powder 99.5%), potassium acetate ($\geq$99.0%), magnesium acetate ($\geq$99.0%), and DTT ($\geq$99.0%) were purchased from Sigma Aldrich.

2.2. Surface Pressure-Area Isotherm

The surface pressure (Π)-Area (*A*) isotherm of h-DPPC was measured using a Langmuir trough (Kibron, Finland) with a maximum area of 166.4 cm^2. The trough was carefully cleaned with ethanol and Milli-Q water before filling it with 120 mL of HKM buffer. Subsequently, the solution of h-DPPC with a concentration of 0.1 mg·mL^{-1} was spread over the subphase using a Hamilton microsyringe (Hamilton Co., Reno, NV, USA) with a precision of 1 µL. After the chloroform was evaporated for about 20 min, the variation of surface pressure during three different compression/expansion cycles was recorded using a Wilhelmy plate made of filter paper and applying a barrier speed of 5 cm^2·min^{-1}. The same cleaning and experiment preparation methods were followed for the NR and ellipsometry measurements. In all the experiments performed in this work, the temperature of the subphase was maintained at 21.0 ± 0.5 °C. Finally, it should be noted here that parameters such as the compression rate, the spreading solvent, the geometry and the dimensions of the trough, the temperature, or the composition of the subphase can affect the resulting isotherm [38]. In view of this, we compare our data with other isotherms performed in a similar way, including a similar compression ratio and spreading solvent.

2.3. Neutron Reflectometry Data Acquisition

Neutron reflectometry (NR) experiments were performed on FIGARO, a time-of-flight reflectometer [27,39,40] at the Institut Laue-Langevin, Grenoble (France). Two different angles of incidence ($\theta_1 = 0.6°$ and $\theta_2 = 3.7°$) and a wavelength resolution of 7% dλ/λ were used, yielding a momentum transfer of 7×10^{-3} Å^{-1} < Q_Z < 0.25 Å^{-1}, normal to the interface, and defined as follows:

$$Q_z = \frac{4\pi}{\lambda} \sin(\vartheta) \tag{1}$$

where λ is the wavelength of the neutron beam. Usually, reflectivity (*R*) is defined as the ratio of the intensity of the neutrons scattered from the air/water interface over the incident intensity of the neutron beam. The measured $R(Q_Z)$ profile is linked to an in plane-averaged scattering length density (SLD) profile perpendicular to the interface, which is a measure of the coherent scattering cross-section of the molecular species that constitutes each interfacial layer. The data were reduced using COSMOS [41]. Data of the samples were normalized to a measurement of pure D_2O.

NR experiments were performed using HKM buffer prepared with a mixture 8:92 *V*/*V* % of D_2O:H_2O as a solvent, generally known as air contrast matched water (ACMW) since its scattering length density is equal to the one of air, i.e., equals to zero. h-DPPC, cm-DPPC (with an SLD of the aliphatic tails equals to zero), and d62-DPPC monolayers were prepared using a Langmuir trough (NIMA, Coventry, UK) with a total area of 354 cm^2. The volume of HKM buffer used to fill the trough was 200 mL. We compressed the monolayer with a barrier speed of 10 cm^2·min^{-1} until Π = 30 mN·m^{-1} and we used the pressure control to keep the pressure constant along the reflectivity measurements.

2.4. NR Data Modeling

The data analysis was performed using AuroreNR software (v5.0) [42]. A two-layer model was used to fit the data, dividing surface-active molecules between polar heads and aliphatic tails (Figure 1). It has been recently demonstrated that using this model results in a better fit of the experimental curves [27]. The fixed parameters used in the fitting procedure (Table 1) are molecular volumes of DPPC heads (V_{m_heads}) and tails (V_{m_tails}) [43,44] and the total scattering length of DPPC heads (Σb_{heads}) and tails (Σb_{tails}). Heads-layer thickness (t_{heads}) was calculated from the V_{m_heads}, and finally, the roughness (r) of the three interfaces (i.e., air/tails-layer, tails-layer/heads-layer, and heads-layer/subphase) was assumed identical following the approach reported by Campbell et al. [27]. Besides, its value was consistent with the presence of capillary waves [45]. A real interface is characterized by a finite roughness, whose minimum value depends on the capillary waves of the subphase [45,46]. Therefore, the change in SLD along the z-axis of a real interface is described here by the SLD profile of the ideal interface modulated by an error function, ERF [47]:

$$\mathrm{ERF}\left(\frac{z - z_0}{\sigma/\sqrt{2}}\right) = \frac{2}{\sqrt{\pi}} \int_0^{\frac{z-z_0}{\sigma/\sqrt{2}}} e^{-t^2} \, dt \tag{2}$$

where z_0 and σ indicate the position and the roughness (respectively) of the interface between the layers.

Table 1. Fixed parameters used for data modeling. The molecular volumes of h-DPPC, cm-DPPC, and d62-DPPC are equal; the only difference between the sample parameters is the value of the total scattering length of the tails Σb_{tail}. One can calculate the scattering length density (SLD) values shown considering b and V_m (SLD = $\Sigma b/V_m$).

Fixed Parameters	h-DPPC	cm-DPPC	d62-DPPC
V_{m_heads} (Å^3)	319	319	319
Σb_{head} (10^{-5} Å)	60.06	60.06	60.06
SLD_{head} (10^{-6} Å^{-2})	1.88	1.88	1.88
V_{m_tails} (Å^3)	825	825	825
Σb_{tail} (10^{-5} Å)	−32.50	0	612.98
SLD_{tails} (10^{-6} Å^{-2})	−0.39	0	7.43
f_{tails} (%)	100	100	100

Experimental data of h-DPPC, cm-DPPC and d62-DPPC were fitted together. Thus, the ambiguity in the interpretation of the sample structure, which may arise from the different sensitivity that the curves exhibit with respect to the different sample components, is significantly reduced. Using this approach, the variables determined through the fitting procedure were solely the surface roughness (r), the thickness of the tails-layer (t_{tails}), and the heads volume fraction (f_{heads}), whose value was constrained to ensure the same surface excess (Γ) of tails Γ_{tails} and heads Γ_{head}, calculated as follows:

$$\Gamma_i = \frac{SLD_i t_i f_i}{\Sigma b_i N_A} \tag{3}$$

where Σb_i and f_i are the total scattering length and the volume fraction of the i-th component (tails or heads), respectively, and N_A is the Avogadro's number. While the tails volume fraction f_{tails} was fixed to unity (i.e., 100%), for the determination of f_{heads} the solvation of the polar headgroups was taken into account. This yields the following equation:

$$f_{heads} = \frac{t_{tails} V_{m\_heads}}{t_{heads} V_{m\_tails}} \tag{4}$$

2.5. Ellipsometry

Ellipsometry experiments were performed on a Picometer Light ellipsometer (Beaglehole Instruments, Kelburn, New Zealand) using a He-Ne laser with $\lambda = 632$ nm. The Langmuir trough used to record the isotherm was coupled with the ellipsometer to measure and control the surface pressure of the lipid monolayer during the measurements of the ellipsometric angles. We studied a DPPC monolayer at the air/buffer interface by measuring the ellipsometric angles as a function of the angle of incidence, AOI, at $\Pi = 30$ mN·m^{-1}. The range of AOI was 45°–70° with a step of 0.5°.

Ellipsometry is a non-destructive optical technique widely used for the study of surfaces and thin films [28,48]. It is based on the determination of the polarization changes that light undergoes when it is reflected at an interface. The reflection coefficients parallel and perpendicular to the plane of incidence, r_p and r_s respectively, are related to the ellipsometric angles Δ, and Ψ. This relationship is known as ellipticity, ρ, and is defined by the following equation:

$$\rho = \frac{r_p}{r_s} = \tan\Psi e^{i\Delta} \tag{5}$$

where ρ is the ellipticity that depends on the AOI, the wavelength of the light and both the thickness and the dielectrical properties of the material on which the reflection of the light beam occurs.

Although the ellipsometric angles are experimentally easily accessible, they do not provide direct access to the refractive index and the thickness of the lipid monolayer. Thus, it is necessary to model the experimental sets of Δ and Ψ vs. AOI to determine d_F and n_F. For the data analysis, we constructed a slab model considering the profile of refractive indices perpendicular to the surface. In contrast to NR, ellipsometry cannot distinguish between heads and tails of lipid molecules, and the different layers are considered as one homogenous layer with negligible roughness (see Figure 1). Therefore, the model used in this work consisted of one slab formed by the lipid monolayer characterized by n_F and d_F.

Once constructed the model, we fitted the data of Δ and Ψ vs. AOI using a numeric nonlinear minimization procedure, specifically, a trust-region reflective algorithm [49]. This method is based on the determination of the ellipsometric angles of the model that minimizes the differences with those experimentally obtained (a more detailed explanation can be found in references [49–51]). For the calculation of Δ and Ψ of the model, we used a power series expansion to the first order of the relative film thickness $(2\pi d_F/\lambda)$ that allows us to relate ρ, i.e., Δ and Ψ, with n_F and d_F as follows:

$$\rho \approx \rho_0 + i\rho'\frac{2\pi d_F}{\lambda} \tag{6}$$

where ρ_0 is the ellipsometric ratio of the ambient/substrate interface and ρ' (Equation (7)) is a linear coefficient defined by the refractive indices of the air and the subphase, n_1 and n_2, respectively, and the incident and transmission angles, α_{inc} and α_{tra}, respectively.

$$\rho' = -2\frac{n_1}{n_2{}^2 - n_1{}^2}\frac{\sin^2(\alpha_{inc})\cos(\alpha_{inc})}{\cos^2(\alpha_{inc} - \alpha_{tra})}\frac{\left(n_F{}^2 - n_1{}^2\right)\left(n_F{}^2 - n_2{}^2\right)}{n_F{}^2} \tag{7}$$

Drude reported this approximation for the first time based on the fact that the terms of a higher order than the first are negligible when the thickness of the film is very small [34,52]. Therefore, the Equations (5)–(7) provide the values of Δ and Ψ for given values of n_F and d_F. Finally, the variation of n_F and d_F allows one to obtain the real parameters of the film as those that minimizes the differences between the calculated ellipsometric angles and those experimentally obtained. The function minimized and used to determine the quality of a given solution is the squared deviation (χ^2) between measured and calculated ellipsometric angles and it is defined by:

$$\chi^2 = \frac{1}{N-M}\sum_{i=1}^{N}\left[\left(\frac{\Delta_{(i)\exp} - \Delta_{(i)\text{model}}}{\delta_{(i)\Delta}}\right)^2 + \left(\frac{\Psi_{(i)\exp} - \Psi_{(i)\text{model}}}{\delta_{(i)\Psi}}\right)^2\right] \tag{8}$$

where N is the number of points, M is the number of parameters determined (i.e., two parameters, d_F and n_F), Δ_{exp} and Δ_{model} correspond to the ellipsometric angle Δ experimentally obtained and the calculated for the model, respectively, and δ_Δ the uncertainty of the i-th experimental Δ or Ψ value.

3. Results

3.1. Π-A Isotherm

Figure 2a shows the Π-*A* isotherm for DPPC in the HKM buffer. DPPC shows a liquid expanded (LE) phase at a very low surface pressure, followed by a minor liquid expanded–liquid condensed (LE–LC) coexistence region at $\Pi \approx 5$ mN·m^{-1}. Further compression yields a LC phase, characterized by a long range molecular order, until it reaches the collapse at $\Pi \approx 54$ mN·m^{-1}. The Π-*A* isotherm does not show a well-defined LE–LC coexistence region as DPPC on water, characterized by a well-defined plateau of coexistence [27,38]. In Figure 2b, we report the corresponding compressional elastic modulus C_s^{-1}, calculated from the surface pressure isotherm following

$$C_s^{-1} = -A\left(\frac{\partial \Pi}{\partial A}\right) \tag{9}$$

where Π represents the surface pressure and A the surface area. The Π-*A* isotherm shown in Figure 2a is quite similar to those previously reported in the literature for DPPC in water, consequently, we interpret them in a similar way [27]. At increasing Π, DPPC molecules are pushed closer and the compressional elastic modulus increases until it reaches a maximum. The low Π region is commonly assigned to a 2D liquid expanded state (LE). A minimum in the compression modulus at $\Pi \approx 5$ mN·m^{-1}, is commonly attributed to the existence of a LE–LC phase transition, which can be more clearly observed than the slight pseudo plateau in the isotherm (Figure 2a). The global maximum value of C_s^{-1} is 130 mN·m^{-1}, which corresponds to the LC phase. The values of C_s^{-1} in this LC phase are smaller than the ones observed for DPPC at the air/water interface, which present a maximum at $C_s^{-1} \approx 230$ mN·m^{-1} [27], indicating a less condensed monolayer. Importantly, Figure 2 shows that a DPPC monolayer in the presence of buffer containing divalent salts exhibits more lateral compressibility due to a less acyl chain compaction, and, therefore, is more permeable [53].

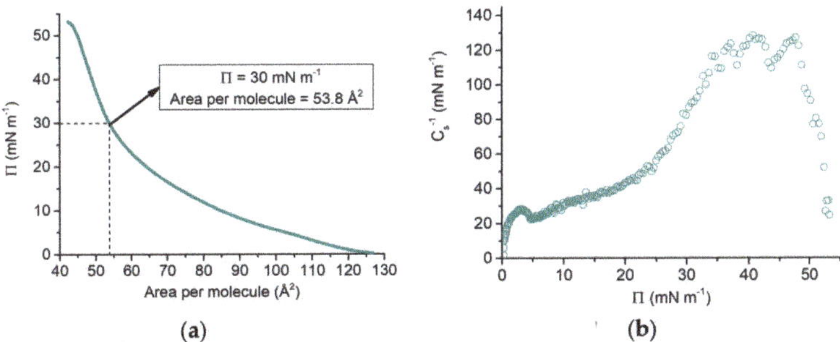

Figure 2. (a) Π-*A* isotherm of a DPPC monolayer at the air/buffer interface and (b) corresponding C_s^{-1} as a function of Π.

3.2. NR Results

Neutron reflectometry measurements were performed to study the structure of DPPC monolayers in the LC phase. In particular, we selected a sample with a surface pressure value of 30 mN·m^{-1} corresponding to an area per molecule of 53.8 Å^2, well above the LE–LC coexistence phase. This guarantees a laterally homogeneous interface on the length scale of the in-plane neutron coherence length, on the order of several microns, and implies that the measured NR can be correlated with the SLD depth profile averaged across the interfacial area delimited by this coherence length. The reflectivity profiles were recorded over the whole Q-range accessible in three isotopic contrasts: h-DPPC, cm-DPPC, and d62-DPPC in ACMW as shown in Figure 3a. As a reference, the measurement of the bare air/D_2O interface is shown also in Figure 3a, including a fit to the data that corresponds to a roughness, r_0, of 2.8 ± 0.1 Å in agreement with the theoretical value expected for thermally excited capillary waves ($\sim \sqrt{k_B T / \gamma_0}$), with γ_0 being the interfacial tension of the bare D_2O interface [45,46].

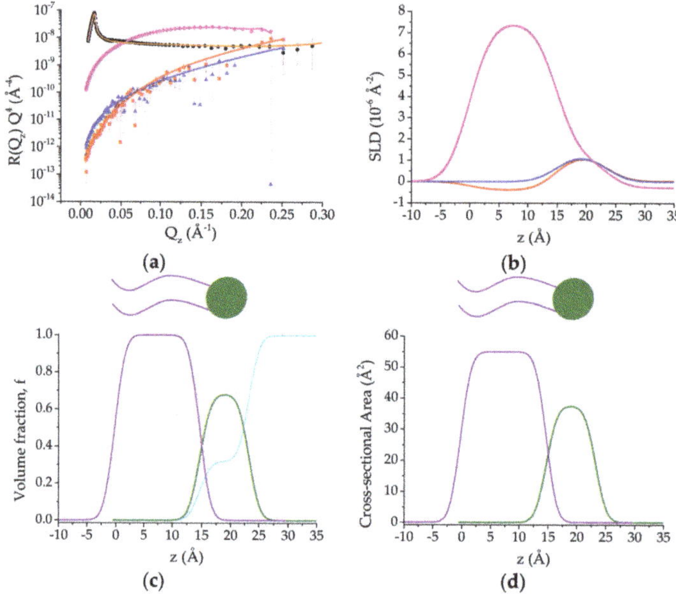

Figure 3. (**a**) Neutron reflectivity data of pure D_2O interface (black circle), h-DPPC (red squares), cm-DPPC (blue triangles), and d62-DPPC (magenta diamonds) monolayers at Π = 30 mN·m^{-1} in HKM-ACMW. The fitting curves of the bare D_2O interface (orange curve), h-DPPC (red), cm-DPPC (blue), and d62-DPPC (magenta) monolayers are shown. The χ^2 value is 2. The figure is displayed on an $R(Q_z)Q_z^4$ scale to highlight the quality of the fits at high Q_z values. (**b**) SLD profiles normal to the interface of monolayers of h-DPPC (red), cm-DPPC (blue), and d62-DPPC (magenta) monolayers at Π = 30 mN·m^{-1}. (**c**) Volume fraction profiles normal to the interface of monolayers to highlight the distribution of tails (violet) and heads (green). (**d**) Cross-sectional area profiles normal to the interface of monolayers to highlight the distribution of tails (violet) and heads (green). Note that the area of the head-groups here does not consider the hydration.

The neutron reflectivity profiles were fitted according to a two-layer model, based on the model recently reported by Campbell et al. [27]. In detail, the model consists of a first layer containing the lipid aliphatic tails in contact with air and, a second one, containing the polar headgroups submerged in the aqueous subphase (see Figure 1a). All parameters used to describe both layers (such as the values of Σb and molecular volumes) are included in Table 1. The best fit of the reflectivity profiles measured is also included in Figure 3a as solid lines. The resulting SLD profiles across the interface are plotted in Figure 3b.

The structural parameters obtained from the fits are summarized in Table 2. The roughness of the three interfaces (air/tails-layer, tails-layer/heads-layer, and heads-layer/subphase) was constrained to be equal. The value obtained from the fitting is 3.0 ± 0.5 Å, which is perfectly consistent with the presence of capillary waves due to thermal fluctuations (usually estimated through the relation: $r \approx r_0 \sqrt{\gamma_0 / (\gamma_0 - \Pi)}$ [54,55]). The value of the thickness of DPPC monolayer is 23.5 Å, with 15.0 Å corresponding to the aliphatic tails in contact with air. Using the parameters from Tables 1 and 2, the variation of the volume fraction, $f_{\mathrm{DPPC}}(z)$, with the distance to the interface, was calculated using the difference of two error functions as follows

$$
f_{\mathrm{DPPC}}(z) = \begin{cases} \frac{1}{2} f_{\mathrm{tails}} \left[\mathrm{ERF}\left(\frac{z}{r/\sqrt{2}}\right) - \mathrm{ERF}\left(\frac{z - t_{\mathrm{tails}}}{r/\sqrt{2}}\right) \right], & 0 \leq z \leq t_{\mathrm{tails}} \\ \frac{1}{2} f_{\mathrm{heads}} \left[\mathrm{ERF}\left(\frac{z}{r/\sqrt{2}}\right) - \mathrm{ERF}\left(\frac{z - t_{\mathrm{heads}}}{r/\sqrt{2}}\right) \right], & t_{\mathrm{tails}} \leq z \leq t_{\mathrm{tails}} + t_{\mathrm{heads}} \end{cases} \tag{10}
$$

The structural information elucidated by NR on DPPC monolayers can be better interpreted, therefore, by the volume fraction profiles and the corresponding cross-sectional area profiles as a function of the distance from the interface, which are shown in Figure 3c,d, respectively. Such as the volume fraction, the cross-sectional area profile is modulated by the same error function and it is calculated as follows:

$$
A_{\mathrm{DPPC}}(z) = \begin{cases} \frac{V_{\mathrm{m\_tails}}}{2\, t_{\mathrm{tails}} f_{\mathrm{tails}}} \left[\mathrm{ERF}\left(\frac{z}{r/\sqrt{2}}\right) - \mathrm{ERF}\left(\frac{z - t_{\mathrm{tails}}}{r/\sqrt{2}}\right) \right], & 0 \leq z \leq t_{\mathrm{tails}} \\ \frac{V_{\mathrm{m\_heads}}}{2\, t_{\mathrm{heads}} f_{\mathrm{heads}}} \left[\mathrm{ERF}\left(\frac{z}{r/\sqrt{2}}\right) - \mathrm{ERF}\left(\frac{z - t_{\mathrm{heads}}}{r/\sqrt{2}}\right) \right], & t_{\mathrm{tails}} \leq z \leq t_{\mathrm{tails}} + t_{\mathrm{heads}} \end{cases} \tag{11}
$$

The values used to calculate the cross-sectional area from Equation (11) are collected from Tables 1 and 2.

Table 2. Parameters resulting from the data modeling. * The hydration degree of the headgroups are considered here.

Fitting Parameters	h-DPPC, cm-DPPC, $\Pi = 30$ mN·m^{-1}
t_{heads} (Å)	8.5
t_{tails} (Å)	15.0 ± 0.5
f_{heads} (%)	68 ± 1
A_{tails} (Å^2)	55.2 ± 0.3
$A_{\mathrm{heads}*}$ (Å^2)	55.2 ± 0.3
Γ_{tails} (µmol·m^{-2})	3.0 ± 0.1
Γ_{heads} (µmol·m^{-2})	3.0 ± 0.1
r (Å)	3.0 ± 0.5

3.3. Ellipsometry

We performed ellipsometry measurements by exploring the variation of the ellipsometric angles Δ and Ψ as a function of the AOI. In Figure 4a,b, we show the variation of Δ and Ψ for the air/water interface, as a reference measurement, the air/buffer interface, and the h-DPPC monolayer ($\Pi = 30$ mN·m^{-1}) spread on the HKM buffer. Firstly, both interfaces air/water and air/buffer yield similar results, which allows us to consider that the refractive index of the buffer does not change to that of water. Nevertheless, the DPPC monolayer at the air/buffer interface shows different values of Δ with respect to the ones obtained for the HKM buffer and water, particularly at values of the angle of incidence close to the Brewster angle. This is the consequence of the change in the state of polarization of the light beam when it interacts with the lipid molecules instead of the bare air/buffer interface. To explain the experimental data, we consider a one-layer optically anisotropic model for the DPPC monolayer (as described in the methods section) with an average refractive index $n_{\mathrm{F}} = \frac{1}{3} n_z + \frac{2}{3} n_x$. Considering the DPPC monolayer in LC as an optically uniaxial system (uniaxially birefringent), the anisotropy can be defined as $\Delta n = n_z - n_x$, being n_x and n_z the refractive indexes of the layer parallel and perpendicular to the interface, respectively [35]. In our first approach, we simultaneously get the values of n_{F} and d_{F} that better fit the experimental data thus yielding the lowest χ^2 value. In detail, we simultaneously fit the variation of Δ and Ψ with AOI shown in Figure 4a,b with different $n_{\mathrm{F}} - d_{\mathrm{F}}$ initial values covering a wide range of n_{F} (from 1.33, corresponding to the bulk phase, to 1.60) and d_{F} (from 0 to 30 Å). Concretely, the combination of 300 values of both parameters resulted in 9×10^5 $n_{\mathrm{F}} - d_{\mathrm{F}}$ pairs of solutions with a given χ^2. This approach allowed us to build a matrix shown as a color-map in Figure 4c. This map presents a clear dark blue area in the region defined by $n_F \in [1.44, 1.60]$ and $d_F \in [10A, 30\text{Å}]$ that correspond to values of $\chi^2 \approx 1$. A priori, it is difficult to select a single pair of values in this area with the minimum χ^2. In the following, we show how we can extract the surface excess and the optical anisotropy of the DPPC monolayer from further analysis of the results shown in Figure 4c and compare with the ones obtained by neutron reflectometry.

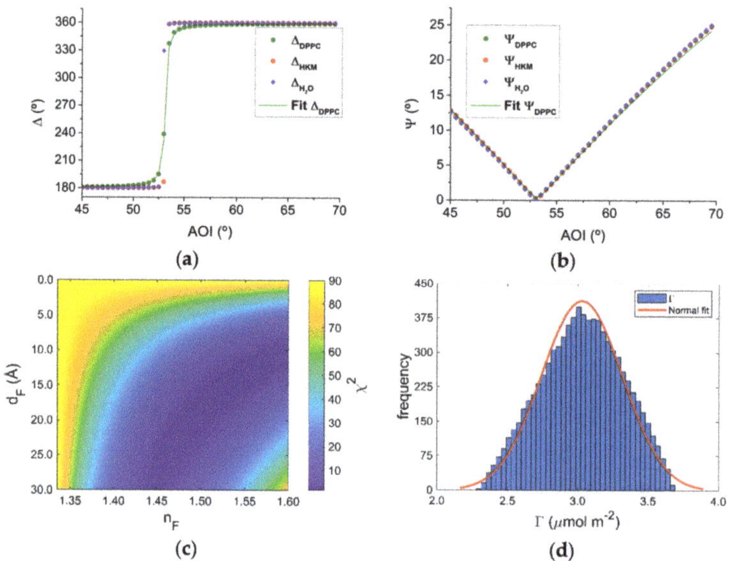

Figure 4. (**a**) Δ and (**b**) Ψ vs. AOI of H_2O (blue diamonds), HKM buffer (red circles), and DPPC (green circles). The green line corresponds to the ellipsometric angles values obtained from the fit that minimizes the χ^2. (**c**) Colormap of the squared deviation χ^2 between measured and calculated Δ and Ψ of a DPPC monolayer at the air/buffer interface at $\Pi = 30$ mN·m^{-1}. Each pixel of the figure represents a value of χ^2 obtained from the fit of Δ and Ψ vs. AOI using the correspondent values of d_F and n_F. (**d**) Normal distribution of Γ values correspondent to the region of lowest χ^2.

Approach 1: Calculation of the surface excess of DPPC monolayer using de Feitjer's equation. We first used the sets of values (n_F, d_F) that yields $\chi^2 \approx 1$ in Figure 4c to calculate the surface excess of DPPC monolayers (Γ) by using de Feitjer's equation [56]:

$$\Gamma = \frac{d_F(n_F - n_{\text{bulk}})}{dn/dc} \tag{12}$$

Here dn/dc is the refractive index increment. In detail, we used a value of 0.138 mL·g^{-1} obtained from reference [22]. n_{bulk} is the refractive index of the bulk phase, here we used a value of 1.335 that corresponds to water taking into account that the presence of buffer does not change the ellipsometry angles as shown in Figure 4a,b. Figure 4d reports the distribution of Γ calculated for the different pairs of (n_F, d_F) values yielding the lowest χ^2. Concretely, we have used all the values that satisfies $\chi^2 \leq 1.5\chi^2_{\text{min}}$, where χ^2_{min} accounts for the best value. The tendency shown by Γ can be modelled by a normal distribution (red line in Figure 4d) obtaining $\Gamma = 3.0 \pm 0.2$ µmol·m^{-2}. This result is in further agreement with the values of Γ_{tails}, and Γ_{heads} obtained by NR.

Approach 2: A new method to calculate the anisotropy of the refractive index n_F. We present here for the first time a novel approach to calculate the anisotropy parameter Δn as well as n_x and n_z, which are the refractive indices corresponding to the aliphatic tails parallel and normal to the surface, respectively. Let us first consider here that the hydrophilic heads-layer has a refractive index close to that of the bulk due to the high level of head group hydration, 32% according to the NR data (see Table 2) and previously demonstrated in references [23,35]. We, therefore, assumed that only the hydrophobic aliphatic chains can be precisely detected by ellipsometry. In this context, using the values of $n_F = 1.547 \pm 0.005$ and $d_F = 13.7 \pm 0.3$ Å that yields the best χ^2 (=1.71, according to data plotted in Figure 4c) together with a value of $d\Delta = \Delta - \Delta_{bulk}$ for a single angle of incidence close to the Brewster conditions, we could calculate the values of n_x and n_z using

$$d\Delta(AOI) = -\frac{4\pi d_F}{\lambda} \frac{n_{air} \sin(AOI) \tan(AOI)}{1 - \frac{n_{air}^2}{n_{water}^2} \tan^2(AOI)}$$
$$\frac{1}{n_{air}^2 - n_{water}^2}\left(n_x^2 - n_{air}^2 - n_{water}^2 + \frac{n_{air}^2 n_{water}^2}{n_z^2}\right) \tag{13}$$

which comes from the separation of the real and imaginary parts in Equation (5) according to the Drude approximation [52], and the definition of the average refractive index

$$n_F = \frac{1}{3}(2n_x + n_z) \tag{14}$$

In particular, we used a value of $d\Delta = 0.1375$ rad that comes from the difference between the value of Δ of the lipid monolayer (188.48°) and the one correspondent to the buffer (180.60°) at an angle of incidence of 52°, for which we had the highest sensitivity as it was close to the Brewster angle (see Figure 4a). Using the values of n_F, d_F and $d\Delta$ in Equations (13) and (14), we obtained $n_x = 1.537$ and $n_z = 1.566$ yielding an anisotropy value, Δn, of 0.029.

4. Discussion

4.1. Effect of HKM Buffer on the DPPC Monolayer Π-A Isotherm

In this work, we studied the optical and structural properties of a DPPC monolayer in the fluid condensed phase ($\Pi = 30$ mN·m^{-1}). Looking at the Π-A isotherm and the corresponding C_s^{-1} as a function of the surface pressure (Figure 2), we obtained information about the different phases of the lipid monolayer and its molecular packaging. The behavior of DPPC at the air/buffer interface is different from the one observed using pure water. Although this is an aspect that merits future work, there are different studies in the literature on the influence of mono and divalent cations on the behavior of DPPC at the air/water interface that agree with our findings: the formation of a liquid-condensed phase with higher lateral compressibility and shifted towards larger molecular areas than the same monolayer at the air/water interface. Complexation effects of mono-, but especially, divalent ions to amphiphilic molecules, such as phospholipids, are known to modify their orientation, morphology and packing. These aspects have been recently evaluated by Adams et al. by studying the influence of highly concentrated salt solutions on the behavior of DPPC monolayers at the air/water interface [15]. The Π-A isotherms reported, together with the Brewster angle microscopy images, showed that the presence of K$^+$, Na$^+$, or Mg^{2+} causes an expansion of the isotherm towards higher values of area per molecule and a decrease in the compressibility modulus, especially in the latter case. This is in full agreement with what we observed here. These authors also demonstrated that the presence of high salt concentration disrupts lipid packing, resulting in an extension of the liquid-expanded phase. Since we observed a slight LE–LC coexistence phase, we thought that we were not in this situation. It is important to mention that the salt concentrations used in that study are higher than the ones that we used. However, the general trend of the expansion of the isotherm and the decrease of the compressibility modulus are indicators that the behavior of DPPC at the air/water interface is affected by the constituents of the HKM buffer. Besides, in this work, there were up to five different species that could affect the organization of DPPC at the interface. Indeed, Bringenzu et al. studied the interaction of a protein with DPPC monolayers using a very similar buffer and they also observed an expansion of the isotherm [57].

4.2. Influence of HKM Buffer on the Structural Parameters of DPPC Monolayers at the LC Phase

By means of NR, we reported a total thickness of the DPPC monolayer of 23.5 Å, in agreement with literature values of DPPC in the condensed phase [20,58]. The thickness of the aliphatic chains was found to be 15 Å by NR. This was close enough to the thickness obtained from ellipsometry (*ca.* 14 Å) corresponding to the best fit. Assuming a length of 19 Å of an aliphatic chain of 16-CH2 groups in trans configuration [59], we obtained a tilting angle of 37°, a value that is higher than the value expected for DPPC at the LC phase (27° at $\Pi \approx 30$ mN·m^{-1}). This agrees with a decrease in the packing density of DPPC monolayers in the presence of divalent salts from the HKM buffer (see Figure 1b). The average refractive index $n_F = 1.547$, found here agrees with previous values reported by Ducharme et al. [23] and Thoma et al. [35] on the DPPC monolayer at the air/water interface. Ducharme et al. rationalized this increase of the refractive index of lipid monolayers to the existence of condensed phases. In addition, the value reported here for the anisotropy was slightly smaller to the one reported by Thoma et al. ($\Delta n = 0.05$). It is well known that an increase in anisotropy is observed for DPPC monolayers going from LE to LC. It is rationalized by the increase of all-trans configurations for the alkyl chains. In our case, the reduced anisotropy found can also be explained by the effect of the buffer on the packing density of the alkyl chains.

The values obtained from the fitting of both NR (Table 2) and ellipsometry yields a surface excess, Γ, or molecular area ($A = 1/(N_A \times \Gamma)$) [60], which is consistent with the one extrapolated from the isotherm (see Figure 2a) at $\Pi = 30$ mN·m^{-1}. The value found of 55.2 Å^2 is slightly larger in comparison with the value for DPPC in water at a similar surface pressure [27]. This has been previously attributed to the effect of the buffer molecules on the monolayer. The calculation of both volume fraction and cross-sectional area profiles with the distance to the interface, (Figure 3c,d, respectively) allows us to rationalize the structure of the DPPC monolayer in real-space, even unraveling the contribution from the aliphatic chains facing the air and the solvated polar headgroups in contact with the buffer. We observed that the solvation of the polar heads by water molecules was about 30% of the volume fraction of the layer (see, cyan curve in Figure 3c) in agreement with previous studies on DPPC/water interfaces [27].

4.3. Combining Neutron Reflectometry and Ellipsometry Experiments

Here, we considered that performing NR and ellipsometry experiments to study lipid monolayers at the air/water interface is convenient due to their complementarity. On one side, the thickness of aliphatic chains as well as the degree of hydration of the polar headgroups can be obtained by NR whilst ellipsometry is more sensitive to the orientation of the aliphatic chains and, therefore, to the physical state of the monolayer. Despite ellipsometry being a very versatile technique, overcoming its limitations in obtaining the refractive index and thickness of monolayers at the air/water interface has required to date complex mathematical calculations (see, for example [34,61]), and/or the parallel use of another technique such as NR. However, we show here an alternative approach to overcome those limitations and the possibility to study optically anisotropic interfacial systems such as phospholipid monolayers in the LC phase (e.g., DPPC). Besides, we report a method to derive the surface excess of the monolayer by the simultaneous modeling of Δ and Ψ as a function of the angle of incidence (Figure 4) that yields a similar value than the one obtained by NR.

5. Conclusions

We firstly reported the Π-*A* isotherm of DPPC at the air/buffer interface. The Π-*A* isotherm presented in this work shows an expansion towards higher molecular areas and a less pronounced phase transition than that using pure water as the subphase. We attributed these differences to the presence of cations (especially Mg^{2+}) and HEPES in the subphase, interacting with the lipid molecules at the interface. We have also shown how, combining NR and ellipsometry, we could get a complete characterization of DPPC monolayers. All of the results obtained in this work are consistent with each other, and can be rationalized due to the presence of the HKM buffer when comparing with the literature data on DPPC thickness, molecular area, and refractive index. Therefore, the use of NR in combination with ellipsometry is proposed as an effective and accurate method for studying the optical properties including the anisotropy in the refractive index and the structure of lipid monolayers in the liquid condensed phase at the air/water interface.

Author Contributions: Conceptualization, methodology, data curation and writing the manuscript by all the authors: J.C.-T., A.S., D.P., and A.M. All authors have read and agreed to the published version of the manuscript.

Funding: This research received no external funding.

Acknowledgments: We thank the ILL for the provision of neutron beam time on FIGARO instrument (10.5291/ILL-DATA.TEST-3104), and the Partnership for Soft and Condensed Matter, PSCM, for the use of the ellipsometer and the Langmuir trough. AS and JC acknowledge a PhD contract from the ILL. The authors are grateful to Richard Campbell for a critical review of this manuscript.

Conflicts of Interest: The authors declare no conflict of interest.

References

1. Lodish, H.; Berk, A.; Zipursky, S.L.; Matsudaira, P.; Baltimore, D.; Darnell, J. Biomembranes: Structural organization and basic functions. In *Molecular Cell Biology*; Freeman, W.H.: New York, NY, USA, 2000; ISBN 978-0-7167-3136-8.

2. Cooper, G.M. Structure of the Plasma Membrane. In *The Cell: A Molecular Approach*; Sunderland, Mass. Sinauer Associates: Washington, DC, USA, 2000.

3. Vacklin, H.P.; Tiberg, F.; Fragneto, G.; Thomas, R.K.; Park, I.S.; Lund, S. Phospholipase A2 hydrolysis of supported phospholipid bilayers: A neutron reflectivity and ellipsometry study. *Biochemistry* **2005**, *44*, 2811–2821. [CrossRef]

4. Bartkowiak, A.; Rojewska, M.; Prochaska, K. Study of mucin interaction with model phospholipid membrane at the air–water interface. *Colloids Surfaces A Physicochem. Eng. Asp.* **2019**, *578*, 123587. [CrossRef]

5. Puglisi, G.; Fresta, M.; Pignatello, R. Synthesis of methotrexate α,γ-bis(amides) and correlation of thermotropic and DPPC biomembrane interaction parameters with their anticancer activity. *Drug Dev. Res.* **1998**, *44*, 62–69. [CrossRef]

6. Caseli, L.; Sakai, A.; de Sousa Mesquita, A.P.; Nader, H.B.; Lopes, C.C.; Nakanishi, W.; Ariga, K. Thermodynamic and morphological properties of trastuzumab regulated by the lipid composition of cell membrane models at the air-water interface. *Biophys. J.* **2020**, *118*, 77a. [CrossRef]

7. Zhao, L.; Feng, S.S.; Go, M.L. Investigation of Molecular interactions between paclitaxel and DPPC by langmuir film balance and differential scanning calorimetry. *J. Pharm. Sci.* **2004**, *93*, 86–98. [CrossRef]

8. Sakai, A.; de Sousa Mesquista, A.P.; Nader, H.B.; Lopes, C.C.; Nakanishi, W.; Ariga, K.; Caseli, L. The lipid composition affects Trastuzumab adsorption at monolayers at the air-water interface. *Chem. Phys. Lipids* **2020**, *227*, 104875. [CrossRef]

9. Deleu, M.; Paquot, M.; Nylander, T. Fengycin interaction with lipid monolayers at the air–aqueous interface—Implications for the effect of fengycin on biological membranes. *J. Colloid Interface Sci.* **2005**, *283*, 358–365. [CrossRef]

10. Simons, K.; Vaz, W.L.C. Model systems, lipid rafts, and cell membranes. *Annu. Rev. Biophys. Biomol. Struct.* **2004**, *33*, 269–295. [CrossRef]

11. Aleskndrany, A.; Sahin, I. The effects of Levothyroxine on the structure and dynamics of DPPC liposome: FTIR and DSC studies. *Biochim. Biophys. Acta Biomembr.* **2020**, 183254. [CrossRef]

12. Ortiz-Collazos, S.; Picciani, P.H.S.; Oliveira, O.N.; Pimentel, A.S.; Edler, K.J. Influence of levofloxacin and clarithromycin on the structure of DPPC monolayers. *Biochim. Biophys. Acta Biomembr.* **2019**, *1861*, 182994. [CrossRef]

13. Vaknin, D.; Kjaer, K.; Als-Nielsen, J.; Lösche, M. Structural properties of phosphatidylcholine in a monolayer at the air/water interface: Neutron reflection study and reexamination of x-ray reflection measurements. *Biophys. J.* **1991**, *59*, 1325–1332. [CrossRef]

14. Didcot, C.; Ox, O. Hydration of DPPC monolayers at the air/water interface and its modulation by the nonionic surfactant C12E4: A neutron reflection study. *Langmuir* **1996**, *11*, 3948–3952.

15. Adams, E.M.; Casper, C.B.; Allen, H.C. Effect of cation enrichment on dipalmitoylphosphatidylcholine (DPPC) monolayers at the air-water interface. *J. Colloid Interface Sci.* **2016**, *478*, 353–364. [CrossRef] [PubMed]

16. Guzmán, E.; Liggieri, L.; Santini, E.; Ferrari, M.; Ravera, F. Mixed DPPC-cholesterol Langmuir monolayers in presence of hydrophilic silica nanoparticles. *Colloids Surfaces B Biointerfaces* **2013**, *105*, 284–293. [CrossRef] [PubMed]

17. Muñoz-López, R.; Guzmán, E.; Velázquez, M.M.; Fernández-Peña, L.; Merchán, M.D.; Maestro, A.; Ortega, F.; Rubio, R.G. Influence of carbon nanosheets on the behavior of 1,2-Dipalmitoyl-sn-glycerol-3-phosphocholine Langmuir monolayers. *Processes* **2020**, *8*, 94. [CrossRef]

18. Kent, M.S.; Murton, J.K.; Sasaki, D.Y.; Satija, S.; Akgun, B.; Nanda, H.; Curtis, J.E.; Majewski, J.; Morgan, C.R.; Engen, J.R. Neutron reflectometry study of the conformation of HIV Nef bound to lipid membranes. *Biophys. J.* **2010**, *99*, 1940–1948. [CrossRef]

19. Ege, C.; Lee, K.Y.C. Insertion of Alzheimer's Aβ40 peptide into lipid monolayers. *Biophys. J.* **2004**, *87*, 1732–1740. [CrossRef]

20. Kienle, D.F.; De Souza, J.V.; Watkins, E.B.; Kuhl, T.L. Thickness and refractive index of DPPC and DPPE monolayers by multiple-beam interferometry. *Anal. Bioanal. Chem.* **2014**, *406*, 4725–4733. [CrossRef]

21. Pusterla, J.M.; Malfatti-Gasperini, A.A.; Puentes-Martinez, X.E.; Cavalcanti, L.P.; Oliveira, R.G. Refractive index and thickness determination in Langmuir monolayers of myelin lipids. *Biochim. Biophys. Acta Biomembr.* **2017**, *1859*, 924–930. [CrossRef]

22. Erbe, A.; Sigel, R. Tilt angle of lipid acyl chains in unilamellar vesicles determined by ellipsometric light scattering. *Eur. Phys. J. E* **2007**, *22*, 303–309. [CrossRef]

23. Ducharme, D.; Max, J.J.; Salesse, C.; Leblanc, R.M. Ellipsometric study of the physical states of phosphatidylcholines at the air-water interface. *J. Phys. Chem.* **1990**, *94*, 1925–1932. [CrossRef]

24. Ehlers, J. *X-ray and Neuron Reflectivity: Principles and Applications*; Springer: Berlin/Heidelberg, Germany, 1999; ISBN 3540661956.

25. Zhou, X.L.; Chen, S.H. Theoretical foundation of X-ray and neutron reflectometry. *Phys. Rep.* **1995**, *257*, 223–348. [CrossRef]

26. Cousin, F.; Chennevière, A. Neutron reflectivity for soft matter. *EPJ Web Conf.* **2018**, *188*, 04001. [CrossRef]

27. Campbell, R.A.; Saaka, Y.; Shao, Y.; Gerelli, Y.; Cubitt, R.; Nazaruk, E.; Matyszewska, D.; Lawrence, M.J. Structure of surfactant and phospholipid monolayers at the air/water interface modeled from neutron reflectivity data. *J. Colloid Interface Sci.* **2018**, *531*, 98–108. [CrossRef] [PubMed]

28. Azzam, R.M.A.; Bashara, N.M. *Ellipsometry and Polarized Light*; North-Holland Publishing Company: New York, NY, USA, 1977; ISBN 0720406943.

29. Motschmann, H.; Teppner, R. Ellipsometry in interface science. *Stud. Interface Sci.* **2001**, *11*, 1–42.

30. Reiter, R.; Motschmann, H.; Knoll, W. Ellipsometric Characterization of Streptavidin Binding to Biotin-Functionalized Lipid Monolayers at the Water/Air Interface. *Langmuir* **1993**, *9*, 2430–2435. [CrossRef]

31. Campbell, R.A.; Tummino, A.; Varga, I.; Milyaeva, O.Y.; Krycki, M.M.; Lin, S.Y.; Laux, V.; Haertlein, M.; Forsyth, V.T.; Noskov, B.A. Adsorption of denatured lysozyme at the air-water interface: Structure and morphology. *Langmuir* **2018**, *34*, 5020–5029. [CrossRef]

32. Tatur, S.; Badia, A. Influence of hydrophobic alkylated gold nanoparticles on the phase behavior of monolayers of DPPC and clinical lung surfactant. *Langmuir* **2012**, *28*, 628–639. [CrossRef]

33. Maestro, A.; Deshmukh, O.S.; Mugele, F.; Langevin, D. Interfacial assembly of surfactant-decorated nanoparticles: On the rheological description of a colloidal 2D glass. *Langmuir* **2015**, *31*, 6289–6297. [CrossRef]

34. Nestler, P.; Helm, C.A. Determination of refractive index and layer thickness of nm-thin films via ellipsometry. *Opt. Express* **2017**, *25*, 321–327. [CrossRef]

35. Thoma, M.; Schwendler, M.; Baltes, H.; Helm, C.A.; Pfohl, T.; Riegler, H.; Mo, H. Ellipsometry and X-ray reflectivity studies on monolayers of phosphatidylethanolamine and phosphatidylcholine in contact with n-dodecane, n-hexadecane, and bicyclohexyl. *Langmuir* **1996**, *12*, 1722–1728. [CrossRef]

36. Richter, R.P.; Brisson, A.R. Following the formation of supported lipid bilayers on Mica: A study combining AFM, QCM-D, and ellipsometry. *Biophys. J.* **2005**, *88*, 3422–3433. [CrossRef]

37. Benjamins, J.; Thuresson, K.; Nylander, T. Ellipsometry studies of nonionic surfactant adsorption at the oil-water interface. *Langmuir* **2005**, *21*, 149–159. [CrossRef]

38. Duncan, S.L.; Larson, R.G. Comparing experimental and simulated pressure-area isotherms for DPPC. *Biophys. J.* **2008**, *94*, 2965–2986. [CrossRef] [PubMed]

39. Campbell, R.A.; Wacklin, H.P.; Sutton, I.; Cubitt, R.; Fragneto, G. FIGARO: The new horizontal neutron reflectometer at the ILL. *Eur. Phys. J. Plus* **2011**, *126*, 1–22. [CrossRef]

40. Braun, L.; Uhlig, M.; von Klitzing, R.; Campbell, R.A. Polymers and surfactants at fluid interfaces studied with specular neutron reflectometry. *Adv. Colloid Interface Sci.* **2017**, *247*, 130–148. [CrossRef]

41. Gutfreund, P.; Saerbeck, T.; Gonzalez, M.A.; Pellegrini, E.; Laver, M.; Dewhurst, C.; Cubitt, R. Towards generalized data reduction on a chopperbased time-of-flight neutron reflectometer. *J. Appl. Crystallogr.* **2018**. [CrossRef]

42. Gerelli, Y. Aurore: New software for neutron reflectivity data analysis. *J. Appl. Crystallogr.* **2016**, *49*, 330–339. [CrossRef]

43. Sun, W.J.; Suter, R.M.; Knewtson, M.A.; Worthington, C.R.; Tristram-Nagle, S.; Zhang, R.; Nagle, J.F. Order and disorder in fully hydrated unoriented bilayers of gel-phase dipalmitoylphosphatidylcholine. *Phys. Rev. E* **1994**, *49*, 4665–4676. [CrossRef]

44. Nagle, J.F.; Tristram-Nagle, S. Structure of lipid bilayers. *Biochim. Biophys. Acta Rev. Biomembr.* **2000**, *1469*, 159–195. [CrossRef]

45. Braslau, A.; Deutsch, M.; Pershan, P.S.; Weiss, A.H.; Als-Nielsen, J.; Bohr, J. Surface roughness of water measured by x-ray reflectivity. *Phys. Rev. Lett.* **1985**, *54*, 114–117. [CrossRef]

46. Sinha, S.K.; Sirota, E.B.; Garoff, S.; Stanley, H.B. X-ray and neutron scattering from rough surfaces. *Phys. Rev. B* **1988**, *38*, 2297–2311. [CrossRef]

47. Maranville, B.B.; Green, A.; Kienzle, P.A. Distributed error-function roughness in refl1d reflectometry fitting program. *arXiv* **2018**, arXiv:1801.04975.

48. Szczepanowicz, K.; Bazylińska, U.; Pietkiewicz, J.; Szyk-Warszyńska, L.; Wilk, K.A.; Warszyński, P. Biocompatible long-sustained release oil-core polyelectrolyte nanocarriers: From controlling physical state and stability to biological impact. *Adv. Colloid Interface Sci.* **2015**, *222*, 678–691. [CrossRef] [PubMed]

49. Conn, A.R.; Scheinberg, K.; Vicente, L.N. *Introduction to Derivative-Free Optimization*; Society for Industrial and Applied Mathematics: Philadelphia, PA, USA, 2009; ISBN 0898716683.

50. Dutta, S. Trust-Region Methods. In *Optimization in Chemical Engineering*; Cambridge University Press: Cambridge, UK, 2016; pp. 74–85.

51. Yuan, Y.X. Recent advances in trust region algorithms. *Math. Program.* **2015**, *151*, 249–281. [CrossRef]

52. Drude, P. *The Theory of Optics*; Longmans, Green, and Co.: New York, NY, USA, 1902; ISBN 0548647755.

53. Nagle, J.F.; Scott, H.L. Lateral compressibility of lipid mono- and bilayers. Theory of membrane permeability. *BBA Biomembr.* **1978**, *513*, 236–243. [CrossRef]

54. Daillant, J.; Bosio, L.; Benattar, J.J.; Meunier, J. Capillary waves and bending elasticity of monolayers on water studied by X-ray reflectivity as a function of surface pressure. *Europhys. Lett.* **1989**, *8*, 458. [CrossRef]

55. Brumm, T.; Naumann, C.; Sackmann, E.; Rennie, A.R.; Thomas, R.K.; Kanellas, D.; Penfold, J.; Bayerl, T.M. Conformational changes of the lecithin headgroup in monolayers at the air/water interface. *Eur. Biophys. J.* **1994**, *23*, 289–295. [CrossRef]

56. De Feijter, J.A.; Benjamins, J.; Veer, F.A. Ellipsometry as a Tool to Study the adsorption behavior of synthetic and biopolymers at the air- water interface. *Biopolymers* **1978**, *17*, 1759–1772. [CrossRef]

57. Bringezu, F.; Majerowicz, M.; Wen, S.; Reuther, G.; Tan, K.T.; Kuhlmann, J.; Waldmann, H.; Huster, D. Membrane binding of a lipidated N-Ras protein studied in lipid monolayers. *Eur. Biophys. J.* **2007**, *36*, 491–498. [CrossRef]

58. Thoma, M.; Möhwald, H. Monolayers of dipalmitoylphosphatidylcholine at the oil-water interface. *Colloids Surfaces A Physicochem. Eng. Asp.* **1995**, *95*, 193–200. [CrossRef]

59. Israelachvili, J.N. *Intermolecular and Surface Forces*, 3rd ed.; Elsevier/Academic Press: Cambridge, MA, USA, 2011; ISBN 9780123751829.

60. Lu, J.R.; Thomas, R.K.; Penfold, J. Surfactant layers at the air/water interface: Structure and composition. *Adv. Colloid Interface Sci.* **2000**, *84*, 143–304. [CrossRef]
61. Barradas, N.P.; Keddie, J.L.; Sackin, R. Bayesian inference analysis of ellipsometry data. *Phys. Rev. E Stat. Physics Plasmas Fluids Relat. Interdiscip. Top.* **1999**, *59*, 6138–6151. [CrossRef] [PubMed]

 *coatings*

Article

Magnetohydrodynamic Boundary Layer Flow of a Viscoelastic Fluid Past a Nonlinear Stretching Sheet in the Presence of Viscous Dissipation Effect

Ahmad Banji Jafar [1,2]**, Sharidan Shafie** [2] **and Imran Ullah** [3,]*

[1] Department of Mathematics, Kebbi State University of Science and Technology, Aliero, P.M.B. 1144 Birnin Kebbi, Kebbi State, Nigeria
[2] Department of Mathematical Sciences, Faculty of Science, Universiti Teknologi Malaysia, 81310 Skudai, Johor Bahru, Malaysia
[3] College of Civil Engineering, National University of Sciences and Technology, 44000 Islamabad, Pakistan
* Correspondence: ullahimran14@gmail.com

Received: 30 May 2019 ; Accepted: 19 July 2019; Published: 2 August 2019

Abstract: This paper numerically investigates the viscous dissipation effect on the boundary layer flow of an electrically-conducting viscoelastic fluid (Walter's B liquid) past a nonlinear stretching sheet. The partial differential equations governing the flow problem are transformed into ordinary differential equations through similarity variables. The transformed equations are then solved using the Keller box method. A careful evaluation of the influence of the pertinent parameters on the velocity field and temperature distributions through various plots is done for the prescribed surface temperature (PST) and prescribed heat flux (PHF) boundary conditions. The computed coefficient of skin friction, the rate of heat transfer (Nusselt number), and the temperature at the wall are also presented in tabular form. It is revealed from this table that the magnitude of the heat transfer is reduced with the increase in the Eckert number Ec, viscoelastic parameter K, and magnetic parameter M for the PST case by about 12%, 20%, and 29%, respectively. Similarly, the temperature at the wall for the PHF case also decreases with the increase in Ec and M by about 8% and 24%, respectively. It is obvious that the application of the PST condition excels at keeping the viscoelastic fluid warmer than the PHF condition. This implies that applying the PHF condition is better for cooling the sheet faster. The temperature at the wall is unchanged with the changes in the pertinent parameters in the PST case, and it is ascertained that the present results are in close agreement with the previous published results.

Keywords: nonlinear stretching sheet; viscoelastic fluid; MHD; viscous dissipation

---

1. Introduction

The existence of the closed-form solution, the simplicity of the mathematical expression, and the numerous applications, such as plastic sheet extrusion, drawing of plastic films, metallic plate cooling, and the glass blowing of the boundary layer flow past a stretching sheet, have received much attention in recent years. In the process of manufacturing the sheets (metal and plastic), it is required that the melt materials from a slit be stretched continuously until the required thickness is achieved. However, the desired final product of the production process largely depends on the rate at which the sheet cools. The rate of cooling is fundamentally influenced by the type of fluid adjacent to the boundary layer surface [1,2]. The rate at which the sheet is stretched and the rate at which it cools are the two major mechanisms that influence the mechanical properties of the desired product. Similarly, the behaviors of the fluid flows and heat transfer induced by elongating or a moving sheet play a vital role in an industrial process [3].

Many researchers have shown keen interest in the study of flow past a stretching sheet, since the work of Crane [4] on the flow past a flat plate. However, Gupta and Gupta [5] observed that the sheet stretching is not always continuous and, hence, may not necessarily conform to the linear speed. For instance, the stretching of the plastic sheet is inextensible and therefore nonlinear. In view of this, Kumaran and Ramanaiah [6], for the first time, presented a note on two-dimensional boundary layer flow past a stretching sheet. They considered the stretching velocity to be quadratic polynomial and obtained a closed-form solution for the problem. Later, the work of Gupta and Gupta [5] was extended by Vajravelu [7] for a nonlinear stretching sheet. He observed that shear stress is an increasing function of the nonlinear stretching sheet parameter. However, an analytical solution of viscous flow past a nonlinear stretching sheet was also solved by Vajravelu and Cannon [8]. They reported that the fluid velocity is a decreasing function of the nonlinear stretching parameter. On the other hand, Cortell [9] numerically extended and studied this problem using the Runge-Kutta method by considering the effect of viscous dissipation with non-isothermal boundary conditions. He reported that an increase in the nonlinear stretching parameter increases the rate of heat transfer. Similarly, Cortell [10] studied the effect of thermal radiation on an induced quiescent fluid past a nonlinear stretching sheet. It was observed from his report that an increase in the radiation parameter reduces the thickness of the thermal boundary, and hence, the heat transfer rate grows.

One of the mechanisms that influence the rate at which heat is being transported in the flow system is the presence of a magnetic field (such as liquid metals, plasma, electrolyte, or salt water) in such flow problem. The concept of such electrically-conducting fluids is known as magnetohydrodynamic (MHD) fluids and has many applications in engineering processes such as MHD power generators, thermal insulators, MHD pumps, and cooling of nuclear reactors [11]. Furthermore, the presence of a magnetic field in the flow problems plays a vital role in controlling the rate of cooling. In view of these important applications of MHD flows past a stretching sheet, many studies [12–14] were conducted by incorporating the magnetic field into the flow problem. Prasad et al. [15] examined the effect of heat generation on the MHD power law flow over a nonlinear stretching sheet. Their results showed that an increase in the power-law index parameter increased the momentum boundary layer thickness and reduced the thickness of the thermal boundary layer. On the other hand, Ullah et al. [16] analyzed the effects of a chemical reaction in the presence of heat generation/absorption and thermal radiation with convective boundary conditions on an unsteady mixed convection flow of Casson fluid over a nonlinear stretching sheet. In the same vein, Ullah et al. [17] used the Keller–Box numerical scheme method to study the effect of a chemical reaction on an electrically-conducting Casson fluid flow past a nonlinear stretching sheet. Furthermore, Hayat et al. [18] analyzed the magnetohydrodynamic Walters' B nanofluid past a nonlinear stretching sheet. They discovered that the rate of heat transfer and the thermal field are enhanced with the increase in temperature ratio.

A viscoelastic fluid is one of the classes of a non-Newtonian fluids that possesses double effect properties (i.e., heat transfer reduction and drag reduction properties), in addition to its properties of exhibiting both viscosity and elasticity, thereby leading to its numerous applications in polymer industries, for instance paper production, production of glass fiber, extrusion processes, thinning and annealing of copper wires, and the production of artificial fibers and plastic film [19,20]. One of the pioneering works of the study of the viscoelastic fluid past a stretching surface was investigated by Rajagopal et al. [21]. They observed from their results that there is a decrease in the skin friction coefficient with the increase in the viscoelastic parameter. Later, Dandapat and Gupta [22] extended this work by including heat transfer in the flow problem. Furthermore, Cortell [23] analytically investigated the influence of a magnetic field on viscoelastic fluid induced by a stretching sheet. His results showed that the velocity boundary layer thickness is thicker in second-grade fluid compared to that of Walter's B liquid and observed that the viscoelasticity parameter influences both the viscoelastic fluids.

MHD viscoelastic fluid flow over a stretching sheet plays a vital role in chemical engineering, the metallurgy industry, the polymer extrusion process, the manufacturing of plastic sheets, the drawing of wires and plastic films, the cooling of metallic sheets, and petroleum engineering [23]. However,

Andersson [24] studied MHD viscoelastic fluid flow over a stretching sheet. He reported from his findings that both viscoelastic and magnetic parameters have the same effect in the fluid flow problem. The radiative effect on MHD viscoelastic fluid flow past a stretching sheet was examined by Char [25]. Likewise, Prasad et al. [26] examined the behavior of an electrically-conducting viscoelastic fluid and heat transfer over a stretching sheet. They noticed that an increase in the magnetic parameter leads to a significant decrease in the wall temperature profile and the velocity gradient. Moreover, an exact solution of a viscoelastic fluid past a stretching sheet with a heat source and viscous dissipation was studied by Abel et al. [27]. Later, Abel et al. [28] examined MHD viscoelastic fluid flow past a stretching sheet and found that the surface temperature of the flow diminishes with the increase in the viscoelasticity. Furthermore, the flow of a viscoelastic fluid induced by a nonlinear stretching sheet using the optimal homotopy analysis method was studied by Mustapha [29].

Besides the importance of a magnetic field in the boundary layer flow and heat transfer is the presence of viscous dissipation in the energy equation. Viscous dissipation plays a significant role similar to that of the energy source, which changes the distribution of the temperature and thereby the rate of heat transfer. This process finds its applications in the flow of oil products through ducts and in polymer processing. The viscous dissipation effect and variable surface temperature on viscous flow past a stretching sheet were examined by Cortell [9]. He explained that the temperature of the fluid rises with an increase in the Eckert number. In the same vein, Abel et al. [30] analyzed the Ohmic and viscous dissipation effect on the MHD boundary layer flow of a viscoelastic fluid past a linear stretching sheet. They observed from their study that the fluid temperature in both the PST and PHF cases amplified with the increase in viscous dissipation parameter. The effects of viscous dissipation and thermal radiation on two-dimensional viscous flow over a nonlinear stretching sheet was addressed numerically through a similarity solution by Cortell [31]. He showed that an increase in thermal radiation and Eckert number leads to the increase in the temperature distribution. A Casson fluid flow over a nonlinear stretching sheet was studied by Medikare et al. [32]. The effects of the heat source/sink and viscous dissipation on magnetohydrodynamic non-Newtonian fluid flow in the presence of Cattaneo-Christov heat flux was examined by Ramandevi et al. [33]. Recently, the effect of joule heating and viscous dissipation on an electrically-conducting tangent hyperbolic nanofluid was examined by Atif et al. [34]. It was reported in their work that the Eckert number and slip parameter enhanced the thermal and concentration fields.

From the above reviews, it is perceived that no consideration has been given to an electrically-conducting viscoelastic fluid over a nonlinear stretching sheet with the viscous dissipation effect. This provides the enthusiasm for the present work, in which the effects of power law surface temperature and power law surface heat flux on the characteristics of the heat transfer of an MHD viscoelastic fluid past a nonlinear stretching sheet in the presence of viscous dissipation are investigated. The unconditionally stable Keller box method was employed in solving the transformed ordinary differential equation by considering non-isothermal boundary conditions. This boundary condition is applicable in an engineering process where the temperature in not constant.

2. Constitutive Equation

The rheological equation of state, which is also known as the constitutive equation, describes the relationship between strain, stress, and their time dependence. According to Newtons's law of viscosity "the stress is often proportional to the strain rate",

$$\tau = \mu \frac{\partial u}{\partial y} \tag{1}$$

where, τ, μ, and $\frac{\partial u}{\partial y}$ respectively represent the shear stress, the dynamic viscosity, and the velocity gradient or rate of the strain. Thus, any fluid that does not obey Newtons's law of viscosity is termed as a non-Newtonian fluid. Examples of such fluids are a Casson fluid, a viscoelastic fluid, a power law fluid, and many more. There is no single constitutive equation that describes the behaviors of non-Newtonian fluids due their diverse nature. However, amongst the numerous non-Newtonian fluids' models is Walter's B viscoelastic fluid model. This model has a constitutive equation of the form [35,36]:

$$\mathbf{T} = -p\mathbf{I} + 2\mu_0\mathbf{e} - 2k_0\frac{\delta\mathbf{e}'}{\delta t} \tag{2}$$

The Cauchy stress tensor $\mathbf{T}$ is expressed in terms of scalar pressure p, identity tensor $\mathbf{I}$, dynamic viscosity μ_0, short memory coefficient k_0, and the convected differentiation of the strain rate $\frac{\delta\mathbf{e}'}{\delta t}$, while the strain rate tensor $\mathbf{e}$ is defined in terms of the velocity vector $\mathbf{u}$ as:

$$\mathbf{e} = \nabla\mathbf{u} + (\nabla\mathbf{u})^T \tag{3}$$

It is convenient to represent (3) in the form of matrix suffix notation as:

$$e_{ij} = \frac{1}{2}\left(\frac{\partial u_i}{\partial x_j} + \frac{\partial u_j}{\partial x_i}\right) \tag{4}$$

where i and j can take values of one and two for two-dimensional flows, and $\frac{\delta\mathbf{e}'}{\delta t}$ is expressed as:

$$\frac{\delta\mathbf{e}'}{\delta t} = \frac{\delta\mathbf{e}}{\delta t} + \mathbf{u}\cdot\nabla\mathbf{e} - \mathbf{e}\cdot\nabla\mathbf{u} - (\nabla\cdot\mathbf{u})^T\cdot\mathbf{e} \tag{5}$$

Here, $\mathbf{u}$ denotes the velocity vector. Furthermore, the generalized constitutive equation for Walter's B viscoelastic fluid can be expressed as:

$$\mathbf{T} = \begin{pmatrix} \tau_{xx} & \tau_{xy} & 0 \\ \tau_{yx} & \tau_{yy} & 0 \end{pmatrix} = -p\mathbf{I} + 2\mu_0\mathbf{e} - 2k_0\left[\frac{\delta\mathbf{e}}{\delta t} + \mathbf{u}\cdot\nabla\mathbf{e} - \mathbf{e}\cdot\nabla\mathbf{u} - (\nabla\cdot\mathbf{u})^T\cdot\mathbf{e}\right] \tag{6}$$

Lastly, the Cauchy equation of motion is given by (see Jaluria [37]):

$$\rho\frac{D\mathbf{u}}{Dt} = \nabla\cdot\mathbf{T} + \mathbf{F} \tag{7}$$

where ρ is the fluid density, $\frac{D\mathbf{u}}{Dt}$ is material derivative, $\mathbf{T}$ is the Cauchy stress tensor and $\mathbf{F}=(F_x, F_y, 0)$ is the body force. Following Ahmad et al. [38], the body force $\mathbf{F}$ can be expressed as $\mathbf{F} = \rho\mathbf{g} + \mathbf{J}\times\mathbf{B}$, where $\mathbf{g}$ is the gravitational field, $\mathbf{J}$ is the current density, $\mathbf{B} = (0, B, 0)$ is the magnetic force and $\mathbf{J}\times\mathbf{B}$ is the Lorentz force and is simplified as

$$\mathbf{F} = \sigma B^2\mathbf{u} \tag{8}$$

where σ is the electrical conductivity.

3. Flow Problem Formulation

A two-dimensional electrically-conducting viscoelastic Walter's B fluid flow in the presence of viscous dissipation over a vertical nonlinear stretching sheet at $y = 0$ is considered in this problem. The sheet was assumed to vary nonlinearly with the velocity $U_w = ax^n$, where n represents the nonlinear stretching parameter and $a(> 0)$ is the stretching rate constant. This fluid flow obeys the constitutive Equation (6). The surface of the sheet was held at power law surface temperature $T_w = T_\infty + bx^{(2n-1)}$ and power law heat flux $-k\frac{\partial T}{\partial y} = cx^{2n-1}$, where n is the parameter for surface temperature, $b(> 0)$ and $c(> 0)$ are constants, and T_∞ is the ambient temperature of the viscoelastic fluid. The variable magnetic field $B(x) = B_0x^{(2n-1)}$ was applied normal to the sheet as shown in Figure 1 with xy-Cartesian coordinates in the horizontal and vertical direction. The steady two-dimensional continuity and Cauchy momentum equations are [36,37,39,40]:

$$\frac{\partial u}{\partial x} + \frac{\partial v}{\partial y} = 0 \tag{9}$$

$$\rho\left(u\frac{\partial u}{\partial x} + v\frac{\partial u}{\partial y}\right) = -\frac{\partial p}{\partial x} + \frac{\partial \tau_{xx}}{\partial x} + \frac{\partial \tau_{xy}}{\partial y} + F_x \tag{10}$$

$$\rho\left(u\frac{\partial v}{\partial x} + v\frac{\partial v}{\partial y}\right) = -\frac{\partial p}{\partial y} + \frac{\partial \tau_{yx}}{\partial x} + \frac{\partial \tau_{yy}}{\partial y} + F_y \tag{11}$$

where τ_{xx}, τ_{xy}, τ_{yx}, and τ_{yy} are the components of the stress matrix, $\frac{\partial \tau_{xx}}{\partial y}$ and $\frac{\partial \tau_{yy}}{\partial y}$ are elastic terms, while $\frac{\partial \tau_{xy}}{\partial y}$ and $\frac{\partial \tau_{yy}}{\partial y}$ are viscous terms.

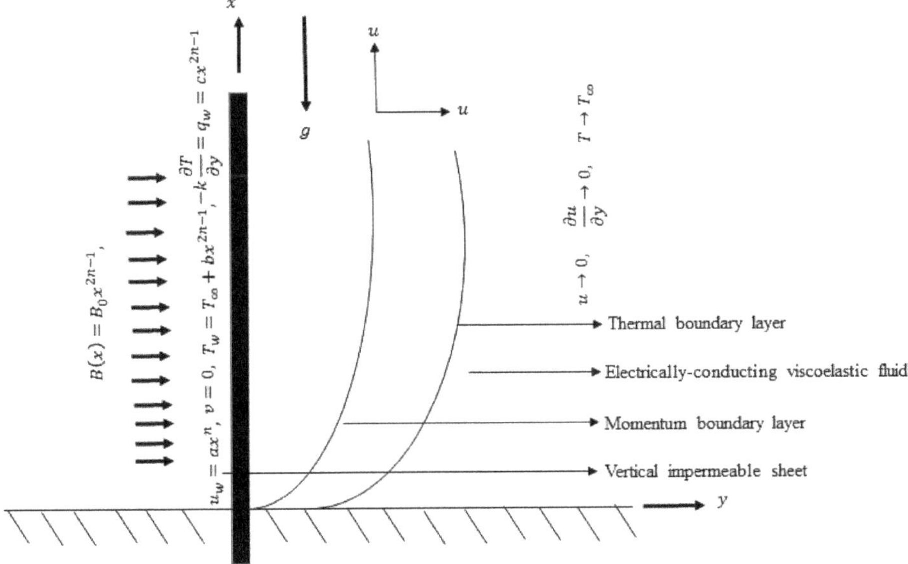

Figure 1. A schematic diagram showing the flow geometry.

To simplify Equations (10) and (11), we need to find $\frac{\partial \tau_{xx}}{\partial y}$, $\frac{\partial \tau_{yy}}{\partial y}$, $\frac{\partial \tau_{xy}}{\partial y}$, and $\frac{\partial \tau_{yy}}{\partial y}$ through Equation (6) and substitute it back into (10) and (11) to get:

The *x* momentum equation:

$$\rho\left(u\frac{\partial u}{\partial x} + v\frac{\partial u}{\partial y}\right) = -\frac{\partial p}{\partial x} + \mu_0\left(\frac{\partial^2 u}{\partial x^2} + \frac{\partial u^2}{\partial y^2}\right) - k_0\left(u\frac{\partial^3 u}{\partial x^3} + u\frac{\partial^3 u}{\partial x\partial y^2} + v\frac{\partial^3 u}{\partial x^2\partial y} + v\frac{\partial^3 u}{\partial y^3} + 3\frac{\partial u}{\partial x}\frac{\partial^2 v}{\partial x\partial y}\right)$$

$$+ k_0\left(\frac{\partial u}{\partial x}\frac{\partial^2 u}{\partial y^2} - \frac{\partial u}{\partial y}\frac{\partial^2 u}{\partial x\partial y} - \frac{\partial u}{\partial y}\frac{\partial^2 v}{\partial x^2} - 2\frac{\partial v}{\partial x}\frac{\partial^2 u}{\partial y\partial x} + F_x\right) \tag{12}$$

The *y* momentum equation:

$$\rho\left(u\frac{\partial v}{\partial x} + v\frac{\partial v}{\partial y}\right) = -\frac{\partial p}{\partial y} + \mu_0\left(\frac{\partial^2 v}{\partial x^2} + \frac{\partial v^2}{\partial y^2}\right) - k_0\left(u\frac{\partial^3 v}{\partial x^3} + u\frac{\partial^3 v}{\partial x\partial y^2} + v\frac{\partial^3 v}{\partial y\partial x^2} + v\frac{\partial^3 v}{\partial y^3} + 3\frac{\partial v}{\partial y}\frac{\partial^2 u}{\partial x\partial y}\right)$$

$$k_0\left(+\frac{\partial v}{\partial y}\frac{\partial^2 v}{\partial y\partial x^2} - \frac{\partial v}{\partial x}\frac{\partial^2 v}{\partial x\partial y\partial x} - \frac{\partial v}{\partial x}\frac{\partial^2 u}{\partial x\partial y^2} - 2\frac{\partial u}{\partial y}\frac{\partial^2 v}{\partial x\partial y} + F_y\right) \tag{13}$$

Following boundary layer theory, we can assume an order of magnitude approach on each term of Equations (12) and (13) as (see Beard and Walter [35]):

$$O(\dot{u}) = 1, \ O(x) = 1, \ O(v) = \delta, \ O(v) = \delta, \ O(\rho) = 1, \ O(k_0) = \delta^2, \ O(\mu_0) = \delta^2, \text{ and } O(B^2) = \delta^2 \quad (14)$$

Simplify boundary layer Equations (12) and (13) with these orders of magnitude to obtain:

$$\rho\left(u\frac{\partial v}{\partial x} + v\frac{\partial v}{\partial y}\right) = -\frac{\partial p}{\partial x} + v\frac{\partial^2 u}{\partial y^2} - k_0\left\{u\frac{\partial^3 u}{\partial x\partial y^2} + \frac{\partial u}{\partial x}\frac{\partial^3 u}{\partial y^2} - \frac{\partial u}{\partial y}\frac{\partial^2 u}{\partial y\partial x\partial y} + v\frac{\partial^3 u}{\partial y^3}\right\} + \sigma B^2 u \quad (15)$$

$$\frac{\partial p}{\partial y} = 0 \quad (16)$$

Equation (15) gives the momentum boundary layer equation for the viscoelastic fluid with the magnetic field. Since the vertical plate is directed along the x-axis, the pressure gradient $\frac{\partial p}{\partial x} = 0$. Hence, the momentum boundary layer governing equations for two-dimensional electrically-conducting viscoelastic fluid are given by:

$$u\frac{\partial v}{\partial x} + v\frac{\partial v}{\partial y} = v\frac{\partial^2 u}{\partial y^2} - \frac{k_0}{\rho}\left\{u\frac{\partial^3 u}{\partial x\partial y^2} + \frac{\partial u}{\partial x}\frac{\partial^3 u}{\partial y^2} - \frac{\partial u}{\partial y}\frac{\partial^2 u}{\partial y\partial x\partial y} + v\frac{\partial^3 u}{\partial y^3}\right\} + \frac{\sigma}{\rho}B^2(x)u \quad (17)$$

subject to the following flow boundary conditions:

$$u = U_w(x) = ax^n, \quad v = 0, \qquad \text{at } y = 0$$

and:

$$u \to 0, \quad \frac{\partial u}{\partial y} \to 0, \qquad \text{as } y \to \infty \quad (18)$$

where x is parallel along the sheet, y is the direction perpendicular to the sheet, u and v are the horizontal and vertical velocity in the xy-direction, respectively, v is the kinematic viscosity, and k_0 is the coefficient of the viscoelasticity, while all other physical parameters are as defined above.

To reduce the complexity of the system of the governing equations into the system of ordinary differential equations, a similarity transformation in the following form is introduced as (see Vajravelu [7]),

$$\eta = \sqrt{\frac{(n+1)u}{2vx}}y \qquad \text{and} \qquad \psi(x,y) = \sqrt{\frac{2vux}{(n+1)}}f(\eta) \quad (19)$$

where $\psi(x,y)$ denotes the stream function and is defined by:

$$u = \frac{\partial \psi}{\partial y} \qquad \text{and} \qquad v = -\frac{\partial \psi}{\partial x} \quad (20)$$

Substituting Equations (19) and (20) into Equations (17) and (18) yields dimensionless ordinary differential equations:

$$f''' + ff'' - \left(\frac{2n}{n+1}\right)f'^2 f' - K\left\{(3n-1)f'f''' - \left(\frac{3n-1}{2}\right)f''^2 - \left(\frac{n+1}{2}\right)ff^{iv}\right\} - M\left(\frac{2}{n+1}\right) = 0 \quad (21)$$

subject to the following dimensionless boundary conditions:

$$f(\eta) = 0, \qquad f'(\eta) = 1 \qquad \text{at} \qquad \eta = 0$$

$$f'(\eta) \to 0, \qquad f''(\eta) \to 0 \qquad \text{as} \qquad \eta \to \infty \quad (22)$$

where the prime represents the derivative of f with respect to η and the dimensionless quantities in these equations are nonlinear sheet parameter n, viscoelastic parameter K, and magnetic parameter M, which are defined as:

$$K = \frac{ax^{n-1}}{\rho v} \qquad \text{and} \qquad M = \frac{\sigma B_0^2}{\rho a} \tag{23}$$

The physical quantity of interest is the coefficient of skin friction C_f at the stretched surface and defined as:

$$C_f = \frac{\tau_w}{\rho u_w^2} \tag{24}$$

where τ_w is the wall shear stress from the plate and is given by:

$$\tau_w = \mu_0 \left(\frac{\partial u}{\partial y}\right)_{y=0} - k_0 \left(u\frac{\partial^2 u}{\partial u \partial y} + v\frac{\partial^2 u}{\partial y^2} - 2\frac{\partial u}{\partial y}\frac{\partial v}{\partial y}\right)_{y=0} \tag{25}$$

4. Heat Transfer Analysis

For the analysis of heat transfer, two cases of the heating process are considered, i.e.,

(i)   Prescribed surface temperature (PST)
(ii)   Prescribed heat flux (PHF)

These non-isothermal conditions are applicable in industrial and engineering processes where the temperature is not constant. The energy equation with the viscous dissipation term is given by (see Cortell [9]):

$$u\frac{\partial T}{\partial x} + v\frac{\partial T}{\partial y} = \alpha\frac{\partial^2 T}{\partial y^2} + \frac{v}{C_p}\left(\frac{\partial u}{\partial y}\right)^2 \tag{26}$$

where T, α, and ρC_p are respectively the temperature, thermal diffusivity, and specific heat capacity of the fluid at constant pressure, while all other physical parameters are as defined above. The thermal boundary conditions depend on the type of heating process under consideration. The momentum Equation (15) and the energy Equation (26) are decoupled and thereby solved sequentially.

4.1. Case I: Prescribed Surface Temperature

For this case, the thermal boundary conditions are:

$$T = T_w = T_\infty + bx^{2n-1} \qquad \text{at } y = 0$$

and:

$$T \to \infty \qquad \text{as } y \to \infty \tag{27}$$

where n stands for wall temperature parameter, and when $n = \frac{1}{2}$, we have the isothermal boundary condition. The dimensionless temperature $\theta(\eta)$ is assumed to be of the form:

$$\theta(\eta) = \frac{T - T_\infty}{T_w - T_\infty} \tag{28}$$

Substituting Equations (19), (20), (27), and (28) into Equation (26) gives:

$$\theta'' + Pr\left\{f\theta' - \left(\frac{2(2n-1)}{n+1}\right)f'\theta + E_c(f'')^2\right\} = 0 \tag{29}$$

also, the boundary conditions (27) become:

$$\theta(\eta) = 1 \qquad \text{at} \qquad \eta = 0$$

$$\theta(\eta) \to 0 \qquad \text{as} \qquad \eta \to \infty \tag{30}$$

Here, Pr and Ec represent the Prandtl and Eckert numbers, respectively, defined as:

$$Pr = \frac{v}{\alpha} \qquad \text{and} \qquad EC = \frac{U^2 w}{c_p(T - T_\infty)} \tag{31}$$

The physical quantity of interests here is the local Nusselt number Nu_x, which is defined by:

$$Nu_x = \frac{x q_w}{\alpha}(T_w - T_\infty) \tag{32}$$

where q_w is the wall heat flux from the plate and is expressed as:

$$q_w = -k\left(\frac{\partial T}{\partial y}\right)_{y=0} \tag{33}$$

4.2. Case II: Prescribed Heat Flux

For this case, the thermal boundary conditions are:

$$-k\frac{\partial T}{\partial y} = q_w = cx^{2n-1} \qquad \text{at} \ \ y = 0$$

and:

$$T \to \infty \qquad \text{as} \ \ y \to \infty \tag{34}$$

where n stands for the surface temperature parameter. The dimensionless temperature $g(\eta)$ is assumed to be of the form:

$$g(\eta) = \frac{T - T_\infty}{T_w - T_\infty}, \qquad \text{where} \qquad T_w - T_\infty = \frac{c}{k}x^{2n-1}\sqrt{\frac{2vx}{(n+1)u}} \tag{35}$$

Substitute Equations (19), (20), (34), and (35) into Equation (26) to get dimensionless energy equation for PHF as:

$$g'' + Pr\left\{fg' - \left(\frac{2(2n-1)}{n+1}\right)f'g + E_c(f'')^2\right\} = 0 \tag{36}$$

similarly, the corresponding boundary conditions (34) give:

$$g'(\eta) = -1 \qquad \text{at} \qquad \eta = 0$$

$$g(\eta) = 0 \qquad \text{as} \qquad \eta \to \infty \tag{37}$$

and all other physical parameters are in analogy with those mentioned in case 1, except for the constant c, which will be replaced in PHF.

5. Method of the Solution

An implicit finite difference scheme known as the Keller box method is used in solving the ordinary differential Equations (21), (29), and (36) with their respective boundary conditions (22), (30), and (32). This method is very accurate, useful, efficient, and unconditionally stable. More explanations of this method can be found in [41,42], and it involves four basic steps:

(i)   the transformed ordinary equations are expressed in the system of first order equations in η.
(ii) the resulting first order system of equations is written as finite difference equations using the central difference method about the mid-point.

(iii)   the resulting finite difference equations are linearized by Newton's method.

(iv)   the linearized system of equations is written in matrix vector form and then solved using the block tri-diagonal elimination method.

The step size $\Delta\eta = 0.02$ was used to achieve the numerical solution, and the procedures were repeated until the convergence to a specified accuracy was achieved.

6. Results and Discussion

In order to examine the effects of nonlinear stretching parameter n, viscoelastic parameter K, magnetic parameter M, Eckert number Ec, and Prandtl number Pr, on boundary flow of an electrically-conducting viscoelastic fluid past a nonlinear stretching sheet, results from the graphs for velocity and temperature profiles, as well as the numerical results of the skin friction coefficient and heat transfer rate (reduced Nusselt number) for the PST and PHF cases are illustrated in tables. To validate the accuracy of the present numerical method, we computed and compared the present results for skin friction coefficient $f''(0)$ with different values of n and K with that of Vajravelu [7] and Arnold et al. [43] as displayed in Table 1, heat transfer rate $\theta'(0)$ (for the PST case), and the surface temperature $g(0)$ (for the PHF case) for different values of Pr with that of Arnold et al. [43] when $Ec = 0.1$ and $K = 0.02$, as shown in Table 2. The current results demonstrated a close agreement with the previous results under some certain conditions. It is also observed from Table 2 that the rate of heat transfer increased with an increase in Pr for the PST case. This is because fluid with higher Pr has a comparatively lower thermal conductivity, which decreases the conduction, which in turn increases the variation. This phenomenon reduces the thickness of the thermal boundary layer and increases the heat transfer at the surface. Similarly, the surface temperature reduced with the increase in the Pr for the PHF case. This implies that an increase in Pr has a cooling effect on the surface.

Table 3 is introduced to analyze the influence of some physical parameters for $M = 0$ and $M = 5$ on the coefficient of skin friction $f''(0)$, and it is observed from this table that an increase in n and K increased the coefficient of skin friction $f''(0)$ significantly. This behavior is also true with the inclusion of a magnetic field M into the flow problem. In Table 4, the variation of the rate of heat transfer $\theta'(0)$ and the temperature at the wall $g(0)$ with different values $M, K, Ec,$ and Pr for $n = 2$ in both PST- and PHF cases is recorded. It is clear that the magnitude of the heat transfer rate decreased with an increase in $M, K,$ and Ec and increased significantly with the increase in the value of Prandtl number Pr for the PST case. Similarly, the temperature at the wall increased with the increase in Ec and M and reduced with respect to the increase in K. The effect of K was to increase the heat transfer rate for the PST case and the temperature at the surface for the PHF case, while the effect of Ec and M was to amplify the heat transfer rate and diminish the temperature at the surface for both the PST- and PHF cases. This implies that the thickness of the thermal boundary layer reduced as M and Ec increased, which led to the higher heat transfer rate at the surface, thereby enhancing the temperature $g(\eta)$. Similarly, an increase in Pr led to the reduction in the thickness of the thermal boundary layer in both cases.

The effects of different dimensionless parameters, such as nonlinear stretching parameter n, viscoelastic parameter K, magnetic parameter M, Eckert number Ec, and Prandtl number Pr, for prescribed surface temperature (PST) and prescribed heat flux (PHF) on the electrically-conducting viscoelastic fluid and heat transfer are depicted in Figures 1–13.

Figures 2–4 illustrate the behaviors of velocity and temperature profiles for various values of nonlinear stretching sheet parameter n with and without a magnetic field in the PST and PHF cases. An increase in n decreased the velocity profile, which led to the increase in the coefficient of skin friction. The decrease was high in the presence of a magnetic field, which physically shows the influence of a magnetic field in an electrically-conducting viscoelastic fluid to produce an opposing force known as Lorentz force. This force has the ability of slowing down the fluid flow in the layer region. It is also noticed from Figure 2 that an increase in M decreased the velocity profile. Figures 3 and 4 illustrate that the temperature profiles for PST- and PHF cases increased with the increase in the nonlinear stretching sheet parameter n. This implies that the rate of heat transfer reduced with the increase in n. This phenomenon showed that the thickness of the momentum boundary layer became thinner and thermal boundary thickness became thicker with an increase in n. The same behavior of this trend was reported by Vajravelu [7].

Table 1. Comparison of local skin friction $f''(0)$ at the wall for the present results and that of Vajravelu [7] when $K = R = Q = Ec = 0$ with that of Arnold et al. [43] for various values of K when $M = Ec = 0$.

n	K	$\dfrac{f''(0)}{\text{Vajravelu [7]}}$	$\dfrac{f''(0)}{\text{Present Results}}$	$\dfrac{-f''(0)}{\text{Arnold et al. [43]}}$	$\dfrac{-f''(0)}{\text{Present Results}}$
0	–	1.0000	0.9999	–	–
5	–	1.1945	1.1946	–	–
10	–	1.2348	1.2346	–	–
–	0	–	–	−1.0000	−1.0004
–	0.1	–	–	−1.0041	−1.0052
–	0.3	–	–	−1.1952	−1.1971
–	0.5	–	–	−1.4142	−1.4147

Table 2. Comparison of the numerical results of the heat transfer rate $-\theta'(0)$ and the surface temperature $g(0)$ with the published results of Arnold et al. [43] for the PST- and PHF cases, respectively, for different values of Pr with $Ec = 1.0$ and $K = 0.02$.

Pr	$-\theta'(0)$		$g(0)$	
Pr	Arnolds et al. [43]	Present Results	Arnolds et al. [43]	Present Results
1	0.988	1.000	1.009	1.000
5	2.236	2.229	0.624	0.622
10	3.080	3.056	0.564	0.567
100	8.787	8.786	0.503	0.506

Table 3. Variation of the numerical results for skin friction $-f''(0)$ with different values of $M, n,$ and K.

M	n	K	$-f''(0)$
0	1	0.4	1.0590
0	2	0.4	2.1084
0	3	0.4	6.1704
5	1	0.4	2.4888
5	2	0.4	3.3021
5	3	0.4	6.1704
0	2	0.4	2.1084
0	2	2.0	10.2509
0	2	4.0	12.6666
5	2	0.4	3.3021
5	2	2.0	10.1751
5	2	4.0	12.3794

The effect of viscoelastic parameter K on the velocity and temperature profiles is illustrated in Figures 5–7 in the presence and absence of a magnetic field. The momentum boundary layer thickness decreased with the increase in the value of K, as shown in Figure 5. This implies that tensile stress enhances the viscoelasticity through a large value of K, which has a tendency to increase the adherence to the surface of the momentum boundary layer, thereby reducing the velocity of the fluid. In the same vein, the temperature profile increased with the increase in the value of K in both PST- and PHF cases. Physically, an increase in the viscoelastic normal stress leads to the increase in the thermal boundary layer thickness. This process was the same in both the PST- and PHF cases. However, it is revealed by these figures that the inclusion of a magnetic field increased the thermal boundary thickness a bit higher compared to the case where $M = 0$. This result coincides with the results reported by Hayat et al. [18].

Table 4. Variation of the numerical values for the rate of heat transfer $-\theta'(0)$ and the wall temperature $g(0)$ for PST- and PHF cases for different values of $n, M, K, Ec,$ and Pr.

				$-\theta'(0)$		$g(0)$	
n	M	K	Ec	$Pr = 0.7$	$Pr = 7$	$Pr = 0.7$	$Pr = 7$
2	0	0.4	0.1	0.968208	3.388100	0.828806	0.201817
2	0	0.4	0.5	0.846103	2.595794	0.893924	0.295620
2	0	0.4	1.0	0.693472	1.605411	0.975321	0.412874
2	0	2.0	0.1	0.772763	2.402988	0.760442	0.183771
2	0	2.0	0.5	0.451428	−0.365513	0.791460	0.224805
2	0	2.0	1.0	0.049759	−3.826139	0.830232	0.276098
2	5	0.4	0.1	0.681828	2.842824	1.078284	0.266512
2	5	0.4	0.5	0.418812	0.959411	1.281700	0.563798
2	5	0.4	1.0	0.090042	−1.345851	1.535970	0.935405
2	5	2.0	0.1	0.563905	2.067077	0.875843	0.213427
2	5	2.0	0.5	0.160478	−1.345851	0.965558	0.343080
2	5	2.0	1.0	−3.343807	−5.612011	1.077701	0.505146

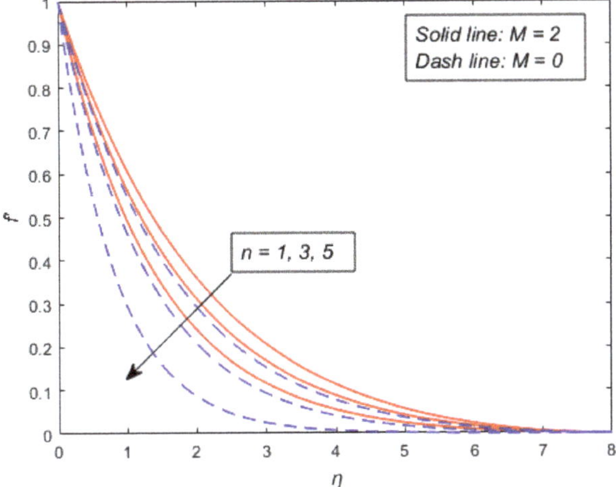

Figure 2. Variation of the velocity profile for different values of nonlinear stretching sheet parameter n and magnetic parameter M for $K = 1, Pr = 0.71,$ and $Ec = 0.1$.

The influence of the magnetic field M on the temperature profiles is shown in Figures 8 and 9. It is observed in these figures that the temperature profile increased with the increase in M. Physically, the presence of a magnetic field in an electrically-conducting fluid generates a drag-like body force, which always acts against flow, and as a result, the fluid flow decelerates. This force is known as Lorentz force and has the ability to oppose the fluid motion. The temperature of the fluid is also enhanced due to the resistance offered by this force. This shows the thermal boundary layer thickness was reduced in both the PST- and PHF cases. However, the decrease was more pronounced when $Pr = 0.71$ as compared to when $Pr = 7$.

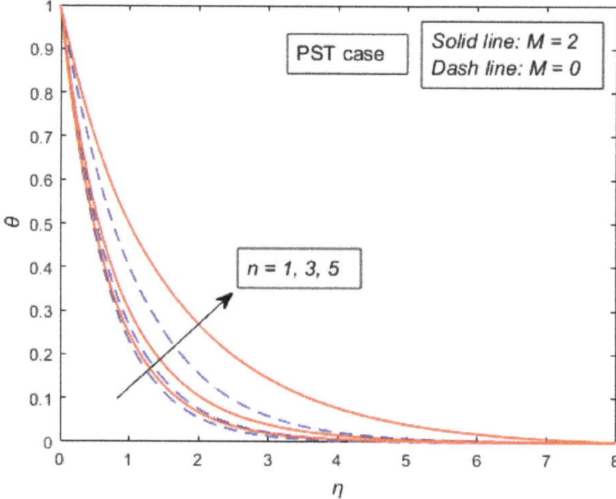

Figure 3. Variation of the temperature profile for different values of nonlinear stretching sheet parameter n and magnetic parameter M for the prescribed surface temperature (PST) case when $K = 1, Pr = 0.71$, and $Ec = 0.1$.

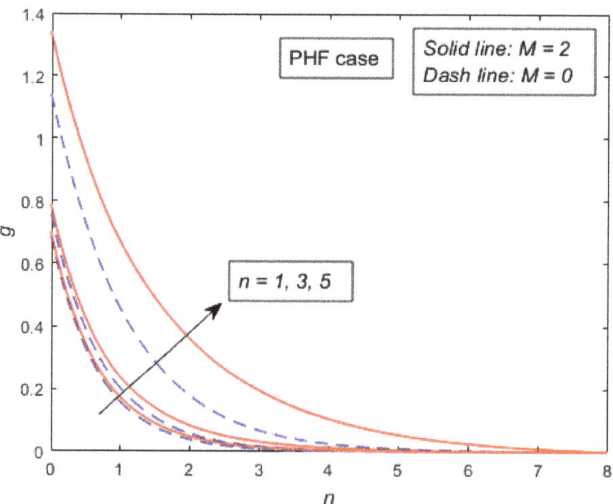

Figure 4. Variation of the temperature profile for different values of nonlinear stretching sheet parameter n and magnetic parameter M for the prescribed heat flux (PHF) case when $K = 1, Pr = 0.71$, and $Ec = 0.1$.

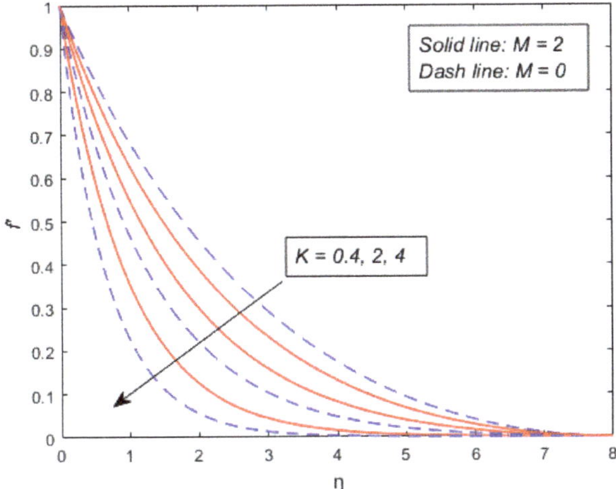

Figure 5. Variation of the velocity profile for different values of viscoelastic parameter K and magnetic parameter M for $n = 3$, $Pr = 0.71$, and $Ec = 0.1$.

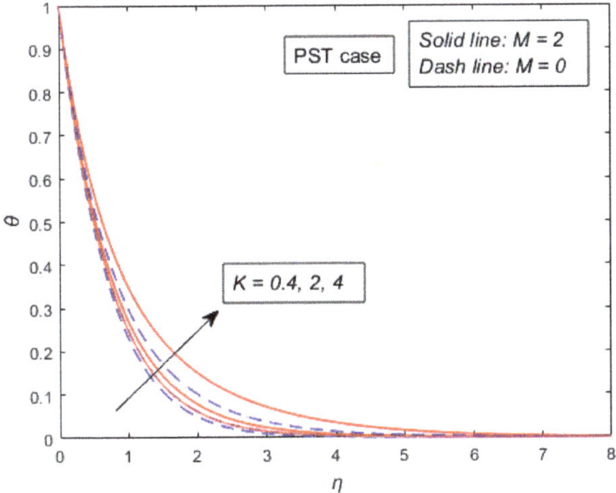

Figure 6. Variation of the temperature profile for different values of viscoelastic parameter K and magnetic parameter M for the PST case when $n = 3$, $Pr = 0.71$, and $Ec = 0.1$.

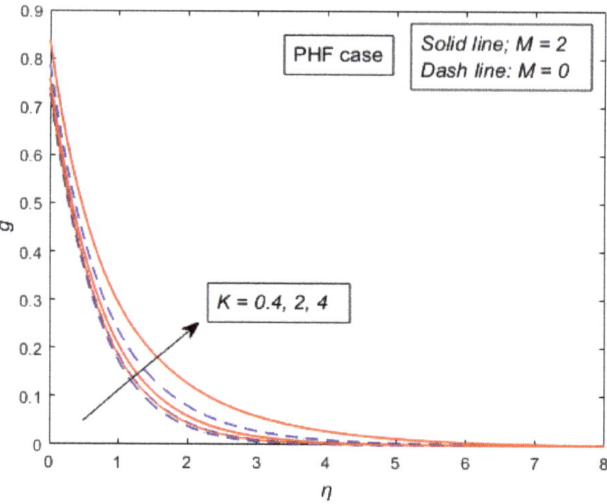

Figure 7. Variation of the temperature profile for different values of viscoelastic parameter K and magnetic parameter M for the PHF case when $n = 3$, $Pr = 0.71$, and $Ec = 0.1$.

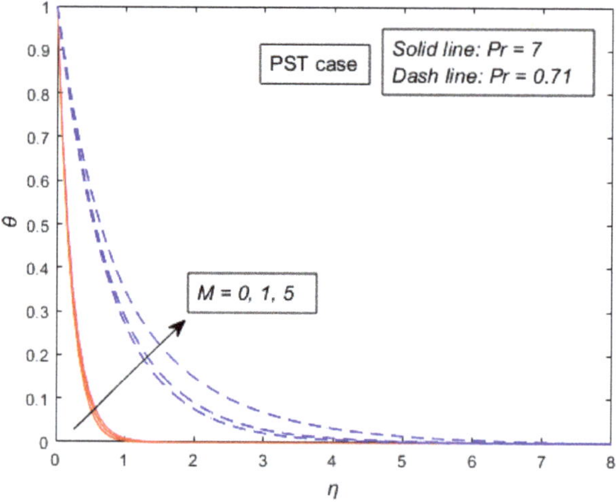

Figure 8. Variation of the temperature profile for different values of magnetic parameter M and Prandtl number Pr for the PST case when $n = 3$, $K = 1$, and $Ec = 0.1$.

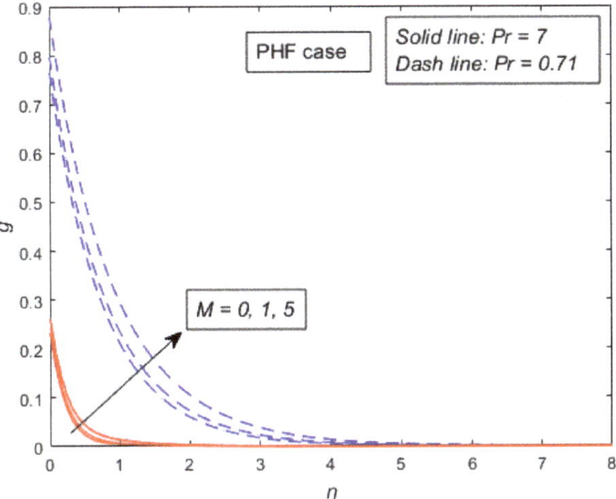

Figure 9. Variation of the temperature profile for different values of magnetic parameter M and Prandtl number Pr for the PHF case when $n = 3, K = 1$, and $Ec = 0.1$.

Figures 10 and 11 depict the effects of Eckert number Ec on the temperature profiles in the PST- and PHF cases in the presence of a magnetic field. It is noticed that an increase in the value of the Eckert number amplified the temperature profiles in both cases. The enhancement was higher for a small value of the Prandtl number. The influence of augmenting Ec is to improve the temperature distribution in the boundary layer region, thereby increasing the thickness of the thermal boundary layer. This is because heat energy is stored in the fluid due to frictional heating, which arises due to the presence of viscous dissipation. A large value of Ec produces more heat in the fluid. We can therefore infer that an increase in Ec improved the temperature of the fluid at any point for the PST- and PHF cases. However, the temperature distribution remained the same at the surface with the variation of the Eckert number for the PST case. A similar behavior was reported by [27].

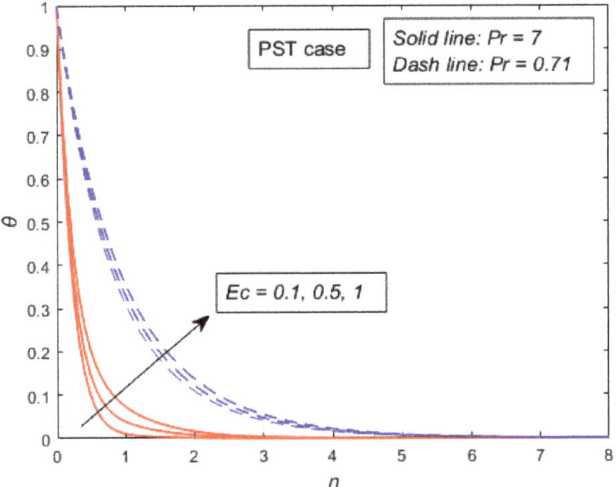

Figure 10. Variation of the temperature profile for different values of Eckert number Ec and Prandtl number Pr for the PST case when $n = 3, K = 1$, and $M = 2$.

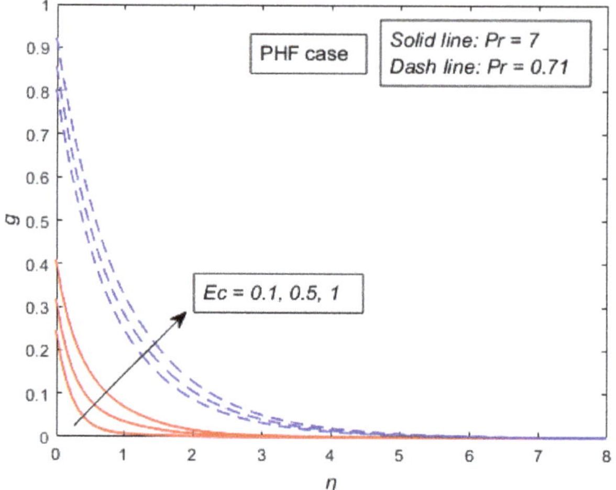

Figure 11. Variation of the temperature profile for different values of Eckert number Ec and Prandtl number Pr for the PHF case when $n = 3, K = 1$, and $M = 0$.

Figures 12 and 13 are the plots of $\theta(\eta)$ and $g(\eta)$ with respect to η for various values of Pr in the presence and absence of the Eckert number. It is clearly evident from these figures that an increase in Pr decreases the temperature profiles in the PST and PHF cases, respectively. This implies that the thickness of the thermal boundary layer is augmented with the decrease in the value of the Prandtl number. This is in close agreement with the results of Arnold et al. [43]. Physically, an increase in the Prandtl number will cause heat transfer enhancement, and this is consistent with the fact that the thickness of the thermal boundary reduces with an increase in Pr.

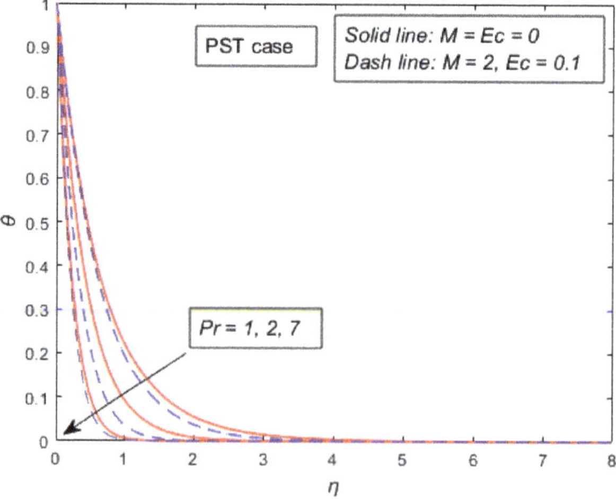

Figure 12. Variation of the temperature profile for different values of Prandtl number Pr, Eckert number Ec, and magnetic parameter M on the temperature profile for the PST case when $n = 3$ and $K = 1$.

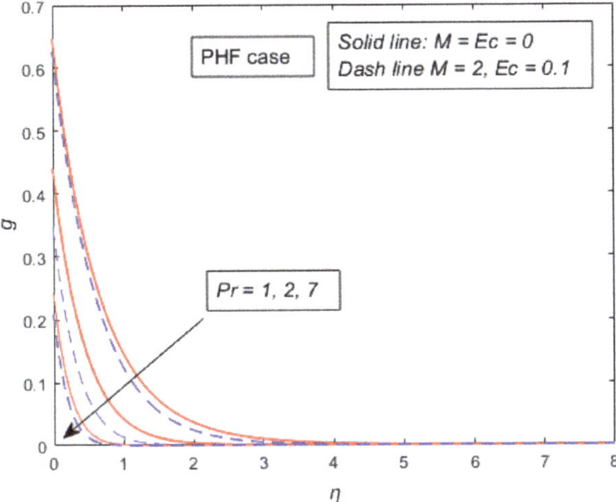

Figure 13. Variation of the temperature profile for different values of Prandtl number Pr, Eckert number Ec, and magnetic parameter M on the temperature profile for the PHF case when $n = 3$ and $K = 1$.

7. Conclusions

The numerical solutions for the two-dimensional boundary layer flow of a viscoelastic fluid and heat transfer past a nonlinear stretching sheet in the presence of magnetic field M and viscous dissipation Ec for prescribed surface temperature (PST) and prescribed heat flux (PHF) were studied in this paper. The dimensional equations that govern the flow problem were reduced into nonlinear ordinary differential equations with the aid of an appropriate similarity transformation for nonlinear stretching sheet. The reduced equations were then solved numerically by the Keller box method. The effects of nonlinear stretching parameter n, viscoelastic parameter K, magnetic parameter M, Eckert number Ec, and Prandtl number Pr on the viscoelastic fluid and the characteristics of heat transfer were examined. The results for different values of n, K, and Pr were compared with the earlier results reported by Vajravelu [7] and Arnold et al. [43] for skin friction coefficient $f''(0)$, heat transfer rate $\theta'(0)$, and surface temperature $g(0)$ in the case of PST and PHF, respectively. The summary of the present results is shown below

- The effect of nonlinear stretching parameter n was to reduce the thickness of the momentum boundary layer and increase the thermal boundary layer thickness in the PST and PHF cases. However, the rate of increase was higher in the PST case.
- The increase in the value of the magnetic parameter resulted in decreasing the velocity profile and increasing the temperature profiles in both cases.
- The influence of n, M, and K was to increase the coefficient of skin friction with its increase.
- The magnitude of the rate of heat transfer reduced with an increase in M, K, and Ec, thereby increasing the temperature of the fluid for the PST case. However, the decrease was more pronounced with a higher value of Pr. Similarly, the temperature at the wall also decreased with the increase in these governing parameters for the PHF case.
- The effect of Pr was to reduce the thickness of the thermal boundary layer and the rate of heat transfer at the surface for the PST case.
- The surface temperature reduced with the increase in Pr for the PHF case. This implies that an increase the Prandtl number had a cooling effect on the surface.

Author Contributions: A.B.J. modeled the problem and solve it; S.S. and I.U contributed to the discussion of the problem and writing the manuscript. All the authors read and approved the final version of the manuscript.

Funding: This work was funded by the Ministry of Education (MOE) and Research Management Centre-UTM, Universiti Teknologi Malaysia (UTM) for the financial support through vote numbers 5F004, 07G70, 07G72, 07G76 and 07G77.

Acknowledgments: The authors would like to thank the Ministry of Education (MOE) and Research Management Centre-UTM, Universiti Teknologi Malaysia (UTM) for this research, the anonymous reviewers and the Editor-in-Chief for the constructive comments that led to the fruitful improvement in the manuscript.

Conflicts of Interest: The authors declare no conflict of interest.

References

1. Mastroberardino, A.; Mahabaleshwar, U.S. Mixed convection in viscoelastic flow due to a stretching sheet in a porous medium. *J. Porous Media* **2013**, *16*, 483–500. [CrossRef]
2. Nayak, M.K. Chemical reaction effect on MHD viscoelastic fluid over a stretching sheet through porous medium. *Meccanica* **2016**, *51*, 1699–1711. [CrossRef]
3. Mukhopadhyay, S.; Gorla, R.S. Effects of partial slip on boundary layer flow past a permeable exponential stretching sheet in presence of thermal radiation. *Heat Mass Transf.* **2012**, *48*, 1773–11781. [CrossRef]
4. Crane, L.J. Flow past a stretching plate. *Zeitschrift für Angewandte Mathematik und Physik ZAMP* **1970**, *21*, 645–647. [CrossRef]
5. Gupta, P.S.; Gupta, A.S. Heat and mass transfer on a stretching sheet with suction or blowing. *Can. J. Chem. Eng.* **1977**, *55*, 744–746. [CrossRef]
6. Kumaran, V.; Ramanaiah, G. A note on the flow over a stretching sheet. *Acta Mech.* **1996**, *116*, 229–233. [CrossRef]
7. Vajravelu, K. Viscous flow over a nonlinearly stretching sheet. *Appl. Math. Comput.* **2001**, *124*, 281–288. [CrossRef]
8. Vajravelu, K.; Cannon, J.R. Fluid flow over a nonlinearly stretching sheet. *Appl. Math. Comput.* **2006**, *181*, 609–618. [CrossRef]
9. Cortell, R. Viscous flow and heat transfer over a nonlinearly stretching sheet. *Appl. Math. Comput.* **2007**, *184*, 864–873. [CrossRef]
10. Cortell, R. Fluid flow and radiative nonlinear heat transfer over a stretching sheet. *J. King Saud Univ. Sci.* **2014**, *26*, 161–167. [CrossRef]
11. Pal, D.; Mandal, G.; Vajravelu, K. MHD convection–dissipation heat transfer over a non-linear stretching and shrinking sheets in nanofluids with thermal radiation. *Int. J. Heat Mass Trans.* **2013**, *65*, 481–490. [CrossRef]
12. Sarpkaya, T. Flow of non-Newtonian fluids in a magnetic field. *AIChE J.* **1961**, *7*, 324–328. [CrossRef]
13. Andersson, H.I.; Bech, K.H.; Dandapat, B.S. Magnetohydrodynamic flow of a power-law fluid over a stretching sheet. *Int. J. Non-Linear Mech.* **1992**, *27*, 929–936. [CrossRef]
14. Cortell, R.A. note on magnetohydrodynamic flow of a power-law fluid over a stretching sheet. *Appl. Math. Comput.* **2005**, *168*, 557–566. [CrossRef]
15. Prasad, K.V.; Santhi, S.R.; Datti, P.S. Non-Newtonian power-law fluid flow and heat transfer over a non-linearly stretching surface. *Appl. Math.* **2012**, *3*, 425–435. [CrossRef]
16. Ullah, I.; Bhattacharyya, K.; Shafie, S.; Khan, I. Unsteady MHD mixed convection slip flow of Casson fluid over nonlinearly stretching sheet embedded in a porous medium with chemical reaction, thermal radiation, heat generation/absorption and convective boundary conditions. *PLoS ONE* **2016**, *24*, e0165348. [CrossRef] [PubMed]
17. Ullah, I.; Shafie, S.; Khan, I. MHD free convection flow of Casson fluid over a permeable nonlinearly stretching sheet with chemical reaction. *Malays. J. Fundam. Appl. Sci.* **2017**, *13*, 263–270. [CrossRef]
18. Hayat, T.; Qayyum, S.; Alsaedi, A.; Ahmad, B. Magnetohydrodynamic (MHD) nonlinear convective flow of Walters-B nanofluid over a nonlinear stretching sheet with variable thickness. *Int. J. Heat Mass Transf.* **2017**, *110*, 506–514. [CrossRef]

19. Javed, T.; Ahmad, H.; Ghaffari, A. Influence of radiation on vertical wavy surface with constant heat flux: Using Keller box scheme. *Alex. Eng. J.* **2016**, *5*, 2221–2228. [CrossRef]

20. Mahapatra, T.R.; Nandy, S.K.; Vajravelu, K.; Van Gorder, R.A. Stability analysis of fluid flow over a nonlinearly stretching sheet. *Arch. Appl. Mech.* **2011**, *81*, 1087–1091. [CrossRef]

21. Rajagopal, K.R.; Na, T.Y.; Gupta, A.S. Flow of a viscoelastic fluid over a stretching sheet. *Rheol. Acta* **1984**, *23*, 213–215. [CrossRef]

22. Dandapat, B.S.; Gupta, A.S. Flow and heat transfer in a viscoelastic fluid over a stretching sheet. *Int. J. Non-Linear Mech.* **1989**, *24*, 215–219. [CrossRef]

23. Cortell, R. A novel analytic solution of MHD flow for two classes of visco-elastic fluid over a sheet stretched with non-linearly (quadratic) velocity. *Meccanica* **2013**, *48*, 2299–2310. [CrossRef]

24. Andersson, H.I. MHD flow of a viscoelastic fluid past a stretching surface. *Acta Mech.* **1992**, *95*, 227–230. [CrossRef]

25. Char, M.I. Heat and mass transfer in a hydromagnetic flow of the viscoelastic fluid over a stretching sheet. *J. Math. Anal. Appl.* **1994**, *186*, 674–689. [CrossRef]

26. Prasad, K.V.; Pal, D.; Umesh, V.; Rao, N.P. The effect of variable viscosity on MHD viscoelastic fluid flow and heat transfer over a stretching sheet. *Commun. Nonlinear Sci. Numer. Simul.* **2010**, *15*, 331–344. [CrossRef]

27. Abel, M.S.; Siddheshwar, P.G.; Nandeppanavar, M.M. Heat transfer in a viscoelastic boundary layer flow over a stretching sheet with viscous dissipation and non-uniform heat source. *Int. J. Heat Mass Transf.* **2007**, *50*, 960–966. [CrossRef]

28. Abel, M.S.; Mahesha, N. Heat transfer in MHD viscoelastic fluid flow over a stretching sheet with variable thermal conductivity, non-uniform heat source and radiation. *Appl. Math. Modell.* **2008**, *32*, 1965–1983. [CrossRef]

29. Mustafa, M. Viscoelastic flow and heat transfer over a non-linearly stretching sheet: OHAM Solution. *J. Appl. Fluid Mech.* **2016**, *9*, 1321–1328. [CrossRef]

30. Abel, M.S.; Sanjayanand, E.; Nandeppanavar, M.M. Viscoelastic MHD flow and heat transfer over a stretching sheet with viscous and ohmic dissipations. *Commun. Nonlinear Sci. Numer. Simul.* **2008**, *13*, 1808–1821. [CrossRef]

31. Cortell, R. Effects of viscous dissipation and radiation on the thermal boundary layer over a nonlinearly stretching sheet. *Phys. Lett. A* **2008**, *28*, 631–636. [CrossRef]

32. Medikare, M.; Joga, S.; Chidem, K.K. MHD stagnation point flow of a Casson fluid over a nonlinearly stretching sheet with viscous dissipation. *Am. J. Comput. Math.* **2016**, *23*, 37–48. [CrossRef]

33. Ramandevi, B.; Reddy, J.R.; Sugunamma, V.; Sandeep, N. Combined influence of viscous dissipation and non-uniform heat source/sink on MHD non-Newtonian fluid flow with Cattaneo-Christov heat flux. *Alex. Eng. J.* **2018**, *57*, 1009–1018. [CrossRef]

34. Atif, S.M.; Hussain, S.; Sagheer, M. Effect of viscous dissipation and Joule heating on MHD radiative tangent hyperbolic nanofluid with convective and slip conditions. *J. Braz. Soc. Mech. Sci. Eng.* **2019**, *41*, 189. [CrossRef]

35. Beard, D.W.; Walters, K. Elastico-viscous boundary-layer flows I. Two-dimensional flow near a stagnation point. *Math. Proc. Camb. Philos. Soc.* **1964**, *60*, 667–674. [CrossRef]

36. Madani Tonekaboni, S.A.; Abkar, R.; Khoeilar, R. On the study of viscoelastic Walters' B fluid in boundary layer flows. *Math. Probl. Eng.* **2012**, *2012*, 861508. [CrossRef]

37. Jaluria, Y. *Natural Convection: Heat and Mass Transfer*; Pergamon: Oxford, UK, 1980; Volume 5.

38. Ahmed, N.; Sharma, D.; Deka, H. MHD mixed convection and mass transfer from an infinite vertical porous plate with chemical reaction in presence of a heat source. *Appl. Math. Sci.* **2012**, *6*, 1011–1020.

39. Xu, H.; Liao, S.J. Series solutions of unsteady magnetohydrodynamic flows of non-Newtonian fluids caused by an impulsively stretching plate. *J. Non-Newton. Fluid Mech.* **2005**, *29*, 46–55. [CrossRef]

40. Kasim, A.M. Convective Boundary Layer Flow of Viscoelastic Fluid. Ph.D. Thesis, Universiti Teknologi Malaysia, Skudai, Johor, Malaysia, 2014.

41. Cebeci, T.; Bradshaw, P. *Physical and Computational Aspects of Convective Heat Transfer*; Springer Science Business Media: New York, NY, USA, 2012.

42. Sarif, N.M.; Salleh, M.Z.; Nazar, R. Numerical solution of flow and heat transfer over a stretching sheet with Newtonian heating using the Keller box method. *Procedia Eng.* **2013**, *53*, 542–554. [CrossRef]
43. Arnold, J.C.; Asir, A.A.; Somasundaram, S.; Christopher, T. Heat transfer in a viscoelastic boundary layer flow over a stretching sheet. *Int. J. Heat Mass Transf.* **2010**, *53*, 1112–1118. [CrossRef]

Article

Thin Film Flow of Couple Stress Magneto-Hydrodynamics Nanofluid with Convective Heat over an Inclined Exponentially Rotating Stretched Surface

Asifa Tassaddiq [1,*], **Ibni Amin** [2], **Meshal Shutaywi** [3], **Zahir Shah** [4], **Farhad Ali** [5], **Saeed Islam** [2,6,7,*] **and Asad Ullah** [8]

[1]   College of Computer and Information Sciences Majmaah University, Al Majmaah 11952, Saudi Arabia
[2]   Department of Mathematics, Abdul Wali Khan University Mardan, Khyber 23200, Pakhtunkhwa, Pakistan; iameen96@gmail.com
[3]   Department of Mathematics College of Science and Arts, King Abdul-Aziz University, Rabigh 21911, Saudi Arabia; mshutaywi@kau.edu.sa
[4]   Center of Excellence in Theoretical and Computational Science (TaCS-CoE), SCL 802 Fixed Point Laboratory, King Mongkut's University of Technology Thonburi (KMUTT), Bangkok 10140, Thailand; Zahir.sha@kmutt.ac.th
[5]   Department of Mathematics, City University of Science and Information Technology, Peshawar 25000, KPK, Pakistan; farhadali@cusit.edu.pk
[6]   Informetrics Research Group, Ton Duc Thang University, Ho Chi Minh City 70000, Vietnam
[7]   Faculty of Mathematics and Statistics, Ton Duc Thang University, Ho Chi Minh City 70000, Vietnam
[8]   Institute of Numerical Sciences, Kohat University of Science and Technology, KUST, Khyber 26000, Pakhtunkhwa, Pakistan; asad.ullah@kust.edu.pk
*   Correspondence: a.tassaddiq@mu.edu.sa (A.T.); saeed.islam@tdtu.edu.vn (S.I.)

Received: 20 February 2020; Accepted: 30 March 2020; Published: 1 April 2020

Abstract: In this article a couple stress magneto-hydrodynamic (MHD) nanofluid thin film flow over an exponential stretching sheet with joule heating and viscous dissipation is considered. Similarity transformations were used to obtain a non-linear coupled system of ordinary differential equations (ODEs) from a system of constitutive partial differential equations (PDEs). The system of ordinary differential equations of couple stress magneto-hydrodynamic (MHD) nanofluid flow was solved using the well-known Homotopy Analysis Method (HAM). Nusselt and Sherwood numbers were demonstrated in dimensionless forms. At zero Prandtl number the velocity profile was analytically described. Furthermore, the impact of different parameters over different state variables are presented with the help of graphs. Dimensionless numbers like magnetic parameter M, Brownian motion parameter Nb, Prandtl number Pr, thermophoretic parameter Nt, Schmidt number Sc, and rotation parameter S were analyzed over the velocity, temperature, and concentration profiles. It was observed that the magnetic parameter M increases the axial, radial, drainage, and induced profiles. It was also apparent that Nu reduces with greater values of Pr. On increasing values of the Brownian motion parameter the concentration profile declines, while the thermophoresis parameter increases.

Keywords: MHD; nanofluid; stretching surface; rotating fluid; Homotopy Analysis Method (HAM)

---

1. Introduction

The word nanofluid denotes a mixture of nanoparticles and base fluids. Usually nanoparticles contain metals such as silver, copper, aluminum, nitrides like silicon nitride, carbides such as silicon carbides, oxides e.g., aluminum oxide and nonmetals such as graphite. The usual liquids are water, oil, and ethylene glycol. The combination of nanoparticles with a base liquid greatly helps to develop the thermal qualities of the vile liquid. Choi et al. [1] introduced the term nanofluid and heat transfer features of vile fluids and studied the thermal conductivity enhancement. Wang et al. [2] investigated convective physiognomies of vile fluid and they found that these fluids are enriched by adding metal and non-metal atoms to them. Heat transfer enhancement and thermal conductivity variation of fluids by the addition of copper nanoparticles were studied by Eastman et al. [3,4]. Murshed et al. [5] found the thermal conductivity of vile liquid increases by adding sphere-shaped nanoparticles. Maïga et al. [6], scrutinized nanofluid flow in a uniformly heated tube with heat transfer. Bianco et al. [7] studied the implication between nanoparticles and liquid matrix in two-phase flow; further, nanofluids involuntary convection in circular tubes was deliberated. Buongiorno [8] introduced the two-phase model for convective transport in nanofluids. The single-phase model was studied by Tiwari et al. [9]. After these two models, several investigators considered nanofluids thermal attraction to investigate the actual fluid characteristics discussed in references [10–12]. Thin film Darcy–Forchheimer nanofluid flow with Joule dissipation and MHD effect were scrutinized by Jawad et al. [13]. Rotating flow in the existence of aqueous suspensions with the effect of non-linear thermal radiation was investigated by Jawad et al. [14]. Bhatti et al. [15] scrutinized Jeffrey nanofluid with immediate effects of variable magnetic field. MHD non-Newtonian nanofluid flow over a pipe with heat reliant viscosity was scrutinized by Ellahi et al. [16]. A Cu-water nanofluid applying porous media with a micro-channel heat sink was investigated by Hatami et al. [17]. Laminar nanofluid flow with heat transfer between rotating disks was studied by Hatami et al. [18]. Recently, Shah et al. [19–22] considered Hall current and thermal radiations of the nanofluid flow through a rotating system. Non-Newtonian fluids are complex in nature and various models have beenconstituted and developed for the purpose of defining the strain rate in these fluids. Recently, Ullah et al. [23] analyzed the Reiner–Philippoff fluid model analytically over a stretching surface. They studied the thermophoresis and Brownian motion impacts over the thin film. The heat transfer enhancement not only depends on the nanoparticles added, but also depends on the nature of the fluid. The couple stress impacts with joule heating and viscous dissipation have been studied and analyzed by different researchers over different surfaces [24–26]. Heat transfer enhancement and its detailed study with engineering applications can be found in the references [27–31].

Various techniques have been used to analyze the problem constituted and modeled in the literature. Soleimani et al. [32] used the finite element method for natural convective nanofluid flow with transfer of heat, in a semi-annular object. They described the turning angle effect on the isotherms, streamlines, and local Nusselt number. Rudraiah et al. [33] studied numerically the natural convection inside a rectangular obstacle subject to magnetic field. They found that the heat transfer reduces with a magnetic field. For the simulation of magnetic drug targeting and ferrofluid flow, Sheikholeslami and Ellahi and Kandelousi [34] considered the lattice Boltzmann method (LBM). They found that both the magnetic parameter and the Reynolds number decrease the coefficient of skin friction. Ramzan et al. [24] found a series solution by using HAM for the flow of 3D nanofluid couple stress with joule heating. Recently, with the impacts of convective condition couple stress 3D MHD nanofluid flow in the presence of Cattaneo–Christov heat flux was explored by Hayat et al. [35,36]. Maxwell boundary layer flow of nanofluid was investigated by Hayat et al. [37]. Malik et al. [38] considered MHD flow through a stretching surface of Erying Powell nanofluid. Nadeem et al. [39] considered a vertical stretching surface and analyzed the flow of Maxwell's liquid with nanoparticles. Raju et al. [40] investigated an MHD nanoliquid flow with free convective heat transfer through a cone. The impact of Lorentz forces and entropy generation for different nanofluids were numerically investigated by Sheikholeslami et al. [41–43]. They also used the control volume finite element method (CVFEM) and some new modified techniques for the analysis of the nanofluid flow through a square cavity by considering shape factors. More detailed studies on nanofluid investigation by considering different models can be found in the references [44–49].

Keeping in view the applications of nanofluid and its role in heat transfer enhancement various analyses have been made by researchers. Shah et al. [50] analyzed the Titanium nanofluid flow over a rotating surface analytically. In their work, they studied the impacts of the Hall current and the magnetic parameter. Considering a similar approach for the problem geometry, in this work an inclined rotating surface is considered and is extended to the couple stress nanofluid MHD flow with convective heat transfer by ignoring the mass flux. Furthermore, the impacts of viscous dissipation and Joule heating are also considered in investigating the overall effects of the fluid parameters. The basic equations for the physical problem are constituted from the geometry of the problem and the assumptions made. The set of PDEs obtained from the fundamental equations of fluid dynamics and further assimilated from the boundary layer theory are transformed to a non-linear ODEs system, by using similarity transformations. The analytical Homotopy Analysis Method [51–54] is applied to solve the set of ODEs.

2. Problem Formulation

Assume a three-dimensional nanofluid thin liquid flow through a steady rotating disk. The disk rotates on its own axis due to the angular velocity (Ω) as shown in Figure 1. With the horizontal axis the inclined disk creates an angle β. The radius of the disk is greater as it is associated to the thickness of the fluid film and therefore the last impact is unobserved. Here T_0 and T_w denote the temperatures of the surface film and the surface of the disk respectively. Similarly, at the film C_0, C_h are the concentrations on the disk surface respectively. The equations for the steady state flow are displayed as follows [50,55,56]:

$$u_x + v_y + w_z = 0 \tag{1}$$

$$uu_x + vu_y + wu_z = \nu u_{zz} - \nu' u_{zzzz} - \frac{\sigma B_0^2 u}{\rho} \tag{2}$$

$$uv_x + vv_y + wv_z = \nu v_{zz} - \nu' v_{zzzz} - \frac{\sigma B_0^2 v}{\rho} \tag{3}$$

$$uw_x + vw_y + ww_z = \nu w_{zz} - \nu' w_{zzzz} - \frac{\sigma B_0^2 w}{\rho} \tag{4}$$

$$uT_x + vT_y + wT_z = \frac{k_{nf}}{\left(\rho c_p\right)_{nf}}(T_{zz}) + \tau\left[D_B C_z T_z + \frac{D_T}{T_0}(T_z)^2\right] \tag{5}$$

$$uC_x + vC_y + wC_z = D_\beta(C_{zz}) + \left(\frac{D_T}{T_0}\right)(T_{zz}) \tag{6}$$

with boundary conditions:

$$
\begin{aligned}
u = -\Omega y, v = \Omega x, w = 0, T = T_w, C = C_h \qquad &\text{at} \qquad z = 0 \\
u_z = v_z = 0, w = 0, T = T_w, C = C_0 \qquad &\text{at} \qquad z = h
\end{aligned}
\tag{7}
$$

using the Similarity transformation [50]:

$$
\begin{aligned}
&u = -\Omega y g(\eta) + \Omega x f'(\eta) + \overline{g} k(\eta) \sin\frac{\beta}{\Omega}\\
&v = \Omega x g(\eta) + \Omega y f'(\eta) + \overline{g} s(\eta) \sin\frac{\beta}{\Omega}\\
&w = -2\sqrt{\Omega \nu_{nf}} f(\eta), T = (T_0 - T_w)\theta(\eta) + T_w\\
&\eta\phi(\eta) = \frac{C-C_w}{C_0-C_w}, \eta = z\sqrt{\frac{\Omega}{\nu_{nf}}}
\end{aligned}
\tag{8}
$$

we obtain:

$$g'' - 2gf' + 2g'f - Kg^{vii} - Mg = 0 \tag{9}$$

$$f''' - f'^2 + g^2 + 2ff'' - Kf^{vii} - Mf' = 0 \tag{10}$$

$$k'' + gs - kf' + 2kf - Kk^{vi} - Mk = 0 \tag{11}$$

$$s'' - kg - sf' + 2s'f - Ks^{vi} - Ms = 0 \tag{12}$$

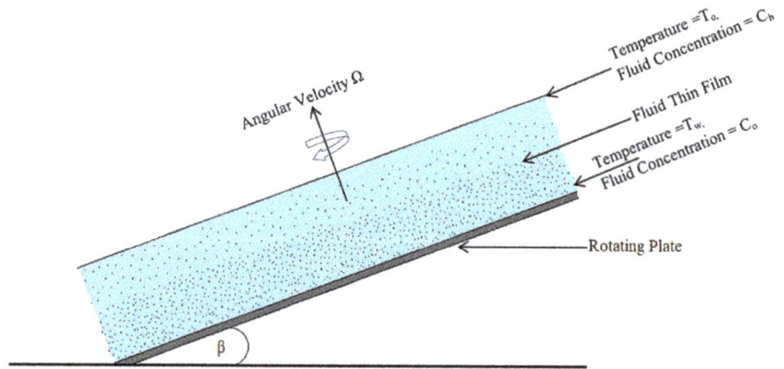

Figure 1. Geometrical description of the problem.

Now if $\theta(\eta)$ and $\phi(\eta)$ depend only on z, then Equations (5) and (6) are reduced to the following forms:

$$\theta'' + 2Pr\theta f + PrNb\eta\phi'\theta' + Pr\Omega Nt\theta'^2 = 0 \tag{13}$$

$$\phi'' + 2Scf\phi' + \frac{Nt}{Nb}\theta'' + \frac{S}{2}\left(\eta\phi' + \eta^2\phi''\right) = 0 \tag{14}$$

$$\begin{aligned}
f(0) &= 0, f'(0) = 0, f''(\delta) = 0 \\
g(0) &= 0, g'(\delta) = 0 \\
k(0) &= 0, k'(\delta) = 0 \\
s(0) &= 0, s'(\delta) = 0 \\
\theta(0) &= 0, \theta(\delta) = 1 \\
\phi(0) &= 0, \phi(\delta) = 1
\end{aligned} \tag{15}$$

while, Pr M, Sc, Nt, K, S and Nb are defined as below [57]:

$$\begin{aligned}
Pr &= \frac{\rho_{cp}\nu_{nf}}{k}, Sc = \frac{\mu}{\rho f D}, Nb = \frac{\tau D_B}{\nu}(C_0 - C_w) \\
Nt &= \frac{\tau D_T}{\nu T_w}(T_0 - T_w), S = \frac{\alpha}{\Omega}, M = \frac{\sigma B_0^2}{\Omega\rho}, K = \frac{\nu'\Omega}{\nu_{nf}^2}
\end{aligned} \tag{16}$$

here, the constant of the normalized thickness is defined as:

$$\delta = h\sqrt{\frac{\Omega}{\upsilon_{nf}}} \tag{17}$$

where the velocity of condensation is written as:

$$f(\delta) = \frac{W}{2\sqrt{\Omega\nu}} = \alpha \tag{18}$$

Direct integration of Equation (4) gives the pressure term. Using Pr $= 0$ and $\theta(\delta) = 1$, we get

$$\theta'(0) = \frac{1}{\delta} \tag{19}$$

For small δ an asymptotic limit is explained by Equation (17). The decrease of $\theta'(0)$ for increasing δ is not monotonic, thus Nu is signified as [57]:

$$Nu = \frac{k_{nf}}{k_f}\frac{(T_z)_w}{(T_0 - T_w)} = A_4\delta\theta'(0) \tag{20}$$

Similarly, the Sherwood number is given by:

$$Sh = \frac{(C_z)_w}{C_0 - C_w} = \delta\phi'(0) \tag{21}$$

3. Solution by HAM

For the solution of three dimensional nanofluid thin layer flows through a steady rotating disk, the optimal approach is used. The obtained Equations (9)–(15) are solved by using HAM. The basic derivation and mechanism is explained below.

The operators $L_{\widehat{f}}, L_{\widehat{\theta}}$ and $L_{\widehat{\phi}}$ are defined as [13,51]:

$$L_{\widehat{f}}(\widehat{f}) = \widehat{f}''', L_{\widehat{k}}(\widehat{k}) = \widehat{k}'', L_{\widehat{g}}(\widehat{g}) = \widehat{g}'',$$
$$L_{\widehat{s}}(\widehat{s}) = \widehat{s}'', L_{\widehat{\theta}}(\widehat{\theta}) = \widehat{\theta}'', L_{\widehat{\phi}}(\widehat{\phi}) = \widehat{\phi}'' \tag{22}$$

where,

$$L_{\widehat{f}}(e_1 + e_2\eta + e_3\eta^2) = 0, L_{\widehat{k}}(e_5 + e_6\eta) = 0, L_{\widehat{g}}(e_7 + e_8\eta) = 0,$$
$$L_{\widehat{s}}(e_9 + e_{10}\eta) = 0, L_{\widehat{\theta}}(e_{11} + e_{12}\eta) = 0, L_{\widehat{\phi}}(e_{13} + e_{14}\eta) = 0 \tag{23}$$

The consistent non-linear operators are reasonably selected as $N_{\widehat{f}}, N_{\widehat{k}}, N_{\widehat{g}}, N_{\widehat{s}}, N_{\widehat{\theta}}$ and $N_{\widehat{\phi}}$, and are recognized in the system [13,22]:

$$N_{\widehat{f}}\left[\widehat{f}(\eta;\zeta), \widehat{g}(\eta;\zeta)\right] = \widehat{f}_{\eta\eta\eta} - \widehat{f}_\eta^2 + \widehat{g}^2 + 2\widehat{f}\widehat{f}_{\eta\eta} - K\widehat{f}_{\eta\eta\eta\eta\eta\eta} - M\widehat{f}_\eta \tag{24}$$

$$N_{\widehat{g}}\left[\widehat{g}(\eta;\zeta), \widehat{f}(\eta;\zeta)\right] = \widehat{g}_{\eta\eta} - 2\widehat{g}\widehat{f}_\eta + 2\widehat{g}_\eta\widehat{f} - K\widehat{g}_{\eta\eta\eta\eta\eta\eta} - M\widehat{g} \tag{25}$$

$$N_{\widehat{g}}\left[\widehat{g}(\eta;\zeta), \widehat{f}(\eta;\zeta)\right] = \widehat{g}_{\eta\eta} - 2\widehat{g}\widehat{f}_\eta + 2\widehat{g}_\eta\widehat{f} - K\widehat{g}_{\eta\eta\eta\eta\eta\eta} - M\widehat{g} \tag{26}$$

$$N_{\widehat{s}}\left[\widehat{s}(\eta;\zeta), \widehat{g}(\eta;\zeta), \widehat{f}(\eta;\zeta), \widehat{k}(\eta;\zeta)\right] = \widehat{s}_{\eta\eta} - \widehat{k}\widehat{g} - \widehat{s}\widehat{f}_\eta + 2\widehat{s}_\eta\widehat{f} - K\widehat{s}_{\eta\eta\eta\eta\eta} - M\widehat{s} \tag{27}$$

$$N_{\widehat{\theta}}\left[\widehat{\theta}(\eta;\zeta), \widehat{f}(\eta;\zeta)\right] = \widehat{\theta}_{\eta\eta} + 2Pr\widehat{\theta}\widehat{f} + PrNb\eta\widehat{\phi}_\eta\widehat{\theta}_\eta + Pr\Omega Nt\widehat{\theta}_\eta^2 \tag{28}$$

$$N_{\widehat{\phi}}\left[\widehat{\phi}(\eta;\zeta), \widehat{f}(\eta;\zeta), \widehat{\theta}(\eta;\zeta)\right] = \widehat{\phi}_{\eta\eta} + 2Sc\widehat{f}\widehat{\phi}_\eta + \frac{Nt}{Nb}\widehat{\theta}_{\eta\eta} + \frac{S}{2}\left(\eta\widehat{\phi}_\eta + \eta^2\widehat{\phi}_{\eta\eta}\right) \tag{29}$$

For Equations (8)–(10) the 0th-order system is written as [21]:

$$(1 - \zeta)L_{\widehat{f}}\left[\widehat{f}(\eta;\zeta) - \widehat{f}_0(\eta)\right] = p\hbar_{\widehat{f}}N_{\widehat{f}}\left[\widehat{f}(\eta;\zeta), \widehat{g}(\eta;\zeta)\right] \tag{30}$$

$$(1 - \zeta)L_{\widehat{k}}\left[\widehat{k}(\eta;\zeta) - \widehat{k}_0(\eta)\right] = p\hbar_{\widehat{k}}N_{\widehat{k}}\left[\widehat{k}(\eta;\zeta), \widehat{g}(\eta;\zeta), \widehat{f}(\eta;\zeta), \widehat{s}(\eta;\zeta)\right] \tag{31}$$

$$(1 - \zeta)L_{\widehat{g}}\left[\widehat{g}(\eta;\zeta) - \widehat{g}_0(\eta)\right] = p\hbar_{\widehat{g}}N_{\widehat{g}}\left[\widehat{f}(\eta;\zeta), \widehat{g}(\eta;\zeta)\right] \tag{32}$$

$$(1-\zeta)L_{\widehat{s}}\left[\widehat{s}\,(\eta;\zeta)-\widehat{s}_0(\eta)\right]=p\hbar_{\widehat{s}}N_{\widehat{s}}\left[\widehat{s}\,(\eta;\zeta),\widehat{g}\,(\eta;\zeta),\widehat{f}\,(\eta;\zeta),\widehat{k}\,(\eta;\zeta)\right] \tag{33}$$

$$(1-\zeta)\,L_{\widehat{\theta}}\left[\widehat{\theta}\,(\eta;\zeta)-\widehat{\theta}_0(\eta)\right]=p\hbar_{\widehat{\theta}}N_{\widehat{\theta}}\left[\widehat{\theta}\,(\eta;\zeta),\widehat{f}\,(\eta;\zeta),\widehat{\phi}(\eta;\zeta)\right] \tag{34}$$

$$(1-\zeta)\,L_{\widehat{\phi}}\left[\widehat{\phi}(\eta;\zeta)-\widehat{\phi}_0(\eta)\right]=p\hbar_{\widehat{\phi}}N_{\widehat{\phi}}\left[\widehat{\phi}(\eta;\zeta),\widehat{f}\,(\eta;\zeta),\widehat{\theta}\,(\eta;\zeta)\right] \tag{35}$$

where the boundary conditions are [22]:

$$\begin{aligned}
&\left.\widehat{f}\,(\eta;\zeta)\right|_{\eta=0}=0,\ \left.\frac{\partial\widehat{f}(\eta;\zeta)}{\partial\eta}\right|_{\eta=0}=0,\ \left.\frac{\partial^2\widehat{f}(\eta;\zeta)}{\partial\eta^2}\right|_{\eta=\delta}=0,\ \left.\widehat{g}\,(\eta;\zeta)\right|_{\eta=0}=0\\[4pt]
&\left.\frac{\partial\widehat{g}(\eta;\zeta)}{\partial\eta}\right|_{\eta=\delta}=0,\ \left.\widehat{k}\,(\eta;\zeta)\right|_{\eta=0}=0,\ \left.\frac{\partial\widehat{k}(\eta;\zeta)}{\partial\eta}\right|_{\eta=\delta}=0,\ \left.\widehat{s}\,(\eta;\zeta)\right|_{\eta=0}=0\\[4pt]
&\left.\frac{\partial\widehat{s}(\eta;\zeta)}{\partial\eta}\right|_{\eta=\delta}=0,\ \left.\widehat{\theta}\,(\eta;\zeta)\right|_{\eta=0}=0,\ \left.\frac{\partial\widehat{\theta}(\eta;\zeta)}{\partial\eta}\right|_{\eta=\delta}=1,\ \left.\widehat{\phi}(\eta;\zeta)\right|_{\eta=0}=0,\ \left.\widehat{\phi}(\eta;\zeta)\right|_{\eta=\delta}=1
\end{aligned} \tag{36}$$

while the embedding constraint is $\zeta\in[0,1]$, to regulate for the solution convergence $\hbar_{\widehat{f}},\hbar_{\widehat{k}},\hbar_{\widehat{g}},\hbar_{\widehat{s}}$, $\hbar_{\widehat{\theta}}$ and $\hbar_{\widehat{\phi}}$ are used. When $\zeta=0$ and $\zeta=1$ we have:

$$\begin{aligned}
&\widehat{f}\,(\eta;1)=\widehat{f}\,(\eta),\ \widehat{k}\,(\eta;1))=\widehat{k}\,(\eta),\ \widehat{g}\,(\eta;1)=\widehat{g}\,(\eta)\\
&\widehat{s}\,(\eta;1)=\widehat{s}\,(\eta),\ \widehat{\theta}\,(\eta;1)=\widehat{\theta}\,(\eta),\ \widehat{\phi}\,(\eta;1)=\widehat{\phi}\,(\eta)
\end{aligned} \tag{37}$$

expand the $\widehat{f}\,(\eta;\zeta),\widehat{k}\,(\eta;\zeta),\widehat{g}\,(\eta;\zeta),\widehat{s}\,(\eta;\zeta),\widehat{\phi}\,(\eta;\zeta)$ and $\widehat{\phi}\,(\eta;\zeta)$ through Taylor's series for $\zeta=0$:

$$\begin{aligned}
&\widehat{f}\,(\eta;\zeta)=\widehat{f}_0(\eta)+\sum_{n=1}^{\infty}\widehat{f}_n(\eta)\zeta^n,\ \widehat{k}\,(\eta;\zeta)=\widehat{k}_0(\eta)+\sum_{n=1}^{\infty}\widehat{k}_n(\eta)\zeta^n\\
&\widehat{g}\,(\eta;\zeta)=\widehat{g}_0(\eta)+\sum_{n=1}^{\infty}\widehat{g}_n(\eta)\zeta^n,\ \widehat{s}\,(\eta;\zeta)=\widehat{s}_0(\eta)+\sum_{n=1}^{\infty}\widehat{s}_n(\eta)\zeta^n\\
&\widehat{\theta}\,(\eta;\zeta)=\widehat{\theta}_0(\eta)+\sum_{n=1}^{\infty}\widehat{\theta}_n(\eta)\zeta^n,\ \widehat{\phi}\,(\eta;\zeta)=\widehat{\phi}_0(\eta)+\sum_{n=1}^{\infty}\widehat{\phi}_n(\eta)\zeta^n
\end{aligned} \tag{38}$$

$$\begin{aligned}
&\widehat{f}_n(\eta)=\frac{1}{n!}\left.\frac{\partial\widehat{f}\,(\eta;\zeta)}{\partial\eta}\right|_{p=0},\ \widehat{k}_n(\eta)=\frac{1}{n!}\left.\frac{\partial\widehat{k}\,(\eta;\zeta)}{\partial\eta}\right|_{p=0},\ \widehat{g}_n(\eta)=\frac{1}{n!}\left.\frac{\partial\widehat{g}\,(\eta;\zeta)}{\partial\eta}\right|_{p=0}\\[4pt]
&\widehat{s}_n(\eta)=\frac{1}{n!}\left.\frac{\partial\widehat{s}\,(\eta;\zeta)}{\partial\eta}\right|_{p=0},\ \widehat{\theta}_n(\eta)=\frac{1}{n!}\left.\frac{\partial\widehat{\theta}\,(\eta;\zeta)}{\partial\eta}\right|_{p=0},\ \widehat{\phi}_n(\eta)=\frac{1}{n!}\left.\frac{\partial\widehat{\phi}\,(\eta;\zeta)}{\partial\eta}\right|_{p=0}
\end{aligned} \tag{39}$$

where the boundary restrictions are:

$$\begin{aligned}
&\widehat{f}\,(0)=0,\ \widehat{f}'\,(0)=0,\ \widehat{f}''\,(\delta)=0,\ \widehat{g}\,(0)=0,\ \widehat{g}'\,(\delta)=0,\ \widehat{k}\,(0)=0,\ \widehat{k}'\,(\delta)=0,\\
&\widehat{s}\,(0)=0,\ \widehat{s}'\,(\delta)=0,\ \widehat{\theta}\,(0)=0,\ \widehat{\theta}'\,(\delta)=1,\ \widehat{\phi}\,(0)=0,\ \widehat{\phi}\,(\delta)=1.
\end{aligned} \tag{40}$$

now define

$$\mathfrak{R}_n^{\widehat{f}}(\eta)=\widehat{f}'''_{n-1}-\widehat{f}'^2_{n-1}+\widehat{g}^2_{n-1}+2\sum_{j=0}^{w-1}\widehat{f}_{w-1-j}\widehat{f}''_j-\sum_{j=0}^{w-1}\widehat{k}_{w-1-j}\widehat{f}^{vii}_j-M\widehat{f}'_{n-1} \tag{41}$$

$$\mathfrak{R}_n^{\widehat{k}}(\eta)=\widehat{k}''_{n-1}+\sum_{j=0}^{w-1}\widehat{g}_{w-1-j}\widehat{s}_j-\sum_{j=0}^{w-1}\widehat{k}_{w-1-j}\widehat{f}'_j+2\sum_{j=0}^{w-1}\widehat{k}_{w-1-j}\widehat{f}_j-K\widehat{k}_{n-1}-M\widehat{k}_{n-1} \tag{42}$$

$$\mathfrak{R}_n^{\widehat{g}}(\eta)=\widehat{g}''_{n-1}-2\sum_{j=0}^{w-1}\widehat{g}_{w-1-j}\widehat{f}_j+2\sum_{j=0}^{w-1}\widehat{g}'_{w-1-j}\widehat{f}_j-K\widehat{g}_{n-1}-M\widehat{g}_{n-1}, \tag{43}$$

$$\mathcal{R}_n^{\widehat{s}}(\eta) = \widehat{s}''_{n-1} - \sum_{j=0}^{w-1} \widehat{k}_{w-1-j}\,\widehat{g}_j - \sum_{j=0}^{w-1} \widehat{s}_{w-1-j}\,\widehat{f}'_j + 2\sum_{j=0}^{w-1} \widehat{s}'_{w-1-j}\,\widehat{f}_j - K\widehat{s}^{vi} - M\widehat{s}_{n-1} \tag{44}$$

$$\mathcal{R}_n^{\widehat{\theta}}(\eta) = \left(\widehat{\theta}''_{n-1}\right) + 2\mathrm{Pr}\sum_{j=0}^{w-1} \widehat{\theta}_{w-1-j}\,\widehat{f}_j + \mathrm{Pr}Nb\eta \sum_{j=0}^{w-1} \widehat{\phi}'_{w-1-j}\,\widehat{\theta}'_j + \mathrm{Pr}\Omega Nt\widehat{\theta}'^2_{n-1} \tag{45}$$

$$\mathcal{R}_n^{\widehat{\phi}}(\eta) = \widehat{\phi}''_{n-1} + 2Sc\sum_{j=0}^{w-1} \widehat{f}_{w-1-j}\,\widehat{\phi}'_j + \frac{Nt}{Nb}\widehat{\theta}''_{n-1} + \frac{S}{2}\left(\eta\widehat{\phi}'_{n-1} + \eta^2\widehat{\phi}''_{n-1}\right) \tag{46}$$

where,

$$\chi_n = \begin{cases} 0, & \text{if } \zeta \leq 1 \\ 1, & \text{if } \zeta > 1 \end{cases} \tag{47}$$

4. Results

A magneto-hydrodynamic nanofluid flow over an exponential stretching sheet with joule heating and viscous dissipation effects was modeled. The reduced modeled Equations (9)–(15) were solved by HAM. The graphical interpretation of the modeled problem is articulated in Figure 1. The impacts of important physical parameters such as M, S, Ω, Sc, Nt, Nb and Pr are discussed with the help of Figures 2–17.

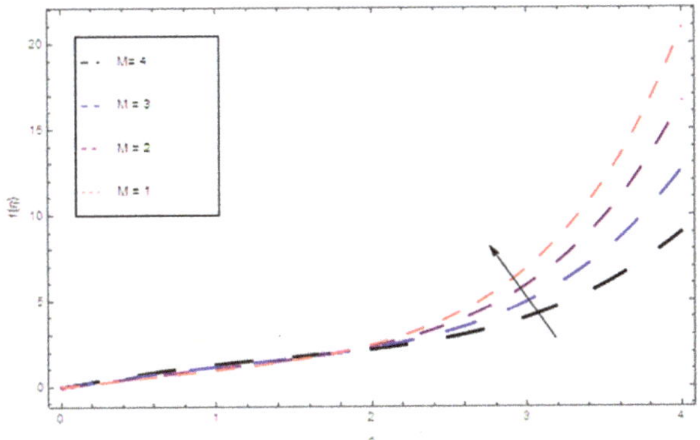

Figure 2. Impact of M on $f(\eta)$ when $k = 1$.

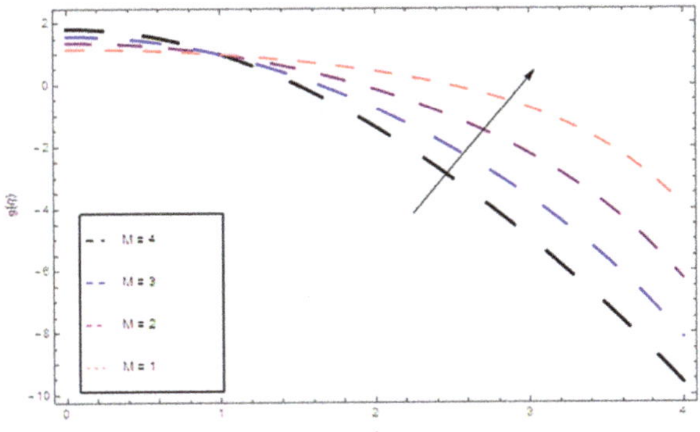

Figure 3. Impact of M on $g(\eta)$ when $k = 2$.

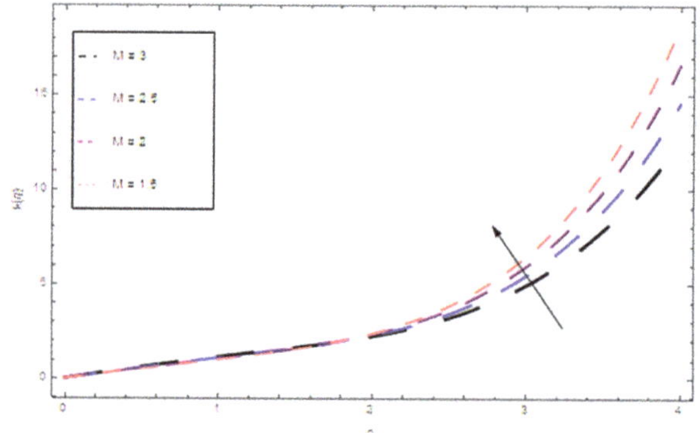

Figure 4. Impact of M on $k(\eta)$ when $k = 1$.

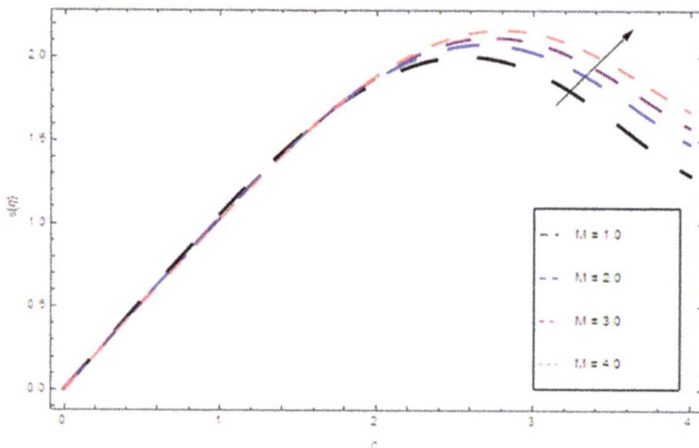

Figure 5. The effect of M on $s(\eta)$ when $k = 0.5$.

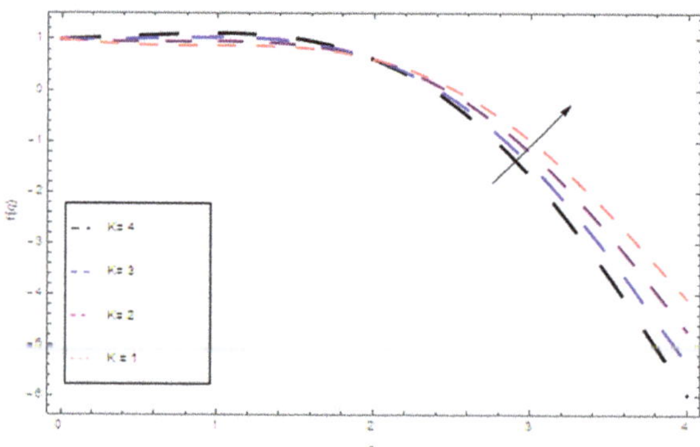

Figure 6. Effect of K on $f(\eta)$ while $M = 0.3$.

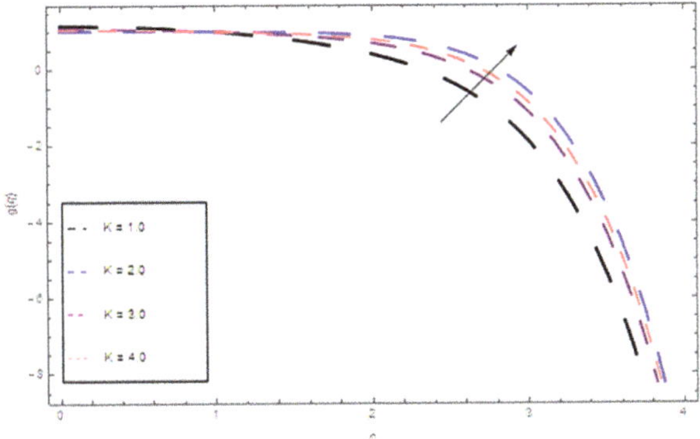

Figure 7. Effect of K on $g(\eta)$ when $M = 0.5$.

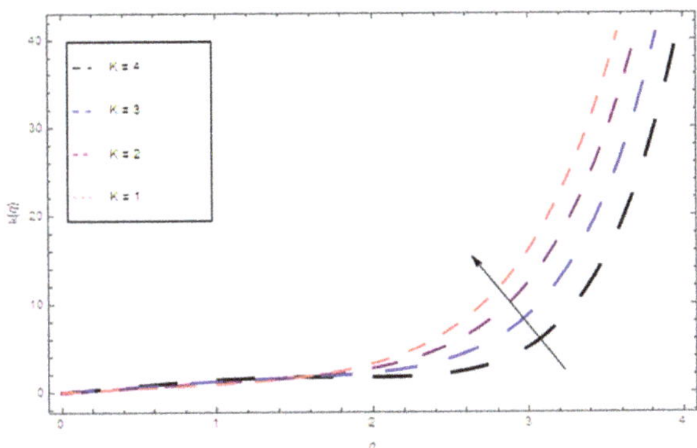

Figure 8. Effect of K on $k(\eta)$ for $M = 0.3$.

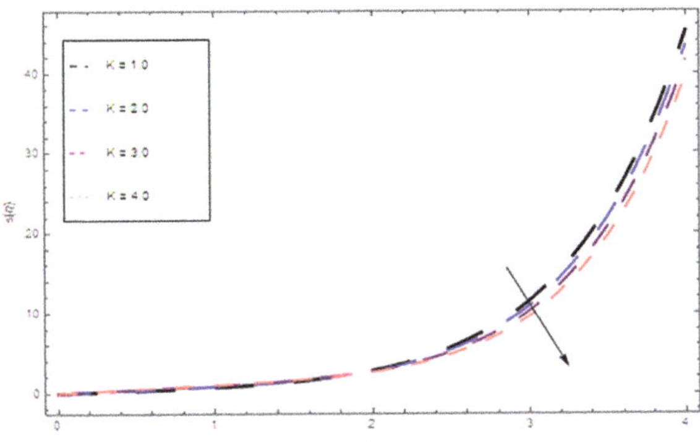

Figure 9. Impact of K on $s(\eta)$ for $M = 0.5$.

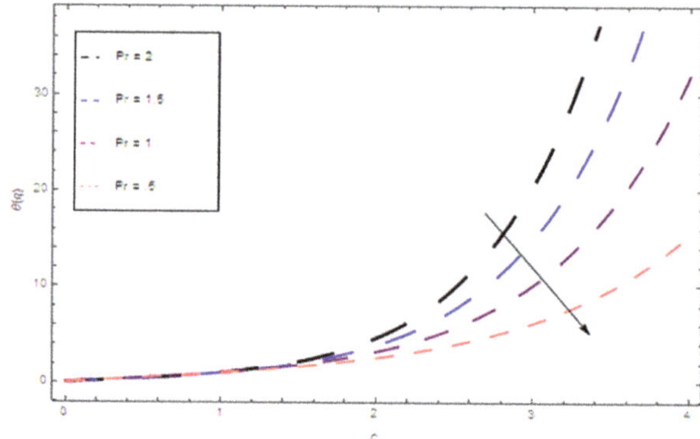

Figure 10. Impact of Pr on $\theta\left(\eta\right)$ for $M = 0.3, \Omega = 0.4, S = 0.5, Nt = 0.8, Sc = 0.7, Nb = 0.6$.

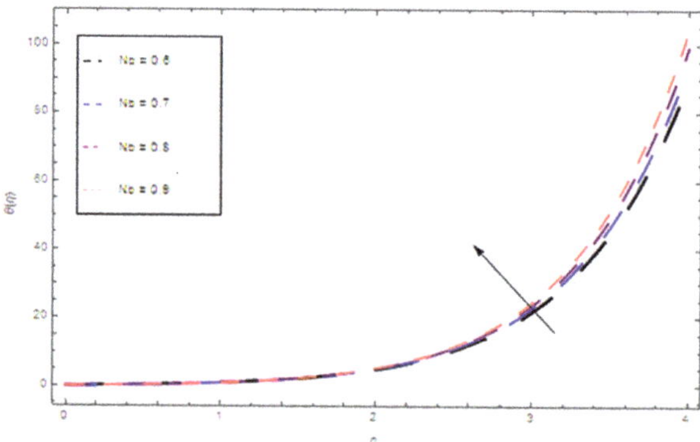

Figure 11. The effect of Nb on $\theta\left(\eta\right)$ when $M = 0.3, \Omega = 1, S = 0.5, \mathrm{Pr} = 2, Sc = 0.7, Nt = 0.5$.

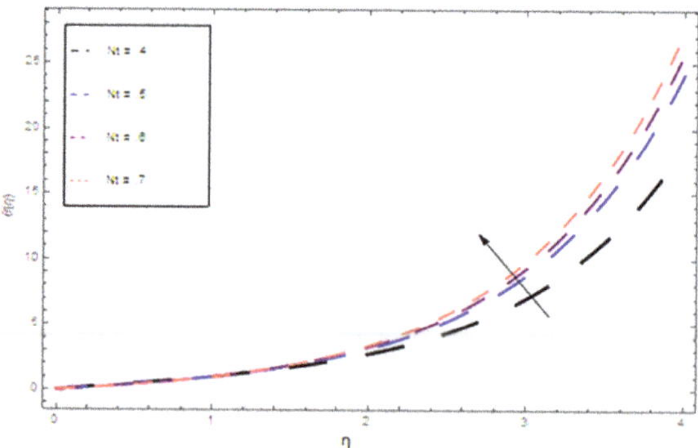

Figure 12. The influence of Nt on $\theta\left(\eta\right)$ when $Sc = 0.7, S = 0.5, \Omega = 1, Nb = 0.6, M = 0.3, \mathrm{Pr} = 0.6$.

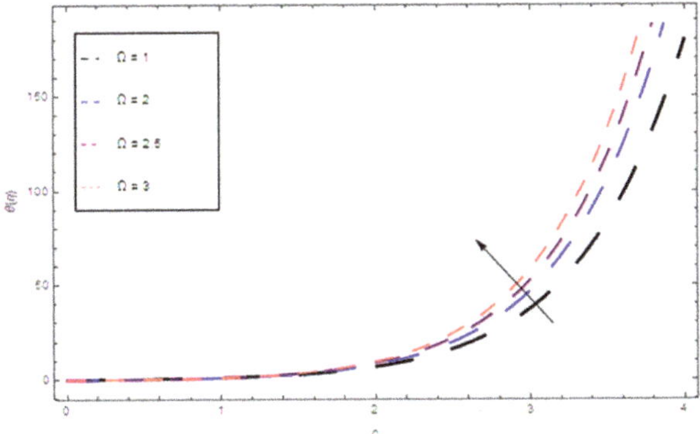

Figure 13. The impact of Ω on $\theta\left(\eta\right)$ while $Nt = 0.8, \Pr = 0.5, S = 0.5, Nb = 0.6, Sc = 0.7, M = 0.3$.

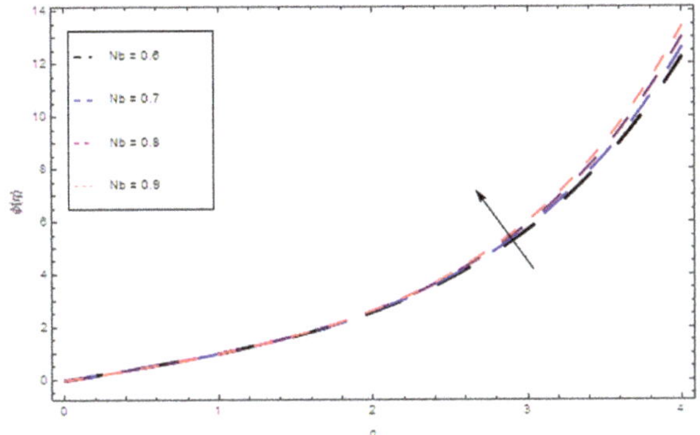

Figure 14. The effect of Nb on $\theta\left(\eta\right)$ when $M = 0.3, Nt = 0.7, S = 0.5, \Pr = 1, Sc = 0.5, \Omega = 1$.

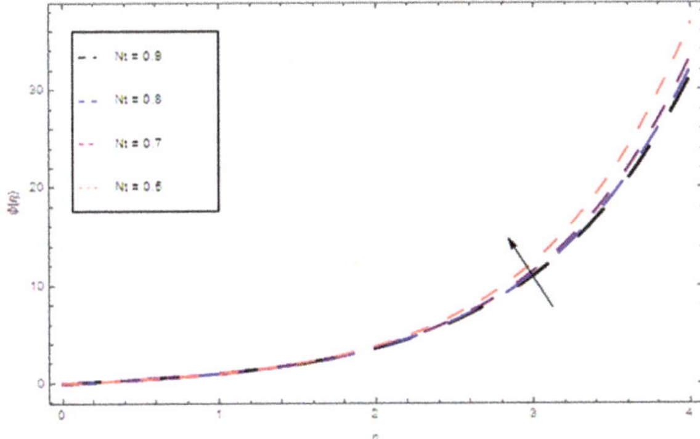

Figure 15. Impact of Nt on $\theta\left(\eta\right)$ for $M = 0.3, Nb = 0.7, S = 0.5, \Pr = 1, Sc = 0.5, \Omega = 1$.

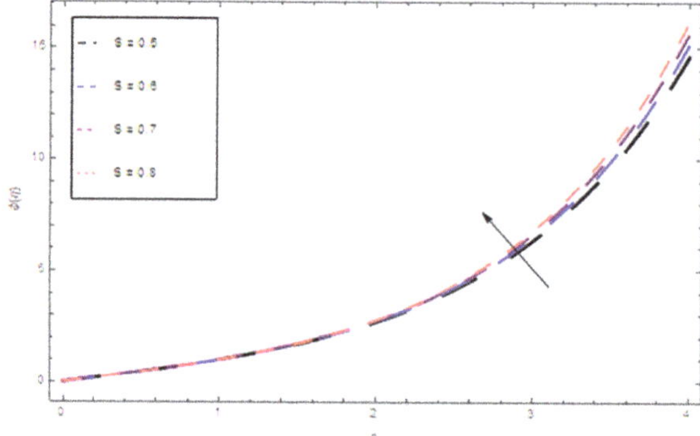

Figure 16. Impact of S over $\theta(\eta)$, when $M = 0.3, Nb = 0.6, Nt = 0.7, \mathrm{Pr} = 2, Sc = 0.5, \Omega = 1$.

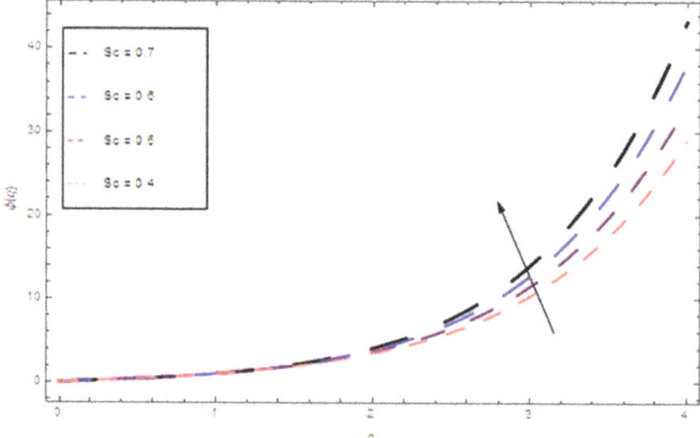

Figure 17. Impact of Sc on $\theta(\eta)$, when $\Omega = 1, \mathrm{Pr} = 1, S = 0.5, Nb = 0.7, Nt = 0.6, M = 0.3$.

5. Discussion

A three-dimensional nanofluid film flow with heat and mass transfer through a steady rotating inclined surface was analyzed. The impact of M, S, Ω, Sc, Nt, Nb, Pr and K were explored for radial velocity $k(\eta)$, axial velocity $f(\eta)$, induced flow $s(\eta)$, and drainage flow $g(\eta)$ respectively. Furthermore, the impact of these parameters was also described for the temperature profile $\theta(\eta)$ and concentration profile $\theta(\eta)$. The impact of the magnetic parameter M over the velocity profiles $f(\eta)$, $k(\eta)$, $g(\eta)$, and $s(\eta)$ are presented in Figures 2–5. The impact of the magnetic parameter on the axial and the radial profiles of the velocity look similar for up to $\eta = 2.0$ as shown in Figures 2 and 4. For greater values of the magnetic parameter these profiles decline sharply and the peak point is at $\eta = 4.0$. Physically, the larger values of the magnetic parameter reduce the rotational parameter, and as a result both the profiles decline. Figures 3 and 5 both show an increasing trend in the drainage and induced flow with the smaller and larger values of the magnetic parameter respectively. For $\eta = 2.3$ all the curves coincide, and after this point the induced velocity jumps in increasing order as shown in Figure 5, while on the other hand smaller values of the magnetic parameter enhance the drainage velocity as presented in Figure 3. The variation in Figures 3 and 5 as a consequence of layer thickness and internal velocity of nanofluid definitely reduce. The direction of this force is perpendicular to both the fields. Also M endorses the ratio of viscous and hydromagnetic body forces, the fluid flow is decreased due to greater values of M which require further hydromagnetic body forces. Lorentz force theory defines that M has a reverse influence on velocity function. Figures 6–9 show the effect of K on $f(\eta)$, $k(\eta)$, $g(\eta)$, and $s(\eta)$. Figure 6 displays the axial velocity which accelerates in response to an increases in the couple stress parameter. This is mainly due to the decrease in friction, which rises from the particle (i.e., base-fluid particles) additives that create a size-dependent influence in couple stress liquids. The influence of K on $k(\eta)$ is demonstrated in Figure 7. The disk rotation increases K, due to which the flow velocity also rises. For $K = 0$ the flow retains Newtonian fluid. Figure 8 displays the influence of K on drainage flow $g(\eta)$. The flow is perceived to rise with rising values of K, since the rise in K causes a reduction in the dynamic viscosity of the liquid and, hence a rise in the molecular distance among the liquid particles. Figure 9 displays the influence of K on induced flow $s(\eta)$. Here a rise in K leads to a rise in the stress among all the coupled fluid atoms, which further leads to a reducttion in the induced velocity. Figure 10 depicts the Pr impact on the temperature profile $\theta(\eta)$ since Pr has an inverse relation to the thermal diffusivity and direct relation to the momentum diffusivity. Greater values of Pr mean that there is robust momentum diffusivity which is associated with the thermal diffusivity and as a result the thermal diffusivity decreases the temperature profile $\theta(\eta)$. These variations almost look similar up to $\eta = 1.8$. Hence, the greater values of Pr drop the boundary layer of heat. The influence of Nb on $\theta(\eta)$ is demonstrated in Figure 11. It is clear that larger values of Nb enhance the thermal boundary layer after $\eta = 2.7$. Physically, when Brownian motion increases the interaction between the particles enhances and as a result the energy transfers rapidly from one point to another, which as a result increase the thermal boundary layer. The impact of the thermophoresis parameter Nt over $\theta(\eta)$ is given in Figure 12. It can be seen that a rise in Nt leads to augment the liquid temperature. Thermophoresis forces generate temperature gradient that further causes a degenerate flow away from the surface. The influence of Ω on $\theta(\eta)$ is demonstrated in Figure 13. The larger values of the rotation parameter Ω enhance the temperature profile $\theta(\eta)$. This is due to the larger values of Ω, the fluid temperature increases. Physically, greater values of Ω increase the kinetic energy, which in consequence increases $\theta(\eta)$. Influence of Nb on $\theta(\eta)$ is demonstrated in Figure 14. For greater values of Nb the concentration profile $\theta(\eta)$ increases. This variation is very effective for greater values of η. In practice, rising values of Nb cause an enhancement of the random motion amongst the nanoparticles, and consequently reduces $\theta(\eta)$ of the liquid. Higher values of Nb decrease the boundary layer thicknesses, which as a result reduces the concentration profile $\Phi(\eta)$. Figure 15 displays the performance of the Nt on $\Phi(\eta)$. Increasing values of Nt push the nanoparticles far away from the hot sheet, and consequently enhance the concentration profile $\Phi(\eta)$. Since Nt varies with the gradient of the temperature of the nanofluids, the kinect energy ncreases of the nanofluids due to the increase of Nt, which further increases $\Phi(\eta)$. The effect of S on $\Phi(\eta)$ is presented in Figure 16. It is seen that $\Phi(\eta)$ varies directly with S. Augmenting S increases the concentration, which as a result raises the kinect energy of the liquid, which further enhances the speed of fluid film. It is evident from the graph that the rising values of S reduce $\Phi(\eta)$. This impact of S over $\Phi(\eta)$ is due to the stretching sheet, steady flow, and the greater concentration, and hence rising S shows a converse influence. The effect of the Schmidt number Sc on the concentration field is presented in Figure 17. The concentration boundary layer reduces due to the rise in Sc. In practice Sc decreases the molecular diffusivity, which reduces the concentration boundary layer. It is observed that the reduction in heat transfer at the sheet leads to a rise in the values of Sc

6. Conclusions

The flow past an exponential stretching sheet of couple stress nanofluid with the impacts of MHD, viscous dissipation, and Joule heating was investigated analytically. The influences of zero mass flux and convective heat condition were also considered. HAM was applied for the solution of the non-linear differential equations. The effects of the numerous constraints on velocity field, temperature profile, and concentration are portrayed by the graphs. The important results of the current study are described below as follows:

- Higher values of S provide coolness to the boundary layer, while Nu increases the profile.
- A larger value of Sc reduces the Sherwood number, which as a result enhances the concentration and kinematic viscosity of the fluid.
- The higher values of the Prandtl number reduce the Nusselt number.

- The higher values of the Brownian motion parameter increase the concentration distribution and decline with increasing values of the thermophoresis parameters.

Author Contributions: Conceptualization, methodology, software, validation, writing—original draft preparation, writing—review and editing, A.T., Z.S., I.A. and M.S.; conceptualization, methodology, software, visualization, writing—review and editing, F.A., A.U. and S.I.; writing—review and editing, visualization, project administration, funding acquisition, investigation, resources, A.T., M.S. and Z.S. All authors have read and agreed to the published version of the manuscript.

Funding: This work is funded by the Deanship of Scientific Research at Majmaah University under Project Number (RGP-2019-28).

Acknowledgments: The authors extend their appreciation to the Deanship of Scientific Research at Majmaah University for funding this work under Project Number (RGP-2019-28).

Conflicts of Interest: The authors declare no conflict of interest.

Nomenclature:

u, v, w	Velocity components in x, y, z directions (m/s)
x, y, z	Space coordinates (m)
T	Temperature of fluid (K)
Tw	Temperature at the surface (K)
T_∞	Ambient temperature (K)
B_0	Uniform magnetic field (Tesla)
C	Concentration of fluid (kg/m^3)
C_w	Concentration at the surface (kg/m^3)
D_B	Brownian diffusion coefficient
D_T	Thermophoresis diffusion coefficient
C_∞	Ambient concentration (kg/m^3)
U_w	Stretching sheet velocity (m/s)
g	Gravity (m/s^2)
α	Thermal diffusivity Coefficient (m^2/s)
β	Angle of inclination
f	Dimensionless stream function
Nt	Thermophoresis parameter
Q	Heat generation/absorption coefficient (W/m^2·K)
Sr	Soret number
δ	Electrical conductivity of the fluid (m/s)
K_0	Chemical reaction coefficient
C_P	Specific heat at constant pressure (J/kg K)
Nb	Brownian motion parameter
Pr	Prandtl number
C_{fx}	Skin friction coefficient
Nu_x	Local Nusselt number
Re	Reynolds number
ρ	Fluid density (kg/m^3)
q_w	Surface heat flux
μ_0	Zero shear viscosity (s^{-1})
μ_∞	Infinite shear viscosity (s^{-1})
ν	Kinematic viscosity (m^2/s)
μ	Dynamic viscosity (N s/m^2)
Φ, θ	Dimensionless concentration and temperature

Subscripts

∞	Condition for away from the surface
w	Condition at the sheet

Superscript

$'$	Derivative with respect to η

References

1. Choi, S.U.; Eastman, J.A. *Enhancing Thermal Conductivity of Fluids with Nanoparticles*; Argonne National Lab.: Lemont, IL, USA, 1995.
2. Wang, X.-Q.; Mujumdar, A.S. Heat transfer characteristics of nanofluids: A Review. *Int. J. Therm. Sci.* **2007**, *46*, 1–19. [CrossRef]
3. Eastman, J.A.; Phillpot, S.R.; Choi, S.U.S.; Keblinski, P. Thermal transport in nanofluids. *Annu. Rev. Mater. Res.* **2004**, *34*, 219–246. [CrossRef]
4. Eastman, J.A.; Choi, S.U.S.; Li, S.; Yu, W.; Thompson, L.J. Anomalously increased effective thermal conductivities of ethylene glycol-based nanofluids containing copper nanoparticles. *Appl. Phys. Lett.* **2001**, *78*, 718–720. [CrossRef]
5. Murshed, S.M.S.; Leong, K.C.; Yang, C. Enhanced thermal conductivity of TiO$_2$—Water based nanofluids. *Int. J. Therm. Sci.* **2005**, *44*, 367–373. [CrossRef]
6. Maïga, S.E.B.; Nguyen, C.T.; Galanis, N.; Roy, G. Heat transfer behaviours of nanofluids in a uniformly heated tube. *Superlattices Microstruct.* **2004**, *35*, 543–557. [CrossRef]
7. Bianco, V.; Chiacchio, F.; Manca, O.; Nardini, S. Numerical investigation of nanofluids forced convection in circular tubes. *Appl. Therm. Eng.* **2009**, *29*, 3632–3642. [CrossRef]
8. Buongiorno, J. Convective transport in nanofluids. *J. Heat Transf.* **2006**, *128*, 240–250. [CrossRef]
9. Tiwari, R.K.; Das, M.K. Heat transfer augmentation in a two-sided lid-driven differentially heated square cavity utilizing nanofluids. *Int. J. Heat Mass Transf.* **2007**, *50*, 2002–2018. [CrossRef]
10. Kasaeian, A.; Daneshazarian, R.M.; Omid, K.; Lioua, C.A.J.; Wongwises, S.; Pop, I. Nanofluid flow and heat transfer in porous media: A review of the latest developments. *Int. J. Heat Mass Transf.* **2017**, *107*, 778–791. [CrossRef]
11. Ramzan, M.; Chung, J.D.; Ullah, N. Radiative magnetohydrodynamic nanofluid flow due to gyrotactic microorganisms with chemical reaction and non-linear thermal radiation. *Int. J. Mech. Sci.* **2017**, *130*, 31–40. [CrossRef]
12. Sheikholeslami, M.; Shehzad, S. RETRACTED: *Magnetohydrodynamic Nanofluid Convective Flow in a Porous Enclosure by Means of LBM*; Elsevier: Amsterdam, The Netherlands, 2017.
13. Jawad, M.; Shah, Z.; Islam, S.; Bonyah, E.; Khan, A.Z. Darcy-Forchheimer flow of MHD nanofluid thin film flow with Joule dissipation and Navier's partial slip. *J. Phys. Commun.* **2018**, *2*, 115014. [CrossRef]
14. Jawad, M.; Shah, Z.; Islam, S.; Majdoubi, J.; Tlili, I.; Khan, W.; Khan, I. Impact of nonlinear thermal radiation and the viscous dissipation effect on the unsteady three-dimensional rotating flow of single-wall carbon nanotubes with aqueous suspensions. *Symmetry* **2019**, *11*, 207. [CrossRef]
15. Bhatti, M.M.; Zeeshan, A.; Ellahi, R. Simultaneous effects of coagulation and variable magnetic field on peristaltically induced motion of Jeffrey nanofluid containing gyrotactic microorganism. *Microvasc. Res.* **2017**, *110*, 32–42. [CrossRef] [PubMed]
16. Ellahi, R. The effects of MHD and temperature dependent viscosity on the flow of non-Newtonian nanofluid in a pipe: Analytical solutions. *Appl. Math. Model.* **2013**, *37*, 1451–1467. [CrossRef]
17. Hatami, M.; Ganji, D. Thermal and flow analysis of microchannel heat sink (MCHS) cooled by Cu–water nanofluid using porous media approach and least square method. *Energy Convers. Manag.* **2014**, *78*, 347–358. [CrossRef]
18. Hatami, M.; Sheikholeslami, M.; Ganji, D. RETRACTED: *Laminar Flow and Heat Transfer of Nanofluid between Contracting and Rotating Disks by Least Square Method*; Elsevier: Amsterdam, The Netherlands, 2014.
19. Shah, Z.; Islam, S.; Ayaz, H.; Khan, S. Radiative heat and mass transfer analysis of micropolar nanofluid flow of Casson fluid between two rotating parallel plates with effects of Hall current. *J. Heat Transf.* **2019**, *141*, 022401. [CrossRef]
20. Shah, Z.; Gul, T.; Islam, S.; Khan, M.A.; Bonyah, E.; Hussain, F.; Mukhtar, F.; Ullah, M. Three dimensional third grade nanofluid flow in a rotating system between parallel plates with Brownian motion and thermophoresis effects. *Results Phys.* **2018**, *10*, 36–45. [CrossRef]
21. Shah, Z.; Alzahrani, E.O.; Alghamdi, W.; Ullah, M.K. Influences of electrical MHD and Hall current on squeezing nanofluid flow inside rotating porous plates with viscous and joule dissipation effects. *J. Therm. Anal. Calorim.* **2020**. [CrossRef]

22. Shah, Z.; Kumam, P.; Deebani, W. Radiative MHD Casson nanofluid flow with activation energy and chemical reaction over past nonlinearly stretching surface through Entropy generation. *Sci. Rep.* **2020**, *10*, 4402. [CrossRef]

23. Ullah, A.; Alzahrani, E.O.; Shah, Z.; Ayaz, M.; Islam, S. Nanofluids thin film flow of Reiner-Philippoff fluid over an unstable stretching surface with Brownian motion and thermophoresis effects. *Coatings* **2019**, *9*, 21. [CrossRef]

24. Ramzan, M. Influence of Newtonian heating on three dimensional MHD flow of couple stress nanofluid with viscous dissipation and joule heating. *PLoS ONE* **2015**, *10*, e0124699. [CrossRef] [PubMed]

25. Hayat, T.; Imtiaz, M.; Alsaedi, A. Melting heat transfer in the MHD flow of Cu–water nanofluid with viscous dissipation and Joule heating. *Adv. Powder Technol.* **2016**, *27*, 1301–1308. [CrossRef]

26. Van Dinh, Q.; Vo, T.Q.; Kim, B. Viscous heating and temperature profiles of liquid water flows in copper nanochannel. *J. Mech. Sci. Technol.* **2019**, *33*, 3257–3263. [CrossRef]

27. Zhao, X.; E., J.; Zhang, Z.; Chen, J.; Liao, G.; Zhang, F.; Leng, E.; Han, D.; Hu, W. A review on heat enhancement in thermal energy conversion and management using Field Synergy Principle. *Appl. Energy* **2020**, *257*, 113995. [CrossRef]

28. Zhao, X.; E., J.; Wu, G.; Deng, Y.; Han, D.; Zhang, B.; Zhang, Z. A review of studies using graphenes in energy conversion, energy storage and heat transfer development. *Energy Convers. Manag.* **2019**, *184*, 581–599. [CrossRef]

29. E., J.; Jin, Y.; Deng, Y.; Zuo, W.; Zhao, X.; Han, D.; Peng, Q.; Zhang, Z. Wetting models and working mechanisms of typical surfaces existing in nature and their application on superhydrophobic surfaces: A review. *Adv. Mater. Interfaces* **2018**, *5*, 1701052. [CrossRef]

30. E., J.; Zhang, Z.; Tu, Z.; Zuo, W.; Han, D.; Jin, Y. Effect analysis on flow and boiling heat transfer performance of cooling water-jacket of bearing in the gasoline engine turbocharger. *Appl. Therm. Eng.* **2018**, *130*, 754–766. [CrossRef]

31. Khare, R.; Keblinski, P.; Yethiraj, A. Molecular dynamics simulations of heat and momentum transfer at a solid–fluid interface: Relationship between thermal and velocity slip. *Int. J. Heat Mass Transf.* **2006**, *49*, 3401–3407. [CrossRef]

32. Soleimani, S.; Sheikhulislami, M.; Ganji, D.D.; Gorji-Bapanday, M. Natural convection heat transfer in a nanofluid filled semi-annulus enclosure. *Int. Commun. Heat Mass Transf.* **2012**, *39*, 565–574. [CrossRef]

33. Rudraiah, N.; Barron, R.M.; Venkatachalappa, M.; Subbaraya, C.K. Effect of a magnetic field on free convection in a rectangular enclosure. *Int. J. Eng. Sci.* **1995**, *33*, 1075–1084. [CrossRef]

34. Kandelousi, M.S.; Ellahi, R. Simulation of ferrofluid flow for magnetic drug targeting using the lattice Boltzmann method. *Zeitschrift für Naturforschung A* **2015**, *70*, 115–124. [CrossRef]

35. Hayat, T.; Muhammad, T.; Shehzad, S.A.; Alsaedi, A. Simultaneous effects of magnetic field and convective condition in three-dimensional flow of couple stress nanofluid with heat generation/absorption. *J. Braz. Soc. Mech. Sci. Eng.* **2017**, *39*, 1165–1176. [CrossRef]

36. Hayat, T.; Muhammad, T.; Alsaedi, A. On three-dimensional flow of couple stress fluid with Cattaneo–Christov heat flux. *Chin. J. Phys.* **2017**, *55*, 930–938. [CrossRef]

37. Hayat, T.; Muhamamd, T.; Shehzad, S.A.; Alsaedi, A. Three-dimensional boundary layer flow of Maxwell nanofluid: Mathematical model. *Appl. Math. Mech.* **2015**, *36*, 747–762. [CrossRef]

38. Malik, M.; khan, I.; Hussain, A.; Salahuddin, T. Mixed convection flow of MHD Eyring-Powell nanofluid over a stretching sheet: A numerical study. *AIP Adv.* **2015**, *5*, 117118. [CrossRef]

39. Nadeem, S.; Haq, R.U.; Khan, Z. Numerical study of MHD boundary layer flow of a Maxwell fluid past a stretching sheet in the presence of nanoparticles. *J. Taiwan Inst. Chem. Eng.* **2014**, *45*, 121–126. [CrossRef]

40. Raju, C.; Sandeep, N.; Malvandi, A. Free convective heat and mass transfer of MHD non-Newtonian nanofluids over a cone in the presence of non-uniform heat source/sink. *J. Mol. Liq.* **2016**, *221*, 108–115. [CrossRef]

41. Sheikholeslami, M.; Rezaeianjouybari, B.; Darzi, M.; Shafee, A.; Li, Z.; Nguyen, T.K. Application of nano-refrigerant for boiling heat transfer enhancement employing an experimental study. *Int. J. Heat Mass Transf.* **2019**, *141*, 974–980. [CrossRef]

42. Sheikholeslami, M.; Jafaryar, M.; Hedayat, M.; Shafee, A.; Nguyen, T.K.; Bakouri, M. Heat transfer and turbulent simulation of nanomaterial due to compound turbulator including irreversibility analysis. *Int. J. Heat Mass Transf.* **2019**, *137*, 1290–1300. [CrossRef]

43. Sheikholeslami, M.; Jafaryar, M.; Shafee, A.; Li, Z.; Haq, R. Heat transfer of nanoparticles employing innovative turbulator considering entropy generation. *Int. J. Heat Mass Transf.* **2019**, *136*, 1233–1240. [CrossRef]

44. Sheikholeslami, M.; Haq, R.; Shafee, A.; Li, Z. Heat transfer behavior of nanoparticle enhanced PCM solidification through an enclosure with V shaped fins. *Int. J. Heat Mass Transf.* **2019**, *130*, 1322–1342. [CrossRef]

45. Sheikholeslami, M. New computational approach for exergy and entropy analysis of nanofluid under the impact of Lorentz force through a porous media. *Comput. Methods Appl. Mech. Eng.* **2019**, *344*, 319–333. [CrossRef]

46. Sheikholeslami, M. Numerical approach for MHD Al_2O_3-water nanofluid transportation inside a permeable medium using innovative computer method. *Comput. Methods Appl. Mech. Eng.* **2019**, *344*, 306–318. [CrossRef]

47. Sheikholeslami, M.; Gerdroodbary, M.B.; Moraadi, R.; Shafee, A.; Li, Z. Application of Neural Network for estimation of heat transfer treatment of Al_2O_3-H_2O nanofluid through a channel. *Comput. Methods Appl. Mech. Eng.* **2019**, *344*, 1–12. [CrossRef]

48. Sheikholeslami, M.; Shah, Z.; Shafee, A.; Khan, I.; Tlili, I. Uniform magnetic force impact on water based nanofluid thermal behavior in a porous enclosure with ellipse shaped obstacle. *Sci. Rep.* **2019**, *9*, 1–11. [CrossRef]

49. Li, Z.; Sheikhulislami, M.; Shah, Z.; Shafee, A.; Al-Qawasmi, A.; Tlili, I. Transient process in a finned triplex tube during phase changing of aluminum oxide enhanced PCM. *Eur. Phys. J. Plus* **2019**, *134*, 173. [CrossRef]

50. Shah, Z.; Ullah, A.; Bonyah, E.; Ayaz, M.; Islam, S.; Khan, I. Hall effect on Titania nanofluids thin film flow and radiative thermal behavior with different base fluids on an inclined rotating surface. *AIP Adv.* **2019**, *9*, 055113. [CrossRef]

51. Kumam, P.; Shah, Z.; Dawar, A.; Rasheed, H.; Islam, S. Entropy generation in MHD radiative flow of CNTs Casson nanofluid in rotating channels with heat source/sink. *Math. Probl. Eng.* **2019**, *2019*, 9158093. [CrossRef]

52. Shah, Z.; Babazadeh, H.; Kumam, P.; Shafee, A.; Thounthong, P. Numerical simulation of magnetohydrodynamic nanofluids under the influence of shape factor and thermal transport in a porous media using CVFEM. *Front. Phys.* **2019**, *7*, 164. [CrossRef]

53. Ameen, I.; Shah, Z.; Islam, S.; Nasir, S.; Khan, W.; Kumam, P.; Thounthong, P. Hall and ion-slip effect on CNTS nanofluid over a porous extending surface through heat generation and absorption. *Entropy* **2019**, *21*, 801. [CrossRef]

54. Vo, D.D.; Shah, Z.; Sheikhulislami, M.; Shafee, A.; Nguyen, T.T. Numerical investigation of MHD nanomaterial convective migration and heat transfer within a sinusoidal porous cavity. *Phys. Scr.* **2019**, *94*, 115225. [CrossRef]

55. Javed, T.; Sajid, M.; Abbas, Z.; Ali, N. Non-similar solution for rotating flow over an exponentially stretching surface. *Int. J. Numer. Methods Heat Fluid Flow* **2011**, *21*, 903–908. [CrossRef]

56. Rosali, H.; Ishak, A.; Nazar, R.; Pop, I. Rotating flow over an exponentially shrinking sheet with suction. *J. Mol. Liq.* **2015**, *211*, 965–969. [CrossRef]

57. Khan, N.S.; Gut, T.; Iskam, S.; Khan, A.; Shah, Z. Brownian motion and thermophoresis effects on MHD mixed convective thin film second-grade nanofluid flow with Hall effect and heat transfer past a stretching sheet. *J. Nanofluids* **2017**, *6*, 812–829. [CrossRef]

Article

Experiment Investigate on the Effectiveness of Flexible Pipes to Isolate Sea-Water Pump Generated Vibration

Yang Yang [1,2,*], Guang Pan [1], Shaoping Yin [2] and Ying Yuan [3]

[1]   School of Marine Science and Technology, Northwestern Polytechnical University, Xi'an 710072, China; panguang601@163.com
[2]   The 705 Research Institute, China Shipbuilding Industry Corporation, Xi'an 710077, China; summertalentyy@163.com
[3]   School of Physics and Optoelectronic Engineering, XIDIAN University, Xi'an 710071, China; yuanying1989@126.com
*   Correspondence: allen-yang1988@mail.nwpu.edu.cn

Received: 5 November 2019; Accepted: 31 December 2019; Published: 3 January 2020

Abstract: Vibration control is important in maintaining the silence of the underwater vehicle. Among the many methods of vibration control, isolation is by far the most efficient approach. However, as one of the major vibration sources in underwater vehicle, the vibration isolation of the sea-water pump has not been well explored. The sea-water pipe is the primary vibration transmit path from the sea-water pump to the housing. In order to realize the vibration isolation of the sea-water pump, the sea-water pipe must have certain flexibility and damping. In this study, scaled model tests were carried out to investigate the isolation effectiveness of flexible pipes in isolated sea-water pump. Specifically, three types of flexible pipes, i.e., double layer metal bellows (DLMB), rubber pipes (RP) and bellows coated rubber (BCR) were designed and tested. Tests were carried out under the operation rotate speeds of the sea-water pump. Our results show that compared with single layer metal bellows (SLMB), the isolation effectiveness of DLMB and BCR were significant and stable in high frequency regions. The optimal pipe can be chosen for different vibration reduction requirements in practical engineering.

Keywords: underwater vehicle; sea-water pump; vibration isolation; flexible pipes

---

1. Introduction

Thermodynamic underwater vehicle power system includes the main engine and multiple auxiliary engines. As shown in Figure 1, the power plant is integrated and supported to the housing through front and rear vibration isolators. The basic function of the vibration isolators is to support the entire power plant, which are called "supporting devices". In addition to vibration isolators, there are also some pipes connected auxiliary engines to the housing, such as sea-water pipes, fuel pipes and oil pipes. The main function of these pipes is to complete the transportation of seawater, fuel and oil, without support functions, collectively referred to as "un supporting devices". The mechanical vibration of power plant during operation transmit to the housing through these supporting and unsupporting devices, and then generate radiated noise in the ocean. The sea-water pump is one of the major vibration and noise sources in the underwater vehicle power system. There are two basic directions to decrease the impact of the sea-water pump generated vibration for the whole vehicle in practical engineering, the first is reducing the vibration of the pump itself, e.g., reduce rotting speed and the optimized structural form. The second is decreasing the vibration transmission in the transfer path. Compared with reducing the vibration of the pump, which needs to sacrifice pump power and increase development cost, the same effectiveness can be achieved from the transfer path without affecting the performance of the product.

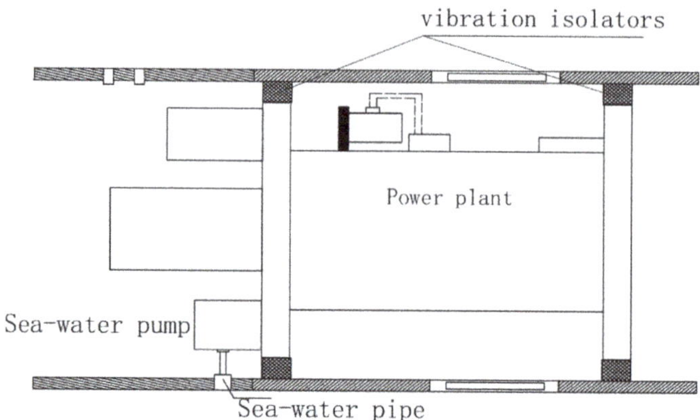

Figure 1. The position of the sea-water pipe in the power system.

The sea-water pipe as the exclusive path to transport the seawater from ocean to the vehicle is the primary path to transmit the mechanical and fluid-excited vibration of sea-water pump to housing during operation. Therefore, the vibration reduction design of the sea-water pipe is of great importance for the silence of underwater vehicle. Single layer metal bellows (SLMB) are widespread applied for its easy processing and long life in practical engineering. However, the pipe could not offer significant and stable vibration reduction for its large stiffness and small damping. In order to reduce fluid induced pulsation, a muffler usually be utilized inside a SLMB. However, the presence of the muffler has an effect on the flow of water in the pipe. As a result, under technically feasible conditions, engineers first consider the flexible connection without a muffler between the sea-water pump and the housing. Due to its good vibration isolation and impact resistance, the flexible pipe can effectively isolate the transmission of mechanical vibration energy of the power equipment to the housing. Besides the function of vibration reduction, the advantage of flexible pipes also includes the role of displacement compensation to avoid a great displacement induced by the equipment force. As the primary transfer path from the sea-water pump to the housing, the design of flexible pipe is much more significant for the vibration reduction of the housing. Nevertheless, subject to the compact internal space of the power plant, the length of the pipe is usually very short. Furthermore, the diameter of the pipe will not be small for the underwater vehicles need enough seawater for cooling and squeezing. Therefore, the pipe commonly owns the feature of short in length and large in diameter, this type of flexible pipe is difficult to design.

Dynamics of fluid-conveying pipes have been well-explored in theoretical [1–10] and experiment research [11,12]. Tan [13] investigated the vibration characteristics of pipes conveying fluid in the super-critical range using Timoshenko beam theory for the first time. FEM [14–16] is also the most popular method up to now. However, experiment methods are the most direct means to evaluate the vibration isolation effectiveness of isolators. Enrique [17] took the experimental method to investigate the two-phase flow-induced vibration in pipes, they found that dynamic pipe response increases with increasing mixture velocity and void fraction, what is more, the hydrodynamic mass parameter is proportional to mixture density. Zhou [18] and Pan [19] focused experimental research on a vibration isolation platform for momentum wheel assembly and laminated rubber bearings to isolate metro generated vibration respectively. Kaiming Bi [20] proposed properly selected viscoelastic materials and constraining layers vibrations of above-ground pipelines can be effectively mitigated.

Double layer metal bellows (DLMB), rubber pipes (RP) and bellows coated rubber (BCR) are widely used for isolation in other engineering. However, the vibration effectiveness of the pipes has not been well explored. Especially for the sea-water pipe installed in underwater vehicles, when the sea-water pump rotates at different speeds, the excitation force and seawater pressure generated by the operation are also different. In the engineering, the working environment of the sea-water pipe is extremely complicated, and the excitation forces are varied. At the same time, in order to meet the assembly and use requirements of the product, the structure of the sea-water pipe is usually complicated, and it is difficult to accurately obtain the dynamic characteristics of the flexible pipes through theoretical calculation. To provide more comprehensive experiment evidence that testifies to the effectiveness of pipes mentioned above for the isolation of sea-water pump generated vibration, a test method was designed and carried out in the laboratory. To simulate the practical working condition, the vibration was induced by a sea-water pump and the pipes were installed between the pump and the simulation holder. In the tests, SLMB, DLMB, RP and BCR were tested respectively. In order to evaluate the isolation effectiveness of flexible pipes at different vibration levels of the sea-water pump, tests were carried out under four different rated operation speed of the sea-water pump, i.e., 1700 r/min, 2000 r/min, 2300 r/min and 2600 r/min. By drawing the energy level transfer coefficients under different frequencies and rotate speeds, a universal approach to evaluating the isolation effectiveness of the sea-water pipe was provided, and the isolation effectiveness of SLMB, DLMB, RP and BCR under various levels of rpm were demonstrated.

2. Design of the Test Specimens

The position of the sea-water pipe in the power system is shown in Figure 1 and the sea-water pipe used in the underwater vehicle is shown in Figure 2. The sea-water pump assembly and the housing were connected by the pipe, which usually is composed by a flange, a straight line part and an elbow. In practice engineering, it is difficult to change the structure form of the flange and elbow restricted by the housing and pump structures. As a result, the straight-line part was the only part of the pipe that could be improved for better isolation effectiveness.

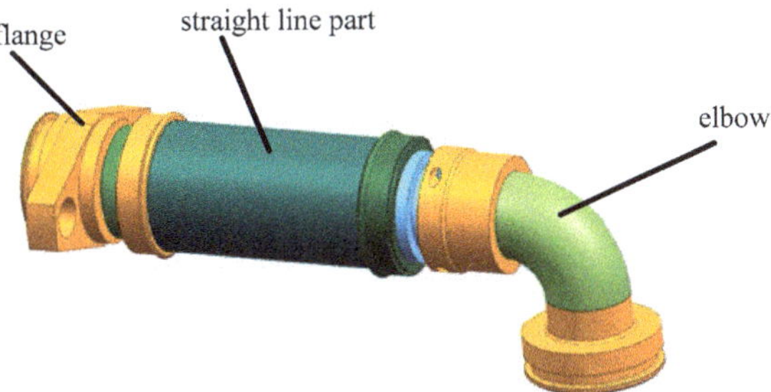

Figure 2. The structure diagram of the sea-water pipe.

To improve the flexibility of pipes and reduce the vibration transport to the housing from the pump, the following direction could be chosen: (i) reduce the pipe stiffness through adjusting the pipe structure; (ii) add high damping rubber sleeve; (iii) replace metal pipes with viscoelasticity pipes and (iv) increase pipes length.

Following the above design principles, three types of flexible pipe were designed and processing, i.e., DLMB, RP and BCR, their structures were shown in Figures 3–5. In the main view of Figure 3, the straight-line part of DLMB consists of a double metal bellows and an external metal braid. The wavy lines in the main view represent the bellows, which owns a wavelength of 3 mm and a thickness of 0.35 mm. The black and white segments represent a protective steel wire braid layer. The black and red ridges each represents a layer of bellows in the partial zoom diagram. The material of the RP was hydrogenated nitrile rubber and the thickness was 3 mm. Figure 5 represents the structure diagram of BCR, which consisted of a bellows and a metal braided layer, with a rubber coating represented by a reticulated line on the outermost layer. Whose thickness was 2 mm. The material parameters of the above three pipes were present in Table 1.

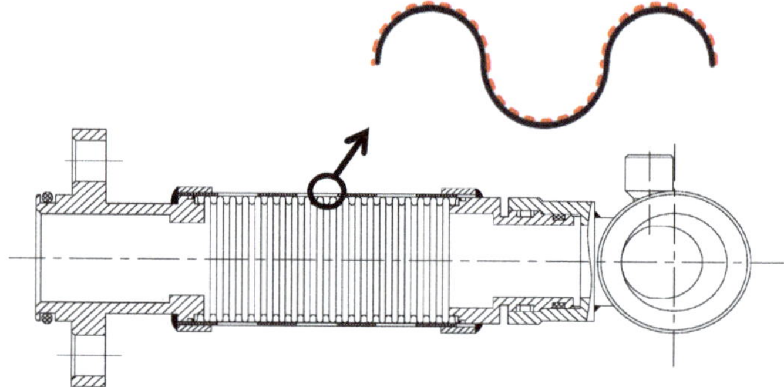

Figure 3. The structure diagram of double layer metal bellows (DLMB).

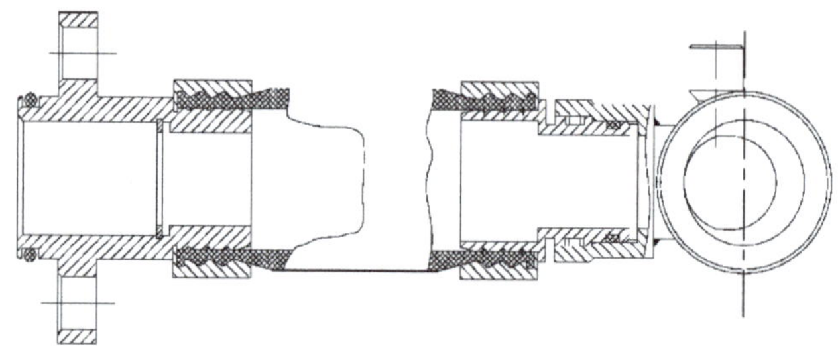

Figure 4. The structure diagram of rubber pipe (RP).

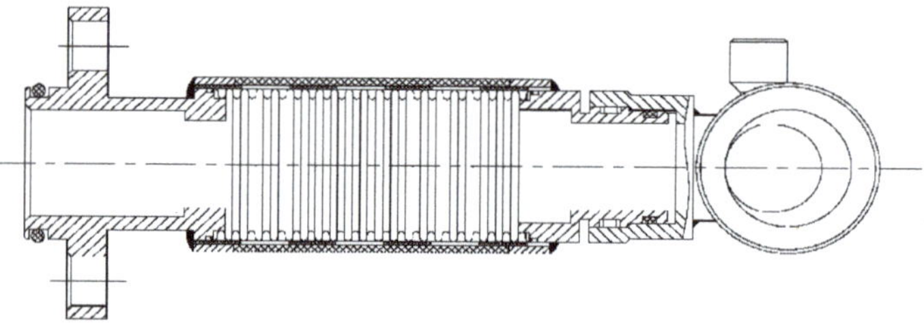

Figure 5. The structure diagram of and bellows coated rubber (BCR).

Table 1. Material parameters.

	Rubber	Stainless Steel
elasticity modulus (pa)	2.8×10^8	1.93×10^{11}
Poisson's ration	0.47	0.30
Material density (kg/m^3)	2200	7980
Material loss factor	5×10^{-3}	1×10^{-4}

Compared with SLMB, DLMB owns the lower flexural stiffness under the same operation load. The vibration energy was consumed by the friction performed between the two-layer metal bellows and reached the aim of reducing the vibration transmission from the sea-water pump to the housing.

Furthermore, there was a layer of metal mesh outside the bellows, which provided adequate axial and radial bearing capacity as well as increases the structural damping.

The material of RP was hydrogenated nitrile rubber, the high damping of the rubber could promote the mechanical impendence of the pipes, the vibration energy could be reduced by the shear deformation of the pipe.

To promote the structural damping of DLMB, a BCR was formed after a layer of rubber was covered on the outer surface of the double-layer.

3. Test Setup

The test setup and photo were shown in Figures 6 and 7. The sea-water pump as the vibrator of the test was riveted in the output shaft of the test bench and driven by the electrical machinery. The front and the end of the tested pipe were connected with the pump and water supply pipe respectively. Furthermore, water supply pipe was supported by the holder, which was a plate with sufficient thickness. The holder was riveted in the workbench with four bolts. The water return pipe was connected with the output of the pump. In the test, the water was transported from the water supply pipe to the pump through the tested pipes and then back to the cistern through water return pipe. There was no connection between the workbench and the electrical machinery. They were respectively fixed on the ground surface. Theoretically, in the test, there were two vibration transmission paths from the sea-water pump to the holder, one was transmitted through the sea-water pipe and the other was transmitted to the workbench through the ground. Compared with the vibration energy transferred through the sea-water pipe, the energy that the second path transfers was sufficiently low. Therefore, we could reasonably ignore the impact of this path, which would not affect the test results of the vibration isolation effectiveness of sea-water pipes.

The surface area of the sea-water pump housing was small and thick in the test. At the other end, the holder owned sufficient thickness. Therefore, in this test, we considered both the sea-water pump and the holder as a mass. In order to accurately evaluate the vibration isolation effectiveness of these pipes, we paid attention to the average vibration reduction effectiveness in multiple directions. As a result, six monitoring points were set in this model. The acceleration of each point was monitored by an acceleration transducer. Three monitoring points were set in different directions on the pump surface and represented the acceleration of the pump. Other monitoring points were on the surface of the holder in three directions and represented the acceleration after the pipe transported. The average responses value of the three directions at both ends of the pipes was utilized to characterize the vibration energy level before and after vibration isolation. The type of acceleration transducers was PCB353B04, the data acquisition system was LMS SCADAS Mobile SCM05, and the LMS.Test.Lab software was used for the acceleration test.

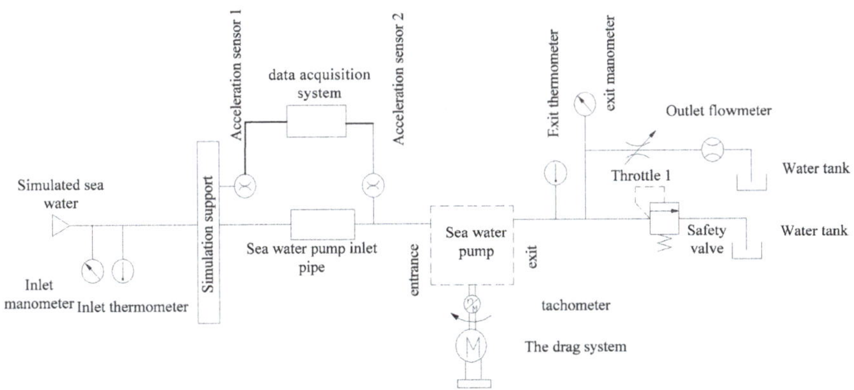

Figure 6. Schematic diagram of the test system.

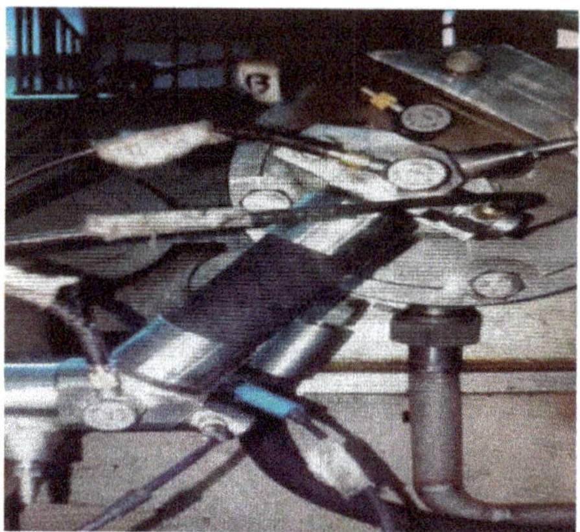

Figure 7. Testing the vibration isolation effectiveness of RP.

In the progress of the test, different vibration energy was induced at different rotate speed of the pump driven by the electrical machinery, which was controlled by the control system of the test bench. In each test condition, the sea-water pump was allowed to run stably for 3 min. After the sea-water pump speed, flow rate and inlet and outlet pressure recorded by the pump platform were stable, time-domain vibration data acquisition was started. The signal acquisition time was 1 min. After the data collection for all working conditions was completed, we performed FFT (Fast Fourier Transform) post-processing, and the selected window function was the Hanning window.

4. Load Scheme

Four stages of tests were considered, the reasons of the operation condition selection included: (i) there were many rated operating speed of underwater vehicle. (ii) The vibration energy level of the pump nonlinear increased with the increasing rotation speed and the tested pipes all owned the strongly nonlinear feature, which was obviously impacted by the vibration energy level and the frequency of the vibrator. Therefore, to accurately assess the vibration reduction effectiveness of the tested pipes in engineering, the following objectives were pursued:

- Measure the isolation effectiveness of various pipes under the 1710 r/min rotation speed of a sea-water pump.
- Measure the isolation effectiveness of various pipes under the 2000 r/min rotation speed of a sea-water pump.
- Measure the isolation effectiveness of various pipes under the 2300 r/min rotation speed of a sea-water pump.
- Measure the isolation effectiveness of various pipes under the 2600 r/min rotation speed of a sea-water pump.

5. Isolation Effectiveness under Design Rpm

Figure 8 showed the results of tests carried out on the four kinds of pipes under 1700 r/min to verify isolation effectiveness. The horizontal axis represented the test frequency, which was from 10 to 10,000 Hz, the dates were recorded at 1–3 octave(s), the curves were combined with center frequency of each 1–3 octave(s), the vertical axis represents the vibration energy level transfer coefficients λ, which was defined as

$$\lambda = \frac{N_2(f)}{N_1(f)} \tag{1}$$

where $N_2(f)$ was the averaged vibration energy level of transducers in all three directions, which represented the response of the holder, and $N_1(f)$ was the averaged vibration energy level of transducers in three directions, which represented the vibration energy of the sea-water pump. If λ was less than unity, it meant that the vibration energy of holder was lower than the sea-water pump, the pipe owned the ability of isolation to mitigate the vibration transport from the pump to the holder, otherwise, the pipe could not be used to alleviate the vibration transport.

The main reasons for choosing to evaluate the vibration isolation effectiveness of flexible pipes in a wide frequency range were as follows:

(a) The vibration spectrum of a sea-water pump presents a typical strong line spectrum characteristic, which mainly includes the shaft frequency and its multiple frequencies, the meshing frequency of the sea-water pump gears and its multiple frequencies related to rotation at a single speed. However, in engineering practice, a sea-water pump usually owns multiple and different operating speeds at different operation time. As a result, a sea-water pump vibration response spectrum owns rich frequency information from tens of hertz to thousands of hertz.

(b) The sea-water pump is one of the main vibration sources of the underwater vehicle. The development of flexible pipes is to reduce the vibration transmission from sea-water pumps to housings and thus reduce the vibration of the underwater vehicle. For the vibration and noise reduction of underwater vehicles, the focus is on its vibration reduction effectiveness in a wide frequency band. Therefore, the selection of wide frequency bands is also due to the need for vibration and noise reduction of underwater vehicles.

(c) Different sea-water pumps have different rated speeds, flexible pipes need to have the ability to isolate the vibration of different sea-water pumps.

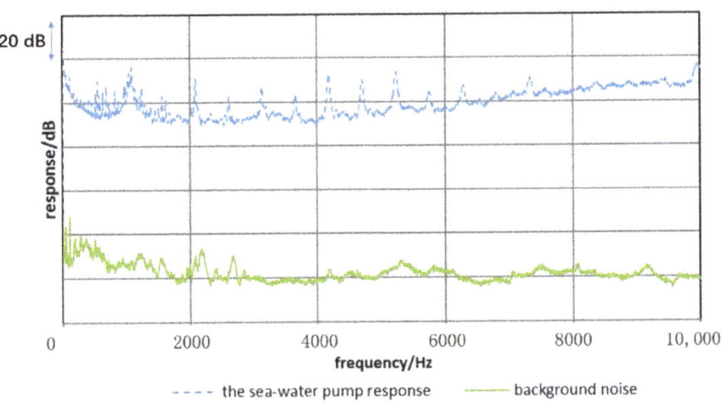

Figure 8. Signal-to-noise ratio of the sea-water pump vibration response at 1710 r/min.

Figure 8 compares the vibration response of the sea-water pump at 1710 r/min with the background noise (the vibration response obtained by the sensors when the sea-water pump is not working). It can be seen from the figure that at the peak of the sea-water pump's response, the signal-to-noise ratio reached 80 dB, and outside of the response peak frequency points, the signal-to-noise ratio at other frequencies had reached more than 40 dB. As the rotation speed increases, the response of the sea-water pump increased. This signal-to-noise ratio was sufficient for evaluating the vibration isolation effectiveness of sea-water pipes.

As shown in Figure 9, for all the pipes, multiple peaks were observed within the test frequency range. However, compared with other pipes, SLMB obviously owned more peaks. What is more, for SLMB, λ remained more than unity in the test frequency range except individual center frequencies, which represented that the pump vibration was magnified to the housing by SLMB in most frequency bands, the pipe offered insignificant vibration reduction.

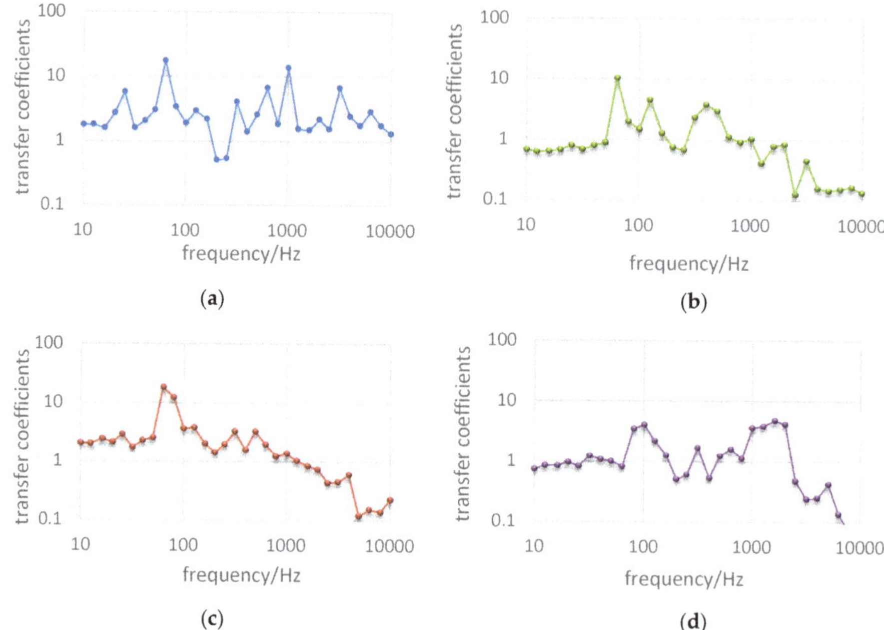

Figure 9. Test results under 1710 r/min: (**a**) single layer metal bellows (SLMB); (**b**) DLMB; (**c**) BCR and (**d**) RP.

The maximum of λ for SLMB, DLMB and BCR within the test frequency range were all observed at the center frequency 63 Hz. It suggested that the natural frequency of the three isolation systems located between 56.2 and 70.8 Hz, which meant their operation stiffness were very similar. The maximum of λ for RP occurred at the center frequency 100 Hz, which represented that the RP owned greater stiffness in operation than other pipes. This was because the structural stiffness of RP was affected by seawater pressure, and the stiffness increased sharply as the inlet pressure of sea-water pump increased. Whereas the maximum of λ for the RP was 4 for its high damping, this was lowest in all pipes.

It can be seen from Figure 9b that the transfer coefficients for DLMB were less than unity below 50 Hz and above 1000 Hz. This suggested that the pump vibration transmitted to the holder was alleviated by the pipe within the above frequency range. Moreover, the transfer coefficients kept oscillation decreasing above the nature frequency, which meant the vibration reduction effectiveness was improved with the increasing frequency.

In Figure 9c, the transfer coefficients for BCR were less than unity above 1250 Hz, and continuously decreased with the rise of frequency. However, the vibration was magnified in other frequency bands. The transfer coefficient for the RP in Figure 9d was less than unity above 2000 Hz and was lower than 0.1 above 6300 Hz, which was the lowest in all pipes.

Figure 10 showed the test results of transfer coefficient curves for the four tested pipes under 2000 r/min. The transfer coefficient for SLMB was greater than unity or close to unity within the test frequency. The transfer coefficient for DLMB was less than unity above 1000 Hz, and continued to reduce. This indicated a stable and positive vibration reduction effectiveness of the DLMB in high frequency. BCR offered good isolation effectiveness above 3150 Hz. The transfer coefficient for RP was less than 1.2 below 63 Hz and lower than 0.1 above 6300 Hz.

It can be seen from the above analysis that the test results of isolation effectiveness of different pipes were basically consistent when the speeds of the sea-water pump were 1710 r/min and 2000 r/min. DLMB owned a stable vibration isolation effectiveness in the frequency band above 1000 Hz, and the vibration isolation effectiveness became better as the frequency increased. BCR also applied stable vibration isolation effectiveness in the high frequency band, but due to its higher damping, its resonance region was wider, and the frequency of vibration isolation was higher than DLMB. The transmission coefficient of RP in the frequency band below 100 Hz was close to 1, which indicated that the vibration of the sea-water pump was not significantly amplified by the RP, and at the same time, RP owned the lowest transfer coefficient in high frequency bands. Unfortunately, SLBM had almost no vibration isolation effectiveness in the test band.

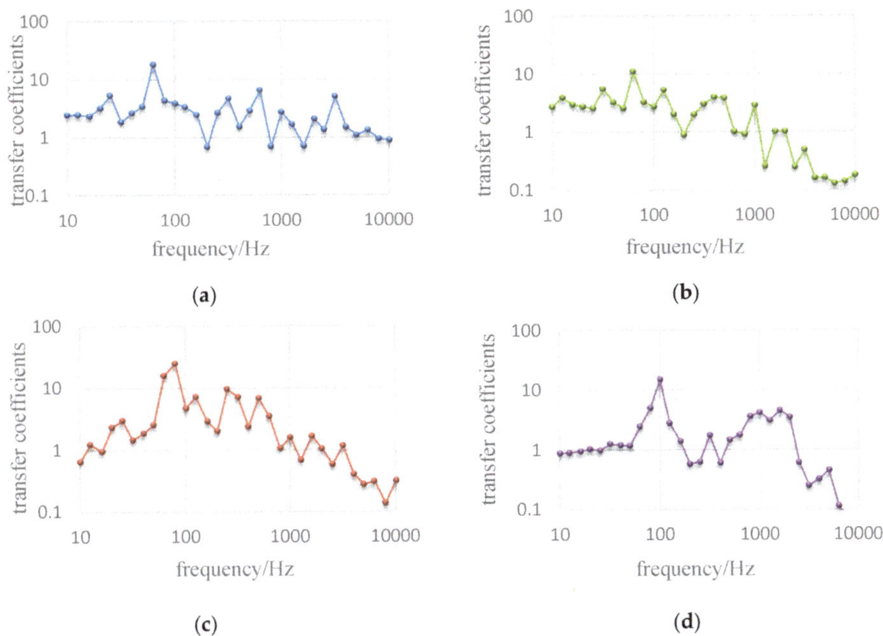

Figure 10. Test results under 2000 r/min: (**a**) SLMB; (**b**) DLMB; (**c**) BCR and (**d**) RP.

In order to find the vibration isolation effectiveness of the pipes at higher sea-water pump speeds, the transfer coefficient for the four pipes at 2300 r/min and 2600 r/min were tested. The results were shown in Figures 11 and 12. Similarly, SLMB offered insignificant isolation effectiveness within the frequency range shown in the figures. DLMB and BCR offered good vibration effectiveness in high frequency. RP always offered the most stable isolation effectiveness below nature frequency, which was the lowest in all pipes, but this phenomenon was more significant at 2600 r/min.

The test results under four different conditions showed some consistency. However, subject to the influences by material nonlinearity and different excitation forces of the sea-water pump at different rotating speeds, there were also some inconsistencies. The frequency at which DLMB began to provide vibration isolation effectiveness had increased at 2300 r/min to 2600 r/min compared with the results under 1710 r/min and 2000 r/min. When the speed of the sea-water pump rose from 2300 to 2600 r/min, the transmission coefficient of BCR was greatly reduced in the frequency band below 200, especially at about 80 Hz, which was also the resonance frequency band of the vibration isolation system, the decrease was most significant. However, at the frequency range of 1000 Hz upwards, the peak of the curve slightly increased. Similar with the test results of the BCR, the transmission coefficient curve of RP was suddenly reduced and owned a transfer coefficient of less than unity in the low frequency band when the speed of the sea-water pump rose from 2300 to 2600 r/min, but the high-frequency vibration isolation effectiveness was not as good as the results at other rotating speeds.

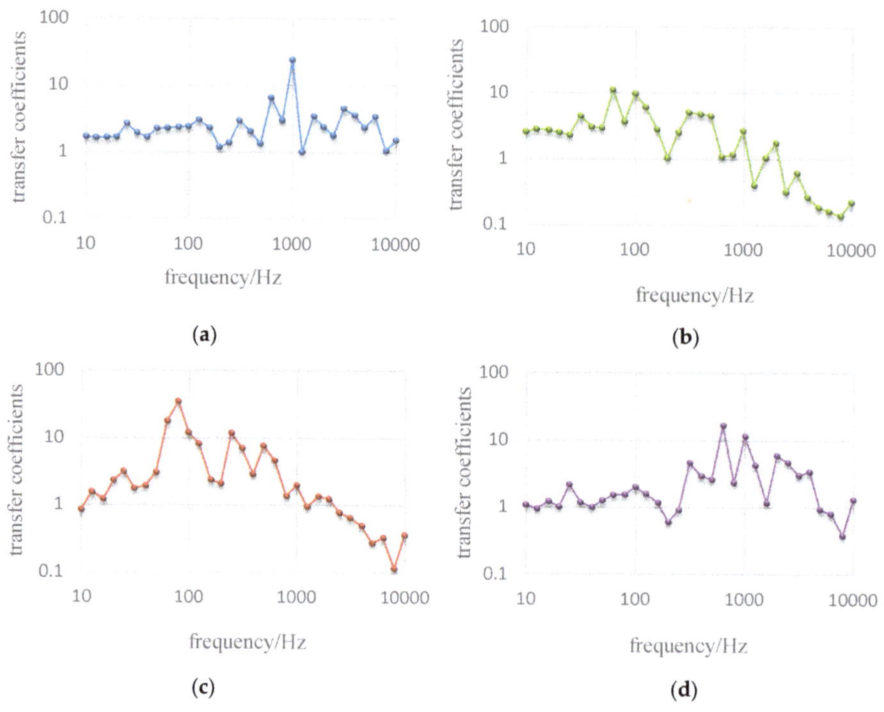

Figure 11. Test results under 2300 r/min: (**a**) SLMB; (**b**) DLMB; (**c**) BCR and (**d**) RP.

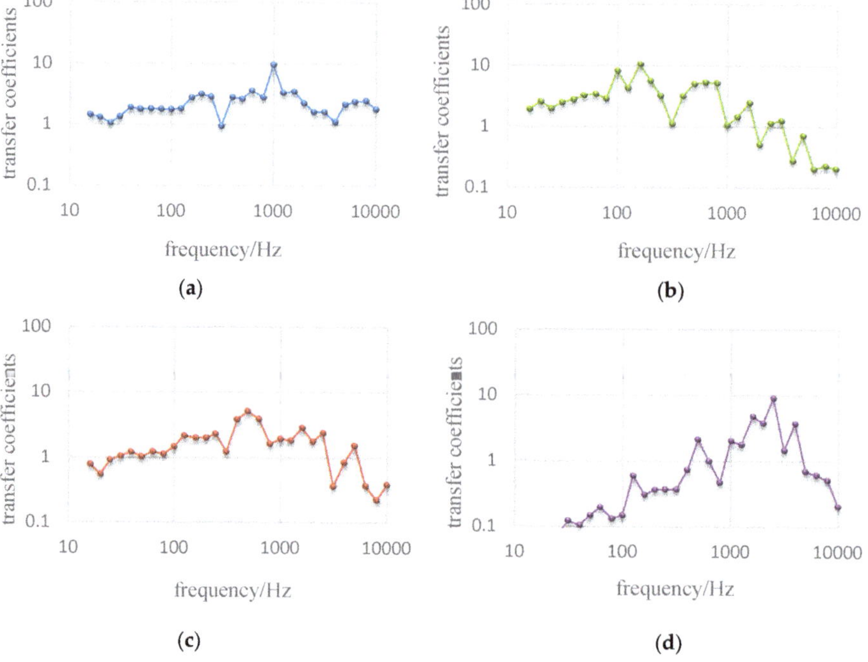

Figure 12. Test results under 2600 r/min: (**a**) SLMB; (**b**) DLMB; (**c**) BCR and (**d**) RP.

It can be seen from the above test results and analysis that none of the flexible pipes had a good vibration damping effectiveness at all sea-water pump speeds and full frequency bands. DLMB owned a stable vibration isolation effectiveness at all speeds in high frequency bands, but had a high transmission coefficient in low frequency bands. The low-frequency vibration reduction effectiveness of BCR and RP were improved, but the high-frequency vibration isolation effectiveness was weakened. This just reflects the complexity of engineering research. In engineering practice, the multi-speed and wide-band vibration responses of sea-water pumps are inevitable, and a vibration isolator cannot take into account the full-band vibration isolation function. In order to effectively reduce the vibration transmission of seawater pumps, it is necessary to use a certain speed and a certain frequency band as input conditions for vibration isolation design under the premise of accurately analyzing the vibration frequency spectrum, so as to design vibration reduction pipes that best satisfies the engineering practice.

6. Conclusions

To investigate the isolation effectiveness of flexible pipes, a scaled model test was carried out. Four types of pipe, i.e., SLMB, DLMB, BCR and RP, were considered. Tests were carried out under 1700 r/min, 2000 r/min, 2300 r/min and 2600 r/min. By calculation the transfer coefficient, the isolation effectiveness was established. The test results under four different conditions showed some consistency as well as some inconsistencies. Nevertheless, DLMB, BCR and RP showed their own advantages in different frequency under all conditions. The primary features are as follows:

- The transfer coefficient of SLMB was closed to or greater than unity within the tested frequency under all conditions, this suggested that the vibration transferred from the sea-water pump the holder in all frequency was increased by the pipe.
- DLMB offered the most significant isolation effectiveness of all pipes at high frequency, this indicated that the vibration of the water pump was alleviated by the pipe. However, the vibration isolation frequency of the DLMB increased as the sea-water pump speed increased.
- Similar to DLMB, BCR also owned stable vibration isolation effectiveness in the high frequency band, but its vibration isolation frequency was slightly higher than DLMB. Compared with other condition, BCR owned a sufficient low transfer coefficient at 2600 r/min.
- The RP kept stabled low transfer coefficient at low frequency in all conditions, especially at 2600 r/min.

Author Contributions: Contributed to Synthesis, Testing, Data Analysis, and Writing the Manuscript: Y.Y. (Yang Yang); Suggested and Supervised the Work: G.P. and S.Y.; Contributed to Revising the Language of the Manuscript: Y.Y. (Ying Yuan). All authors have read and agreed to the published version of the manuscript.

Funding: This research was funded by the Natural Science Basic Research plan in Shaanxi Province of China grant number [2019JQ-071].

References

1. Dai, H.L.; Wang, L.; Qian, Q.; Gan, J. Vibration analysis of three-dimensional pipes conveying fluid with consideration of steady combined force by transfer matrix method. *Appl. Math. Comput.* **2012**, *219*, 2453–2464. [CrossRef]
2. Li, S.; Liu, G.; Kong, W. Vibration analysis of pipes conveying fluid by transfer matrix method. *Nucl. Eng. Des.* **2014**, *266*, 78–88. [CrossRef]
3. Liu, G.; Li, Y. Vibration analysis of liquid-filled pipelines with elastic constraints. *J. Sound Vib.* **2011**, *330*, 3166–3181. [CrossRef]
4. Bochkarev, S.; Lekomtsev, S.; Matveenko, V. Parametric investigation of the stability of coaxial cylindrical shells containing flowing fluid. *Eur. J. Mech. A Solids* **2014**, *47*, 174–181. [CrossRef]
5. Brambley, E.; Peake, N. Stability and acoustics catering in a cylindrical thin shell containing compressible mean flow. *J. Fluid Mech.* **2008**, *602*, 403–426. [CrossRef]

6.	Hu, K.; Wang, Y.; Dai, H.; Wang, L.; Qian, Q. Nonlinear and chaotic vibrations of cantilevered micro pipes conveying fluid based on modified couple stress theory. *Int. J. Eng. Sci.* **2016**, *105*, 93–107. [CrossRef]
7.	Zare, A.; Eghtesad, M.; Daneshmand, F. Numerical investigation and dynamic behavior of pipes conveying fluid based on is geometric analysis. *Ocean Eng.* **2017**, *140*, 388–400. [CrossRef]
8.	Zare, A.; Eghtesad, M.; Daneshmand, F. An isogeometric analysis approach to the stability of curved pipes conveying fluid. *Mar. Struct.* **2018**, *59*, 321–341. [CrossRef]
9.	Zhang, W.M.; Yan, H.; Jiang, H.M.; Hu, K.M.; Peng, Z.K.; Meng, G. Dynamics of suspended micro channel resonators conveying opposite internal fluid flow: Stability, frequency shift and energy dissipation. *J. Sound Vib.* **2016**, *368*, 103–120. [CrossRef]
10.	Amabili, M.; Karagiozis, K.; Païdoussis, M. Effect of geometric imperfections on non-linear stability of circular cylindrical shells conveying fluid. *Int. J. Non-Linear Mech.* **2009**, *44*, 276–289. [CrossRef]
11.	Karagiozis, K.; Païdoussis, M.; Amabili, M.; Misra, A. Nonlinear stability of cylindrical shell subjected to axial flow: Theory and experiments. *J. Sound Vib.* **2008**, *309*, 637–676. [CrossRef]
12.	Mekanik, A.; Païdoussis, M. Unsteady pressure in the annular flow between two concentric cylinders, one of which is oscillating: Experiment and theory. *J. Fluids Struct.* **2007**, *23*, 1029–1046. [CrossRef]
13.	Tan, X.; Mao, X.Y.; Ding, H.; Chen, L.Q. Vibration around non-trivial equilibrium of a supercritical Timoshenko pipe conveying fluid. *J. Sound Vib.* **2018**, *428*, 104–118. [CrossRef]
14.	Alshorbagy, A.E.; Eltaher, M.; Mahmoud, F. Free vibration characteristics of a functionally graded beam by finite element method. *Appl. Math. Model.* **2011**, *35*, 412–425. [CrossRef]
15.	Park, C.H.; Lee, U.; Hong, S.C.; Kim, T.Y. Stability analysis of piping system conveying unsteady flow. *Trans. Korean Soc. Mech. Eng.* **1991**, *15*, 1512–1521.
16.	Salman, H.E.; Sert, C.; Yazicioglu, Y. Computational analysis of high frequency fluid–structure interactions in constricted flow. *Comput. Struct.* **2013**, *122*, 145–154. [CrossRef]
17.	Ortiz-Vidala, L.E.; Mureithib, N.W.; Rodriguez, O.M.H. Vibration response of a pipe subjected to two-phase flow: Analytical formulations and experiments. *Nucl. Eng. Des.* **2017**, *313*, 214–224. [CrossRef]
18.	Zhou, W.; Li, D. Experimental research on a vibration isolation platform for momentum wheel assembly. *J. Sound Vib.* **2013**, *332*, 1157–1171. [CrossRef]
19.	Pan, P.; Shen, S.; Shen, Z.; Gong, R. Experimental investigation on the effectiveness of laminated rubber bearings to isolate metro generated vibration. *Measurement* **2018**, *122*, 554–562. [CrossRef]
20.	Bi, K.; Hao, H. Numerical simulation on the effectiveness of using viscoelastic materials to mitigate seismic induced vibrations of above-ground pipelines. *Eng. Struct.* **2016**, *123*, 1–14. [CrossRef]

Article

Darcy-Forchheimer MHD Hybrid Nanofluid Flow and Heat Transfer Analysis over a Porous Stretching Cylinder

Anwar Saeed [1], Asifa Tassaddiq [2,*], Arshad Khan [3], Muhammad Jawad [1], Wejdan Deebani [4], Zahir Shah [1,5,*] and Saeed Islam [1]

[1] Department of Mathematics, Abdul Wali Khan University, Mardan, Khyber 23200, Pakistan; anwarsaeed769@gmail.com (A.S.); muhammadjawad175@yahoo.com (M.J.); saeedislam@awkum.edu.pk (S.I.)

[2] College of Computer and Information Sciences Majmaah University, Al Majmaah 11952, Saudi Arabia

[3] Institute of Computer Sciences and Information Technology, The University of Agriculture, Peshawar 25000, Pakistan; arshadkhan@aup.edu.pk

[4] Department of Mathematics, College of Science & Arts, King Abdulaziz University, P. O. Box 344, Rabigh 21911, Saudi Arabia; wdeebani@kau.edu.sa

[5] Center of Excellence in Theoretical and Computational Science (TaCS-CoE), SCL 802 Fixed Point Laboratory, Science Laboratory Building, King Mongkut's University of Technology Thonburi (KMUTT), 126 Pracha-Uthit Road, Bang Mod, Thrung Khru, Bangkok 10140, Thailand

* Correspondence: a.tassaddiq@mu.edu.sa (A.T.); zahir.sha@kmutt.ac.th (Z.S.)

Received: 11 February 2020; Accepted: 11 April 2020; Published: 16 April 2020

Abstract: This research work deals with investigation of the thermal characteristics of the Darcy–Forchheimer hydromagnetic hybrid nanofluid (Al_2O_3-Cu/H_2O) flow through a permeable stretching cylinder. The model equations, which consist of continuity, momentum, and energy equations, are converted to a set of coupled ordinary differential equations through similarity variables transformations and appropriate boundary conditions. Brownian motion and Thermophoresis effects are mainly focused in this work. The impacts of some interesting parameters over velocity, temperature, and concentrations profiles are graphically studied. The present study will be helpful in understanding the thermal characteristics of heat transfer liquids.

Keywords: porous media; magnetohydrodynamics; hybrid nanofluid; stretching cylinder; flow characteristics; nanoparticles; convective heat transfer

1. Introduction

The liquid coolants are mainly employed to keep the operating temperature of different equipment in the specified range by transferring heat from them. Presently, the investigations show that the thermal conductivities of different fluids used in liquid coolants are much smaller when compared to those of solid metals. The natural laminar and convective heat energy transfer is a significant process in engineering and industry due to its numerous applications. Currently, a lot of research is in progress to investigate the different heat energy transfer characteristics of the newly developed fluid called "hybrid nanofluid" due to an extensive range of engineering and technological uses, like medical manufacturing, microfluidics, transportation, generator cooling, naval structures, and solar heating. A hybrid nanofluid refers to a combination in a base fluid of two different types of nanoparticles. Therefore, if nanoparticles materials are correctly picked, then they will enhance each other's positive aspects. The metallic nanoparticles, such as aluminum, copper, and zinc, have high thermal conductivities. However, the use of these metallic nanoparticles for nanofluid applications is limited because of their reactivity and stability. On the other hand, ceramic nanoparticles have lower thermal conductivity relative to metal nanoparticles, but they have many desirable possessions, such as chemical inertness and stability. In the case of closed cavities, heat transfer study appeared in several applications and has been widely deliberated in the literature [1–3]. In open cavities, the processes of natural convection give significant results by simulating more complex geometries at the open end. In open cavities natural convection is related to different engineering systems, such as the cooling of electrical equipment, room air conditioning, and solar thermal central receiver systems etc. [4–6]. Skok et al. [4] have undertaken an experimental survey of the open cavity natural convective flow numerically. They found some good agreements between the experimental data and the numerical results. For larger values, their results are very significant. Chan and Tien [7] studied the two-dimensional (2-D) natural convection flow in narrow open cavities and conducted relative studies while using open square cavities with protracted domains. The small thermal conductivities of the base fluids used in the natural convection are showed to be an essential obstacle for increasing the heat energy transformation rate beyond a definite limit. As a ground breaking work, Choi et al. added solid nanoparticles to the base fluid to enhance its thermal conduction, and called it nanofluids [8]. Mahmoudi et al. [9] numerically deliberated the enhancement of natural convective heat energy transfer flow. Sheremet et al. performed the numerical investigation of non-steady natural convection in a permeable open corrugated cavity through a two-phase model [10]. They showed that the Nusselt number reduces by enhancing the number of iterations. Presently, Tassaddiqet al. [11]investigated couple stress magneto-hydrodynamic nanofluid thin film flow over an exponential stretching sheet with joule heating and viscous dissipation. Tahar and Chamkha [12] discussed the flow of a hybrid nanofluid through horizontal and confocal elliptical cylinders with natural convection enhancement. Moreover, it has been determined that the key aim of including the nanomaterials in a transferor liquid is to augment its thermal conductivity. It shall also be noted that stable nanofluids have significant attributes, such as small nanoparticle concentration and higher thermal conductivities. Hence, the majority of the studies in the past have been carried out to attain high thermal conductivity through the use of a single nanomaterial [13–16]. Currently, many articles have considered the topic of hybrid nanofluids [17–20]. Mixed nanomaterials display important chemical and thermo-physical characteristics that do not happen in a single component. Hybrid nanomaterials are mainly categorized into three types [21–23]. In current years, numerous numerical and experimental works that are associated to hybrid nanofluids have been published, and their results display that they are more suitable than the conventional nanofluids. Suresh et al. [24] discussed a copper-alumina nanocomposite powder, which was mixed by using thermochemical method, and prepared a hybrid nanofluid through a two-step procedure. In another study, Suresh et al. [25] examined the heat energy transfer characteristics of hybrid nanofluid (alumina-copper)/water metal nanomaterials and polymer nanomaterials. Nadeem et al. [26] deliberated the MHD Maxwell nanofluid flow through a stretch sheet. Rockney et al. [27] studied the MHD nanofluid flow involving heat energy transfer through two plates. Shehzad et al. [28] investigated a nanofluidic flow by using the Jaffrey fluid model with MHD convective boundary conditions. Mahmood et al. discussed the flow of nanofluids for cooling purposes [29]. Nanofluid flow through a porous medium by incorporating the heat conduction through channels have explored by Fakour et al. [30]. Hatami et al. [31] described the laminar flow of nanofluids through rotating disks. Nadeem et al. [32] investigated the non-orthogonal and nanofluid non-Newtonian flows with heat energy transformation. Sheikholeshlami et al. [33] thoroughly investigated the nanofluid flow through a semi-porous channel. Akbar et al. [34] studied the viscosity and buoyancy impacts during the nanofluid MHD flow over a stretch surface. Fakour et al. [35] have undertaken the nanofluid flow through vertical channels. Maskeen et al. [36] have examined nanofluid flow by using water as base fluid and investigated the enrichment of heat energy transfer through the stretching sheet. Akilu et al. [37] deliberated the flow with the thermo-physical properties of water-based composite nanofluids. Hayat et al. [38] considered the Newtonian nanofluid flow through a cylinder along with the heating impacts. Further study can be read in [39–42].

Ceramic materials like alumina (Al_2O_3) have numerous excellent possessions, such as chemical inertness and good stability. Yet, alumina's thermal conductivity is small when compared to metallic nanoparticles. Metallic nanoparticles, such as copper, have greater thermal conductivity. Yet, reactivity and stability are two significant factors that impede these metallic nanoparticles from being used. The inclusion of small quantities of copper particles in an alumina matrix will significantly increase the thermal possessions without affecting the nanofluid's stability. Jena et al. [43] clarified the synthesis of Cu-Al$_2$O$_3$ nanocomposites while using hydrogen reduction techniques from chemically formulated Cu-Al$_2$O$_3$ mixtures. Niihara [44] and Oh et al. [45,46] revealed the manufacture of Al$_2$O$_3$-Cu nanocomposite made from fine powder mixtures of Al$_2$O$_3$ and CuO nanoscale. The proposed nanocomposites had a new concept of material design, thermal properties, and enhanced mechanical dramatically.

The Darcy–Forchheimer law is the law for porous media flows with Reynolds numbers more than approximately 1 to 10, and the inertial impacts can also become important. Sometimes a term of inertia is applied to the equation of Darcy, defined as the word Forchheimer. The non-linear behavior of the pressure difference vs. flow data can be taken into account by this term. In 1856, Hennery Darcy has developed the flow of a steady fluid through porous media, when he was carrying out his work of moving the water through sand bags. Though his outcome was not very much reliable at that time because he did not deliver a positive consequence. Yet, based on his research, a Dutch scientist, known as Forchheimer, came in 1901. Forchheimer, gave his ideas and expressions more extensive. Forchheimer added the velocity term square in the momentum equations for calculating inertial forces as well as limit layer flow [47]. Muskat later added the name 'Forchheimer term' [48] Pal and Mondal [49] later investigated Darcy–Forchheimer's model over a stretch face, where they defined that the value of the electric field parameter increases with a lessening in the nanoparticles concentration sketch. Ganesh et al. [50] conducted a study on the nanofluid flow of Darcy–Forchheimer MHD over a stretched/shrinking sheet and determined that the temperature value rises with the presence of viscous-dissipation effects. Hayat et al. [51] investigated the Darcy–Forchheimer flow with heat flux and variable thermal conductivity between Cattaneo and Christov. Muhammad et al. [52] an updated Darcy–Forchheimer flow of Maxwell nanofluid model due to convective boundary conditions. They used the pores medium, and found that the concentration and temperature-profile of nanoparticles increased with the value of the porosity parameter. Jawad et al. [53] have been studied MHD Nanofluid Darcy–Forchheimer thin film flow with Navier's partial slip and joule dissipation. Uddin et al. [54] have been deliberated flow with nonlinear thermal Radiation of Darcy–Forchheimer Sisko nanomaterial. Mohamed et al. [55] have studied flow of Carreau nanofluid over a convectively heated nonlinear stretching surface in the presence chemically reactive species. Lahmar et al. [56] have investigated unsteady nanofluid squeezing flow with the effects of an inclined magnetic field and variable thermal conductivity. Mohamed et al. [57] have discussed FEM for blood-based SWCNTs flow with electromagnetic radiation through a circular cylinder in a porous medium. Mohamed et al. [58] have described SQLM for external yield stress effect on three-dimensional (3D) MHD nanofluid flow in a porous medium. Mohamed [59] has studied MHD boundary-layer flow of two-phase nanofluid model over an exponentially stretching sheet with a heat generation and Chemical reaction effect. Mohamed [60] has investigated unsteady flow of water-NPs over a stretching sheet in a saturated porous medium in the stagnation-point region with chemical reaction. Other related study can be read in [61–63].

The aim of this research work is the investigation of the thermal characteristics of the Darcy–Forchheimer hydromagnetic hybrid nanofluid flow through a stretching porous cylinder. The model equations with appropriate boundary conditions are solved analytically. Thermophoresis and Brownian motion impacts are mainly focused in this work. The impacts of modeled parameters over velocity, temperature, and concentrations profiles are graphically studied.

2. Mathematical Formulation

2.1. Flow Modelling

In the current work, we examine the steady and incompressible boundary layer hydro-magnetic flow of alumina-copper/water hybrid nanofluids through a permeable stretch cylinder. An external magnetic field is applied. The hybrid nanofluid flow is affected in the axial direction by the stretching of the elastic cylinder. The Darcy–Forchheimer impact is included in momentum equation. The coordinate system are selected in such a method that fluid flow is started due to elongating cylinder in the axial direction, where (x, r) shows axial and radial directions.

The energy, mass, and momentum conservation laws in boundary layer approximation can be represented as:

$$\frac{\partial}{\partial x}(ru) + \frac{\partial}{\partial x}(rv) = 0 \tag{1}$$

$$\left(u\frac{\partial}{\partial x}u + v\frac{\partial}{\partial r}u\right) = v_{hnf}\frac{1}{r}\frac{\partial}{\partial r}\left(r\frac{\partial u}{\partial r}\right) - \frac{1}{\rho_{hnf}}\left[\sigma_0\beta_0^2 u + (\rho\beta)_{hnf}g(T-T_\infty) - \frac{v_{nf}}{k}u - Fu^2\right], \tag{2}$$

$$\left(u\frac{\partial}{\partial x}T + v\frac{\partial}{\partial r}T\right) = \frac{k_{hnf}}{(\rho c_p)_{hnf}}\left(\frac{\partial}{\partial r}\left(r\frac{\partial}{\partial r}T\right) + \frac{1}{r}\frac{\partial}{\partial r}T\right) + \tau\left(D_B\frac{\partial}{\partial r}C\frac{\partial}{\partial r}T + \frac{D_T}{T_\infty}\left(\frac{\partial}{\partial r}T\right)^2\right), \tag{3}$$

$$\left(u\frac{\partial}{\partial x}C + v\frac{\partial}{\partial r}C\right) = D_B\left(\frac{\partial^2}{\partial r^2}C + \frac{1}{r}\frac{\partial}{\partial r}C\right) + \frac{D_T}{T_\infty}\left(\frac{\partial^2}{\partial r^2}T + \frac{1}{r}\frac{\partial}{\partial r}T\right). \tag{4}$$

Boundary conditions are:

$$\begin{aligned} r = a \quad u &= cx,\ v = 0,\ T = T_w,\ C = C_w, \\ r \to \infty\ u &= 0,\ T = T_\infty,\ C = C_\infty. \end{aligned} \tag{5}$$

We employ the following transformations to convert the model equations into dimensionless form are [36]:

$$\begin{aligned} \eta &= \frac{r^2-a^2}{2a}\sqrt{\frac{c}{v}},\quad u = cxf'(\eta),\quad v = -\frac{a}{r}\sqrt{cv}f(\eta), \\ \theta(\eta) &= \frac{T-T_\infty}{T_w-T_\infty},\ \phi(\eta) = \frac{C-C_\infty}{C_w-C_\infty} \end{aligned} \tag{6}$$

By using Equation (7) in Equations (1) to (6), we get:

$$A_1\left((1+2\eta\gamma)f''' + 2\gamma f''\right) + A_2\left(ff'' - f'^2(1-Fx)\right) - f'(M+k_1A_2) + A_3\lambda\theta = 0 \tag{7}$$

$$(1+2\eta\gamma)\theta'' + \gamma\theta' + Pr(1+2\eta\gamma)\{N_b\theta'\phi' + N_t\theta'^2\} + Prf\theta' = 0 \tag{8}$$

$$(1+2\eta\gamma)\phi'' + \gamma\phi' + \frac{N_t}{N_b}\{(1+2\eta\gamma)\theta'' + \gamma\theta'\} + Scf\phi' = 0 \tag{9}$$

$$f(0) = 0, f'(0) = \theta(0) = 1 \tag{10}$$

$$f(\infty) = \theta(\infty) = \phi(\infty) = 0 \tag{11}$$

We have defined the different parameters in Equations (8)–(10) as; $k_1 = \frac{v_{nf}}{kc}$, is the permeability parameter, $Re = \frac{cx^2}{v}$ is the local Reynolds number, $N_t = \frac{\tau D_T(T_w-T_\infty)x^3}{T_\infty v}$ is the thermophoresis constraint, $Pr = \frac{v\rho c_p}{k}$ is the Prandtl number. $N_b = \frac{\tau D_B(C_w-C_\infty)}{v}$ represents the Brownian motion limitation, $\gamma = \sqrt{\frac{v}{ca^2}}$ is the curvature parameter, and $M = \frac{\sigma_0\beta_0^2}{c\rho_f}$ is the magnetic parameter. The Schmidt number is defined by $Sc = \frac{v}{D_B}$, whereas $F = C_b x/\sqrt{K}$ represents the local inertia parameter.

2.2. Important Physical Quantities

The basic physical quantities of interest are the coefficient of friction (C_f) and Nusselt number (Nu), which are defined by:

$$C_f = \frac{2\tau_w}{\rho_{bf}u_w^2}, \ \tau_w = \mu_{hnf}\left(\frac{\partial}{\partial r}u\right)_{r=a} \tag{12}$$

$$Nu = \frac{xQ_w}{k_{bf}(T_w - T_\infty)}, \ Q_w = -k_{hnf}\left(\frac{\partial}{\partial r}T\right)_{r=a} \tag{13}$$

$$Sh = \frac{xQ_w}{D_B(C_w - C_\infty)}, \ Q_w = -D_B\left(\frac{\partial}{\partial r}C\right)_{r=a} \tag{14}$$

By using the similarity transformations as defined in Equation (7), the dimensionless forms of Equations (13)–(15) are:

$$C_f\sqrt{Re} = \frac{\mu_{hnf}}{\mu_{bf}}f''(0), \ \frac{Nu}{\sqrt{Re}} = -2\left(\frac{k_{hnf}}{k_{bf}}\right)\theta'(0), \ \frac{Sh}{\sqrt{Re}} = -\phi'(0) \tag{15}$$

In Equation (8), the different constant terms appearing are given by:

$$A_1 = \frac{\mu_{hnf}}{\mu_{bf}}, A_2 = \frac{\rho_{hnf}}{\rho_{bf}}, A_3 = \frac{(\rho\beta)_{hnf}}{(\rho\beta)_{bf}} \tag{16}$$

2.3. Hybrid Nanofluid Modelling

The density ρ_{hnf}, viscosity μ_{hnf}, thermal expansion coefficient β_{hnf}, thermal conductivity k_{hnf} of the hybrid nanofluid, and heat capacity $(c_p)_{hnf}$ are given according to [18,19], as:

$$\rho_{hnf} = (1 - \varphi_{np2})\{(1 - \varphi_{np1})\rho_f + \varphi_{np1}\rho_{np1}\} + \varphi_{np2}\rho_{np2}, \tag{17}$$

$$\mu_{hnf} = \mu_{bf}(1 - \varphi_{np1})^{-2.5}(1 - \varphi_{np2})^{-2.5}, \tag{18}$$

$$(\rho c_p)_{hnf} = (1 - \varphi_{np2})\{(1 - \varphi_{np1})(\rho c_p)_f + \varphi_{np1}(\rho c_p)_{np1}\} + \varphi_{np2}(\rho c_p)_{np2}, \tag{19}$$

$$(\rho\beta)_{hnf} = (1 - \varphi_{np2})\{(1 - \varphi_{np1})(\rho\beta)_f + \varphi_{np1}(\rho\beta)_{np1}\} + \varphi_{np2}(\rho\beta)_{np2}, \tag{20}$$

$$\frac{k_{hnf}}{k_{bf}} = \frac{k_{np2} + (n-1)k_{bf} - (n-1)\varphi_{np2}(k_{bf} - k_{np2})}{k_{np2} + (n-1)k_{bf} + \varphi_{np2}(k_{bf} - k_{np2})}, \tag{21}$$

$$\frac{k_{nf}}{k_{bf}} = \frac{(n-1)k_{bf} + k_{np1} - (n-1)\varphi_{np1}(k_{bf} - k_{np1})}{(n-1)k_{bf} + k_{np1} + \varphi_{np1}(k_{bf} - k_{np1})}, \tag{22}$$

The subscripts np_1 and np_2 represent the nanoparticles of Al_2O_3 and Cu, whereas bf and hnf represent the base fluid and hybrid nanofluid. The symbol φ is the fraction of whole volume, which is the combination of two different types of nanoparticles, Alumina–Copper dispersed in the transferor fluid in order to develop the hybrid nanofluid, which is: $\varphi = \varphi_{np1} + \varphi_{np2}$.

2.4. Solution by HAM

For solution process optimal approach is used. Equations (8)–(10) with boundary conditions (11 and 12) are solved by HAM. Basic derivations of the model equations through HAM are specified in details below.

$$L_{\widehat{f}}(\widehat{f}) = \widehat{f}''', L_{\widehat{\theta}}(\widehat{\theta}) = \widehat{\theta}'', L_{\widehat{\phi}}(\widehat{\phi}) = \widehat{\phi}'', \tag{23}$$

Linear operators $L_{\widehat{f}}$, $L_{\widehat{\theta}}$ *and* $L_{\widehat{\phi}}$ are signified as

$$L_{\widehat{f}}(e_1 + e_2\eta + e_3\eta^2) = 0,\ L_{\widehat{\theta}}(e_4 + e_5\eta) = 0,\ L_{\widehat{\phi}}(e_6 + e_7\eta) = 0\,,\tag{24}$$

The consistent non-linear operators are reasonably selected as $N_{\widehat{f}}$, $N_{\widehat{\theta}}$ *and* $N_{\widehat{\phi}}$ and identify as:

$$N_{\widehat{f}}\left[\widehat{f}(\eta;\zeta),\widehat{\theta}(\eta;\zeta)\right] = A_1\left((1+2\eta\gamma)\widehat{f}_{\eta\eta\eta} + 2\gamma\widehat{f}_{\eta\eta}\right) + A_2\left(\widehat{f}\widehat{f}_{\eta\eta} - \widehat{f}_{\eta\eta}^2(1-Fx)\right)$$
$$-\widehat{f}_\eta(M+k_1A_2) + A_3\lambda\widehat{\theta}\,,\tag{25}$$

$$N_{\widehat{\theta}}\left[\widehat{f}(\eta;\zeta),\widehat{\theta}(\eta;\zeta),\widehat{\phi}(\eta;\zeta)\right] = (1+2\eta\gamma)\widehat{\theta}_{\eta\eta} + \gamma\widehat{\theta}_\eta + Pr(1+2\eta\gamma)\{N_b\widehat{\theta}_\eta\widehat{\phi}_\eta + N_t\widehat{\theta}_\eta^2\} + Pr\widehat{f}\widehat{\theta}_\eta\,,\tag{26}$$

$$N_{\widehat{\phi}}\left[\widehat{\phi}(\eta;\zeta),\widehat{f}(\eta;\zeta),\widehat{\theta}(\eta;\zeta)\right] = (1+2\eta\gamma)\widehat{\phi}_{\eta\eta} + \gamma\widehat{\phi}_\eta + \frac{N_t}{N_b}\{(1+2\eta\gamma)\widehat{\theta}_{\eta\eta} + \gamma\widehat{\theta}_\eta\} + Sc\widehat{f}\widehat{\phi}_\eta.\tag{27}$$

For Equations (8)–(10) the 0th-order system is shown as

$$(1-\zeta)L_{\widehat{f}}\left[\widehat{f}(\eta;\zeta) - \widehat{f}_0(\eta)\right] = p\hbar_{\widehat{f}}\,N_{\widehat{f}}\left[\widehat{f}(\eta;\zeta),\widehat{\theta}(\eta;\zeta)\right]\tag{28}$$

$$(1-\zeta)\,L_{\widehat{\theta}}\left[\widehat{\theta}(\eta;\zeta) - \widehat{\theta}_0(\eta)\right] = p\hbar_{\widehat{\theta}}N_{\widehat{\theta}}\left[\widehat{f}(\eta;\zeta),\widehat{\theta}(\eta;\zeta),\widehat{\phi}(\eta;\zeta)\right]\tag{29}$$

$$(1-\zeta)\,L_{\widehat{\phi}}\left[\widehat{\phi}(\eta;\zeta) - \widehat{\phi}_0(\eta)\right] = p\hbar_{\widehat{\phi}}N_{\widehat{\phi}}\left[\widehat{\phi}(\eta;\zeta),\widehat{f}(\eta;\zeta),\widehat{\theta}(\eta;\zeta)\right]\tag{30}$$

Whereas, BCs are

$$\widehat{f}(\eta;\zeta)\Big|_{\eta=0} = 0,\ \ \frac{\partial\widehat{f}(\eta;\zeta)}{\partial\eta}\Big|_{\eta=0} = \widehat{\theta}(\eta;\zeta)\Big|_{\eta=0} = 1$$
$$\widehat{f}(\eta;\zeta)\Big|_{\eta=\infty} = \widehat{\theta}(\eta;\zeta)\Big|_{\eta=\infty} = \widehat{\phi}(\eta;\zeta)\Big|_{\eta=\infty} = 0,\tag{31}$$

While the embedding constraint is $\zeta \in [0,1]$, to regulate for the solution convergence $\hbar_{\widehat{f}}$, $\hbar_{\widehat{\theta}}$ and $\hbar_{\widehat{\phi}}$ are used. When $\zeta = 0$ and $\zeta = 1$, we have:

$$\widehat{f}(\eta;1) = \widehat{f}(\eta),\ \widehat{\theta}(\eta;1) = \widehat{\theta}(\eta)\,,\widehat{\phi}(\eta;1) = \widehat{\phi}(\eta),\tag{32}$$

Expand the $\widehat{f}(\eta;\zeta)$, $\widehat{\theta}(\eta;\zeta)$ and $\widehat{\phi}(\eta;\zeta)$ through Taylor's series for $\zeta = 0$

$$\widehat{f}(\eta;\zeta) = \widehat{f}_0(\eta) + \sum_{n=1}^{\infty}\widehat{f}_n(\eta)\zeta^n$$
$$\widehat{\theta}(\eta;\zeta) = \widehat{\theta}_0(\eta) + \sum_{n=1}^{\infty}\widehat{\theta}_n(\eta)\zeta^n\tag{33}$$
$$\widehat{\phi}(\eta;\zeta) = \widehat{\phi}_0(\eta) + \sum_{n=1}^{\infty}\widehat{\phi}_n(\eta)\zeta^n$$

$$\widehat{f}_n(\eta) = \frac{1}{n!}\frac{\partial\widehat{f}(\eta;\zeta)}{\partial\eta}\Big|_{p=0},\ \widehat{\theta}_n(\eta) = \frac{1}{n!}\frac{\partial\widehat{\theta}(\eta;\zeta)}{\partial\eta}\Big|_{p=0},\ \widehat{\phi}_n(\eta) = \frac{1}{n!}\frac{\partial\widehat{\phi}(\eta;\zeta)}{\partial\eta}\Big|_{p=0}.\tag{34}$$

Whereas, BCs are:

$$\widehat{f}(0) = 0, f'(0) = \widehat{\theta}'(0) = 1$$
$$\widehat{f}(\infty) = \widehat{\theta}(\infty) = \widehat{\phi}(\infty) = 0.\tag{35}$$

Now

$$\mathfrak{R}_n^{\widehat{f}}(\eta) = A_1\left((1+2\eta\gamma)\widehat{f}'''_{n-1} + 2\gamma\widehat{f}''_{n-1}\right) - \left(\sum_{j=0}^{w-1}\widehat{f}_{w-1-j}\widehat{f}''_j - \widehat{f}''^2_{n-1}(1-Fx)\right)$$

$$-\widehat{f}'_{n-1}(M+k_1A_2) + A_3\lambda\widehat{\theta}_{n-1},$$

(36)

$$\mathfrak{R}_n^{\widehat{\theta}}(\eta) = (1+2\eta\gamma)\widehat{\theta}''_{n-1} + \gamma\widehat{\theta}'_{n-1} + \mathrm{Pr}(1+2\eta\gamma)\{N_b\sum_{j=0}^{w-1}\widehat{\theta}'_{w-1-j}\widehat{\phi}_j + N_t\widehat{\theta}'^2_{n-1}\} + \mathrm{Pr}\sum_{j=0}^{w-1}\widehat{\theta}'_{w-1-j}\widehat{f}_j, \quad (37)$$

$$\mathfrak{R}_n^{\widehat{\phi}}(\eta) = (1+2\eta\gamma)\widehat{\phi}''_{n-1} + \frac{N_t}{N_b}\{(1+2\eta\gamma)\widehat{\phi}''_{n-1} + \gamma\widehat{\phi}'_{n-1}\} + \gamma\widehat{\phi}'_{n-1} + Sc\sum_{j=0}^{w-1}\widehat{f}_{w-1-j}\widehat{\phi}'_j, \quad (38)$$

While

$$\chi_n = \begin{cases} 0, & \text{if } \zeta \leq 1 \\ 1, & \text{if } \zeta > 1. \end{cases}$$

(39)

3. Result and Discussion

We discuss in this section the effect produced by the different physical parameters of interest on the velocity profile ($f'(\eta)$), temperature profile ($\theta(\eta)$), and concentration profile ($\phi(\eta)$) during the hybrid nanofluid flow of a boundary layer through a stretching cylinder. Figure 1 describes the geometry of the nanofluid flow. We have plotted the effects that are produced by these different parameters in the Figures 2–14.

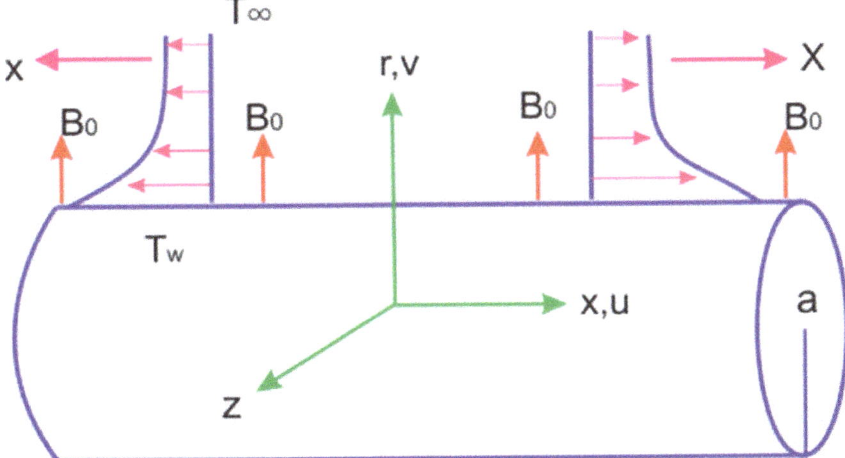

Figure 1. Physical sketch of flow phenomena.

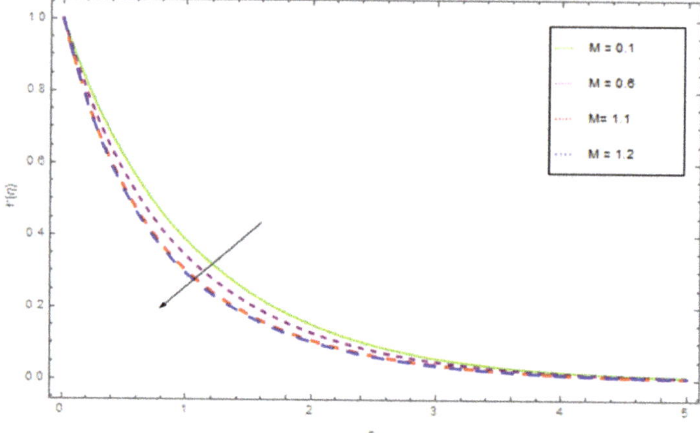

Figure 2. The influence of M on $f'(\eta)$ when $k_1 = 0.2$, $F = 0.1$, $\lambda = 0.3$, $\gamma = 0.4$.

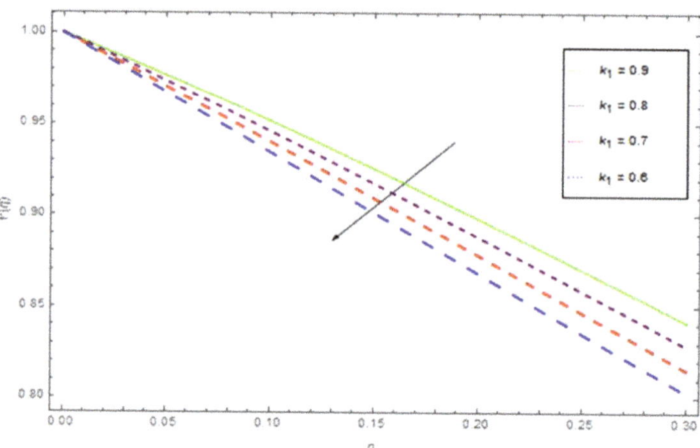

Figure 3. The variation of $f'(\eta)$ with k_1, for $M = 0.6$, $F = 0.5$, $\lambda = 0.3$, $\gamma = 0.4$.

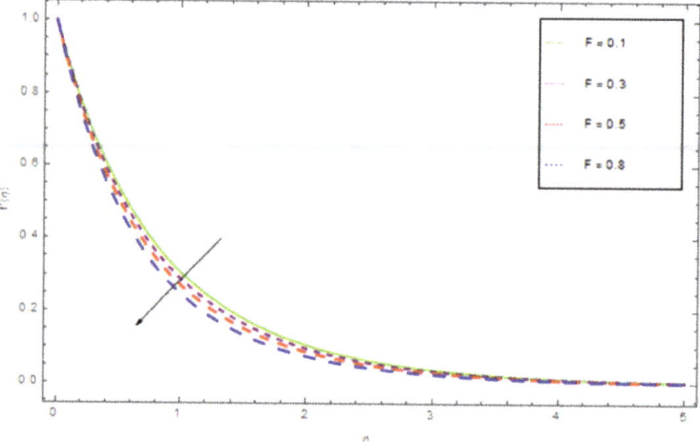

Figure 4. The influence of F on $f'(\eta)$ when $M = 0.5$, $k_1 = 0.4$, $\lambda = 0.2$, $\gamma = 0.6$.

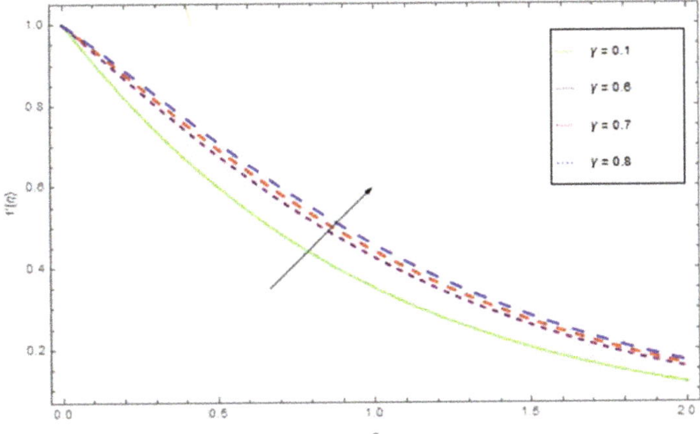

Figure 5. The effect of γ on $f'(\eta)$ when $M = 0.6$, $k_1 = 0.5$, $\lambda = 0.3$, $\gamma = 0.4$.

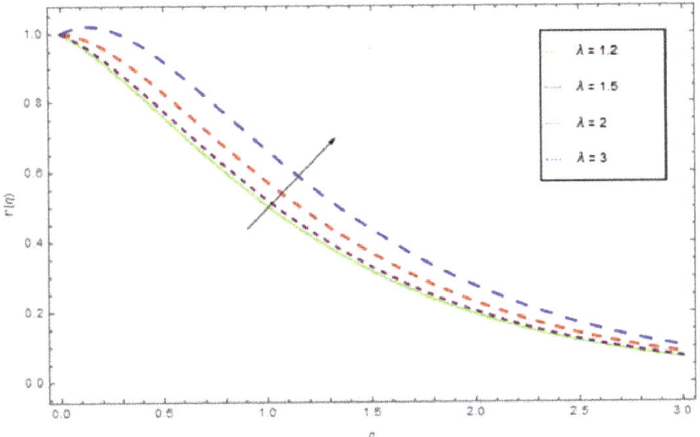

Figure 6. The variation of $f'(\eta)$ with λ, when $M = 0.6$, $k_1 = 0.5$, $\lambda = 0.3$, $\gamma = 0.4$.

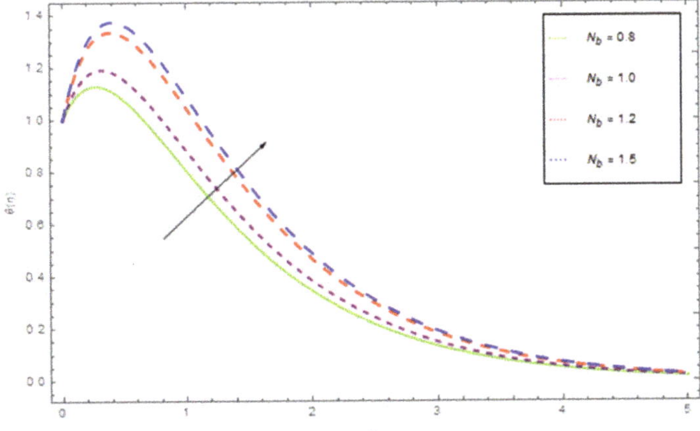

Figure 7. The influence of N_b on $\theta(\eta)$ when $\gamma = 0.8$, $\mathrm{Pr} = 4$, $N_t = 0.5$.

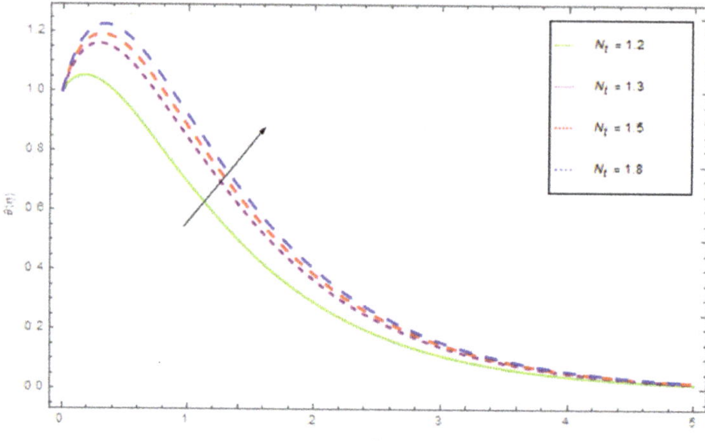

Figure 8. The effect of N_t on $\theta(\eta)$ when $N_b = 0.4$, $\mathrm{Pr} = 4$, $\gamma = 1.2$.

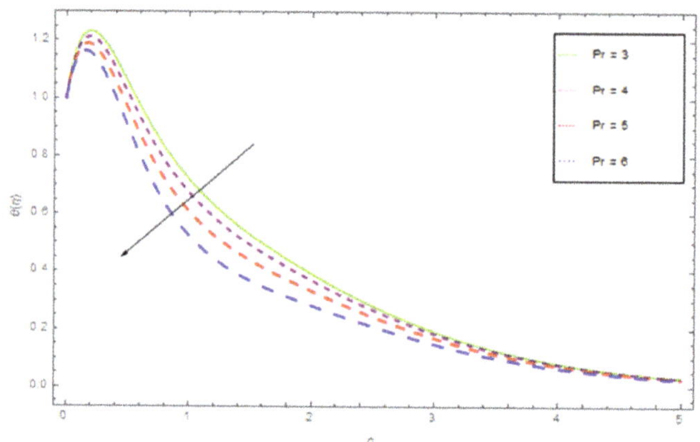

Figure 9. The influence of Pr on $\theta(\eta)$ while $\gamma = 0.4$, $N_b = 0.6$, $N_t = 5$.

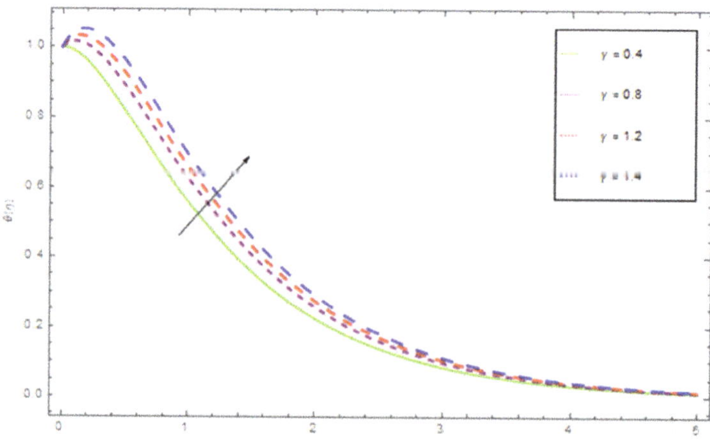

Figure 10. The influence of γ on $\theta(\eta)$ when $N_b = 0.7$, $N_t = 0.5$, $\mathrm{Pr} = 4$.

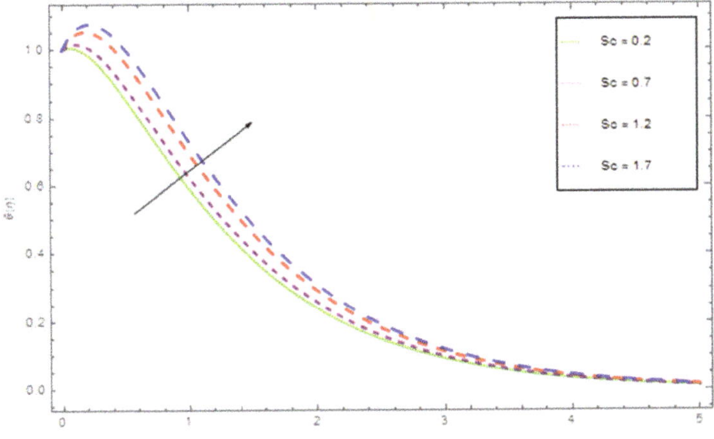

Figure 11. The influence of Sc on $\phi(\eta)$, while $N_t = 0.4$, $\gamma = 0.8$, $N_b = 0.7$.

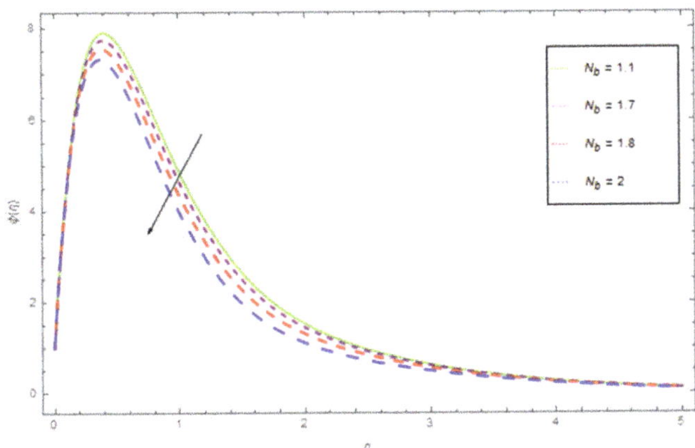

Figure 12. The influence of N_b on $\phi(\eta)$ when $\gamma = 0.4$, $N_t = 0.5$, $Sc = 0.6$.

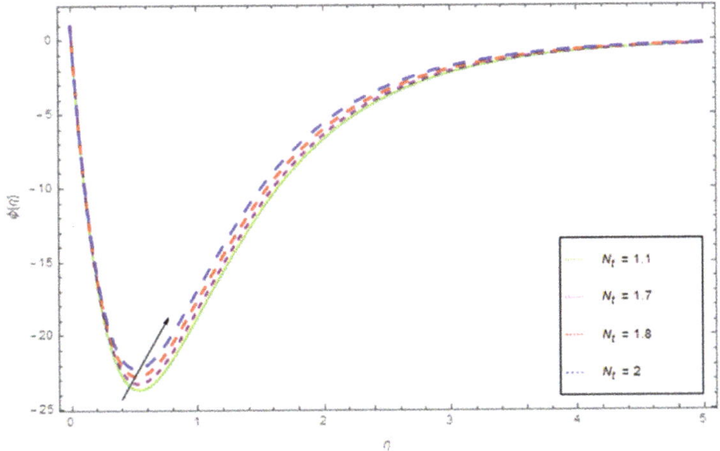

Figure 13. The influence of N_t on $\phi(\eta)$ when $\gamma = 0.4$, $N_t = 0.5$, $Sc = 0.6$.

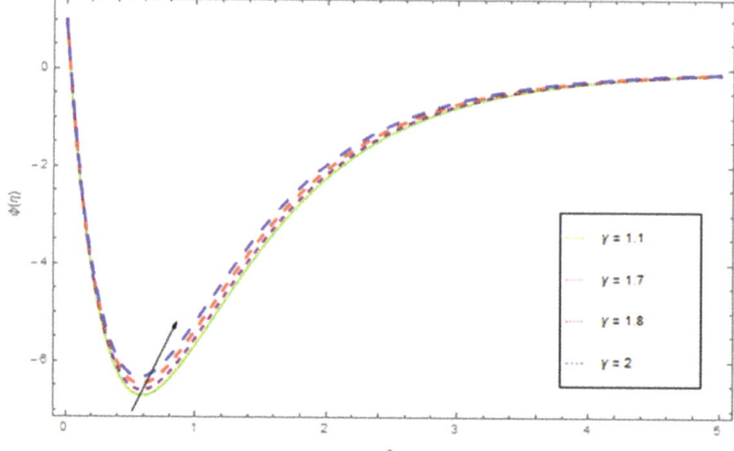

Figure 14. The influence of γ on $\phi(\eta)$ when $Sc = 0.6$, $N_t = 0.5$, $N_b = 0.4$.

3.1. Velocity

Figures 2–6 display the variation of $f'(\eta)$ with respect to the variations in the magnetic parameter M permeability parameter k_1, inertial parameter F, curvature parameter γ, and convection parameter λ. From Figure 2 it is observed that $f'(\eta)$ decreases almost exponentially with η at a given value of the M. The velocity profiler $f'(\eta)$ decreases with the higher values of M. This is because of the Lorentz force, which resists force that acts in the direction opposite to the flow direction. This opposes force slow down the fluid motion. Figure 3 represents the impact of the permeability parameter k_1 over the velocity profile $f'(\eta)$. The higher value of permeability parameter k_1 decreases $f'(\eta)$. Actually, an increasing permeability produces resistance in the flow path and that resistance reduces the fluid flow motion. Figure 4 show the influence of F on velocity profile. Figure 4 illustrates that on rising F, inner nanofluid fluid velocity is diminished, while there is no impact of F on thickness of the fluid. There is hardly an influence of F on free surface velocity, which is obvious from Figure 2. In state of porous gap with larger pores sizes, and porous medium expanded by fluid-solid interaction, which increases the viscous interference. Hence, an increase in F causes a better flow resistance, so the velocity of fluid is reduced. We describe the effect of the curvature parameter γ on the velocity profile $f'(\eta)$ in Figure 5. From Figure 5, we demined that a higher value of curvature parameter γ augmented the fluid flow motion. The influence of convection parameter λ on velocity profile $f'(\eta)$ is illustrated in Figure 7. The increasing behavior for the motion of nanofluid is found with augmentation of convection parameter λ, because of the buoyancy impact.

3.2. Temperature

The variations of N_b (Brownian motion parameter), N_t (thermophoresis parameter), Pr (Prandtl number), and γ (curvature parameter) on temperature profile $\theta(\eta)$ are respectively shown in Figures 7–10. We have shown the effect that is produced by the variation of Brownian motion N_b on $\theta(\eta)$ in Figure 7. Brownian motion is the arbitrary movement of small colloidal particles suspended in a fluid, brought about by the collision of the fluid atoms with the particles. An expansion in the Brownian motion impact yields noteworthy movement of nanoparticles, which offers ascend to the fluid kinetic energy and henceforth temperature increments. It is seen that, at fixed N_b, initially the profile drops to the lowest value at around $\eta = 0.4$, and then rises and reaches to maximum at the largest value of η. By changing N_b to larger values, there is an upshift in the temperature function from about $\eta = 0.4$ to $\eta = 3.0$. This shows that there is an enhancement in the average kinetic energy of the hybrid nanofluid with the higher values of N_b. The influence of the thermophoresis parameter (N_t) on $\theta(\eta)$ is displayed in Figure 8. It is initiated that rise in N_t leads to augmenting both the liquid temperature. The augmented value of N_t Shows stouter thermophoretic force due to temperature gradient, which transfers the nanoparticles from the warm surface to the quiescent fluid. Thermophoresis force is generated by temperature gradient, which fashions a degenerate flow away from the surface. Figure 9 depicts the effects that are produced by the variation of the Prandtl number (Pr) over the temperature function $\theta(\eta)$. Figure 9 is portraying the declining behavior of temperature curves for growing values of Prantl number (Pr), since the thickness of the thermal boundary layer reduced by enhancing (Pr). Figure 10 illustrates the variation of $\theta(\eta)$ with respect to the changing values of the curvature parameter γ. It is apparent that $\theta(\eta)$ increases with increasing values of γ.

3.3. Concentration

The impacts of Sc, N_b, N_t and γ on $\phi(\eta)$ are presented in Figures 11–14. Figure 11 shows the profile of $\phi(\eta)$ for dissimilar values of Sc. We observe that $\phi(\eta)$ increases with the higher values of Sc. The impact of N_b and N_t on the concentration profile is shown in Figures 12 and 13. Figure 12 shows that the concentration profile reduced as the values of N_b reduces. Enhancing thermophoresis process boosts the traveling of particles from higher to lower temperature deference, which results in maximizing the concentration of nanoparticles in the flow regime. Figure 13 show the influence of N_t on the concentration profile. The higher values of N_t corresponds increasing trends in concentration profile and related momentum boundary layer. Figure 14 shows that $\phi(\eta)$ increases with the higher values of γ in quite the same manner.

3.4. Table Discussion

In Table 1, numerical calculation of surface drag force at different value of γ and λ with variation of magnetic parameter M in range of 0 to 1 are shown. It is observed that the surface drag force is augmented in for all value of M when γ is increases. The surface drag force reduces all values of M when λ is segmented. In Table 2, numerical calculation of heat flux at different value of Nb, Nt and Pr with variation of magnetic parameter γ in range of 0 to 0.2 are shown. Increasing Nb, Nt, and Pr heat flux is decreases with variation of γ. The present results are compared with previous published results for validation. Both results are with good agreement. Table 3 shows the physical properties of some nanoparticles.

Table 1. Computed numerical values of skin friction coefficient C_f for numerous values of k, M, and c, when λ, M and γ, $\varphi_{np1} = 2\%$, $\varphi_{np2} = 3\%$.

γ	λ	C_f (at $M=0$) Present Results	C_f (at $M=0$) Results [36]	C_f (at $M=0.5$) Present Results	C_f (at $M=0$) Results [36]	C_f (at $M=1.0$) Present Results	C_f (at $M=0$) Results [36]
0.0	0.5	0.8987	0.8976	1.0227	0.9557	1.1439	1.0110
0.2	-	0.9052	0.9756	1.0292	1.0346	1.15049	1.0908
0.3	-	0.9085	1.0157	1.0325	1.0748	1.1537	1.1311
0.4	-	0.9118	1.0562	1.0358	1.1152	1.1570	1.1716
0.2	0.0	1.0222	1.0296	1.1450	1.0897	1.2650	1.1468
	0.5	0.9052	0.9756	1.0292	1.0346	1.1504	1.0908
	0.8	0.8350	0.9441	0.9597	1.0023	1.0817	1.0580
	1.0	0.7862	0.9233	0.9134	0.9811	1.0359	1.0364

Table 2. Numerical computed values of Nusselt number for numerous values of Pr, Nb, Nt and γ, when $M = 0.5$, $\varphi_{np1} = 2\%$, $\varphi_{np2} = 3\%$.

Pr	Nb	Nt	Nu (at $\gamma = 0$)	Nu (at $\gamma = 0$) Results [36]	Nu (at $\gamma = 0.1$)	Nu (at $\gamma = 0.1$) Results [36]	Nu (at $\gamma = 0.2$)	Nu (at $\gamma = 0.2$) Results [36]
0.7	0.5	0.5	0.1630	0.1653	0.1636	0.0897	0.1642	0.0198
0. 9	-	-	0.1582	0.2296	0.1584	0.1609	0.1586	0.0964
1.2	-	-	0.1512	0.2703	0.1508	0.2060	0.1505	0.1448
1.5	-	-	0.1444	0.3287	0.1435	0.2704	0.4227	0.2139
0.7	0.1	0.5	0.1751	0.1312	0.1764	0.0592	0.1776	0.0225
	0.5	-	0. 1630	0.2086	0.1636	0.1377	0.1642	0.1036
	0.6	-	0.1600	0.2387	0.1604	0.1692	0.1609	0.1372
	0.7	-	0.1570	0.2544	0.1573	0.1861	0.1577	0.1555
0.7	0.5	0.0	0.1769	1.9523	0.1782	1.9867	0.1795	2.0244
	-	0.5	0. 1630	1.5141	0.1636	1.5219	0.1642	1.5334
	-	0.6	0.1603	1.0774	0.1608	1.0588	0.1613	1.0442
	-	0.7	0.15767	0.6423	0.15804	0.5974	0.15844	0.5569

Table 3. Physical properties of water H_2O, alumina (Al_2O_3), and Copper (Cu).

Water, Alumina and Copper Particles	Physical Property				
	$\beta \times 10^{-5}/\text{K}^{-1}$	$cp/\text{Jkg}^{-1}\text{K}^{-1}$	ρ/kgm^{-3}	$k/\text{wm}^{-1}\text{K}^{-1}$	$\mu \times 10^{-3}/\text{kgm}^{-1}\text{s}^{-1}$
Alumina (Al_2O_3)	0.85	765	3970	40	-
Copper (Cu)	1.65	385	8933	400	-
Water (H_2O)	21	4179	997.1	0.613	0.894

4. Conclusions

We have investigated in this work the Darcy–Forchheimer hybrid nanofluid (Al_2O_3-Cu/H_2O) boundary layer hydromagnetic flow over a stretch cylinder. The model equations are converted to a set of coupled ODEs by employing similarity transformations and appropriate boundary conditions. The impact of variation of dissimilar physical parameters over velocity profile, concentration, and temperature profile are graphically studied. We conclude the following:

- The higher value of permeability parameter k_1 decreases $f'(\eta)$.
- Increase in F causes a better flow resistance, so velocity of fluid is reduced.
- The velocity profile decreases exponentially with the increasing of M, while it rises with the rising values of the permeability, inertial, curvature, and convection parameters.
- The augmented value of N_t shows stouter thermophoretic force due to the temperature gradient, which transfers the nanoparticles from the warm surface to the quiescent fluid. Thermophoresis force is generated by temperature gradient, which fashions a degenerate flow away from the surface.
- The temperature profile of the hybrid nanofluid increases almost in the same manner with the increasing values of N_b, Reynold number and convection parameter. The variation is more prominent for the intermediate values of the independent variable η.
- The temperature profile decreases with the higher values of the Prandtl number. The drop in $\theta(\eta)$ is more significant from $\eta = 0.2$ to $\eta = 3.6$. Furthermore, the temperature function drops at a maximum rate for the highest value of Pr, i.e., Pr $= 6.0$.
- The $\phi(\eta)$ increases with the higher values of Sc from $\eta = 0.3$ to $\eta = 2.6$, while it reduces with the rising values of Brownian motion parameter from $\eta = 0.4$ to $\eta = 3.2$.

Author Contributions: Conceptualization, methodology, software, validation, writing—original draft preparation, writing—review and editing, A.S., A.T., Z.S. and M.J.; conceptualization, methodology, software, visualization, writing—review and editing, W.D., A.K. and S.I.; writing—review and editing, visualization, project administration, funding acquisition, investigation, resources, A.T., W.D., A.S. and Z.S. All authors have read and agreed to the published version of the manuscript.

Funding: This work is funded by the Deanship of Scientific Research at Majmaah University under Project Number (RGP-2019-28).

Acknowledgments: The authors extend their appreciation to the Deanship of Scientific Research at Majmaah University for funding this work under Project Number (RGP-2019-28).

Conflicts of Interest: The authors declare no conflict of interest.

Nomenclature

u, v [m/s]	Components of Velocity
x, r [m]	Axial and radial Coordinates
β_0 [kg/s$^2 \cdot$ A]	Strength of Magnetic field
a [m]	Characteristic radius
T [K]	Temperature
f'	Dimensional velocity
θ	Dimensional temperature
Ec	Eckert number
q'''	Hear source/sink
α [m^2/s]	Thermal diffusivity
c_p [J/Kg $\cdot$ K]	Specific heat
ρ [kg/m^3]	Density
F	Local inertial parameter
U_w [m/s]	Stretching velocity
σ^* [$-$]	Stephan-Boltzmann constant
k^* [$-$]	Mean absorption coefficient
ϕ [nm]	Volume fraction of the nanoparticles
η	similarity variable
Pr [$-$]	Prandtl Number
C_f [$-$]	Skin friction coefficient
Nu [$-$]	Nusselt number
Re [$-$]	Local Reynolds number
$\widetilde{k}$	Coefficients of mean absorption
$\widetilde{Q}$	Radiative heat flux
c	Starching rate
k [W/mK]	Thermal conductivity
Sc	Schimid number
ΔT [$-$]	Temperature difference
v [m^2/s]	Kinematic viscosity
τ_w [kg/m $\cdot$ s^2]	Shear stress
γ	Curvature parameter
M [$-$]	Magnetic parameter
Rd [$-$]	Thermal radiation parameter
μ [kg/ms]	Dynamic viscosity
σ [W/mK]	Electrical conductivity
λ	Convection parameter
Subscripts	
bf	Base fluid
nf	Nanofluid
∞	Ambient
hnf	Hybrid nanofluid
W	Wall
P	particles

References

1. Davis, G.d.V. Natural convection of air in a square cavity, a bench mark numerical solution. *Int. J. Numer. Methods Fluids* **1983**, *3*, 249–264. [CrossRef]
2. Fusegi, T.; Hyun, J.M.; Kuwahara, K.; Farouk, B. A numerical study of three dimensional natural convection in a differentially heated cubical enclosure. *Int. J. Heat Mass Transf.* **1991**, *34*, 1543–1557. [CrossRef]
3. Barakos, G.; Mistoulis, E. Natural convection flow in a square cavity revisited: Laminar and turbulent models with wall functions. *Int. J. Numer. Method Heat Fluid Flow* **1994**, *18*, 695–719. [CrossRef]
4. Skok, H.; Ramadhyani, S.; Schoenhals, R.J. Natural convection in a side-facing open cavity. *Int. J. Heat Fluid Flow* **1991**, *12*, 36–45. [CrossRef]
5. Cha, S.S.; Choi, K.J. An interferometric investigation of open-cavity natural convection heat transfer. *Exp. Heat Transf.* **1989**, *2*, 27–40. [CrossRef]
6. Sheikholeslami, M.; Shah, Z.; Shafee, A.; Khan, I.; Tlili, I. Uniform magnetic force impact on water based nanofluid thermal behavior in a porous enclosure with ellipse shaped obstacle. *Sci. Rep.* **2019**, *9*, 1196. [CrossRef]
7. Chan, Y.L.; Tien, C.L. A numerical study of two-dimensional laminar natural convection in shallow open cavities. *Int. J. Heat Mass Transf.* **1985**, *28*, 603–612. [CrossRef]
8. Choi, S.U.S.; Eastman, J.A. Enhancing thermal conductivity of fluids with nanoparticles. In Proceedings of the 1995 International Mechanical Engineering Congress and Exhibition, San Francisco, CA, USA, 12–17 November 1995.
9. Mahmoudi, A.H.; Shahi, M.; Shahedin, A.M.; Hemati, N. Numerical mod-eling of natural convection in an open enclosure with two vertical thin heat sources subjected to a nanofluid. *Int. Commun. Heat Mass Transf.* **2011**, *38*, 110–118. [CrossRef]
10. Sheremet, M.A.; Pop, A.S.I. Unsteady free convection in a porous open wavy cavity filled with a nanofluid using Buongiorno's mathematical model. *Int. Commun. Heat Mass Transf.* **2015**, *67*, 66–72. [CrossRef]
11. Tassaddiq, A.; Amin, I.; Shutaywi, M.; Shah, Z.; Ali, F.; Islam, S.; Ullah, A. Thin Film Flow of Couple Stress Magneto-Hydrodynamics Nanofluid with Convective Heat over an Inclined Exponentially Rotating Stretched Surface. *Coatings* **2020**, *10*, 338. [CrossRef]
12. Tayebi, T.; Chamkha, A.J. Free convection enhancement in an annulus between horizontal confocal elliptical cylinders using hybrid nanofluids. *Numer. Heat Transf. Part A* **2016**, *70*, 1141–1156. [CrossRef]
13. Abbas, T.; Bhatti, M.M.; Ayub, M. Aiding and opposing of mixed convection Casson nanofluid flow with chemical reactions through a porous Riga plate. *Proc. Inst. Mech. Eng. Part E J. Process Mech. Eng.* **2018**, *232*, 519–527. [CrossRef]
14. Shah, Z.; Kumam, P.; Deebani, W. Radiative MHD Casson Nanofluid Flow with Activation energy and chemical reaction over past nonlinearly stretching surface through Entropy generation. *Sci. Rep.* **2020**, *10*, 4402. [CrossRef] [PubMed]
15. Rashidi, S.; Akar, S.; Bovand, M.; Ellahi, R. Volume of fluid model to simulate the nanofluid flow and entropy generation in a single slope solar still. *Renew. Energy* **2018**, *115*, 400–410. [CrossRef]
16. Akbarzadeh, M.; Rashidi, S.; Karimi, N.; Ellahi, R. Convection of heat and thermodynamic irreversibilities in two-phase, turbulent nanofluid flows in solar heaters by corrugated absorber plates. *Adv. Powder Technol.* **2018**, *29*, 2243–2254. [CrossRef]
17. Shah, Z.; Alzahrani, E.O.; Alghamdi, W.; Ullah, M.Z. Influences of electrical MHD and Hall current on squeezing nanofluid flow inside rotating porous plates with viscous and joule dissipation effects. *J. Therm. Anal. Calorim.* **2020**, *140*, 1215–1227. [CrossRef]
18. Takabi, B.; Shokouhmand, H. Effects of Al_2O_3–Cu/water hybrid nanofluid on heat transfer and flow characteristics in turbulent regime. *Int. J. Mod. Phys. C* **2015**, *26*, 1550047. [CrossRef]
19. Shah, Z.; Babazadeh, H.; Kumam, P.; Shafee, A.; Thounthong, P. Numerical simulation of magnetohydrodynamic nanofluids under the influence of shape factor and thermal transport in a porous media using CVFEM. *Front. Phys.* **2019**, *7*, 164. [CrossRef]
20. Minea, A.A. Hybrid nanofluids based on Al_2O_3, TiO_2 and SiO_2: Numerical evaluation of different approaches. *Int. J. Heat Mass Transf.* **2017**, *104*, 852–860. [CrossRef]
21. Taylor, P.K.M.; Della, R.J.; Huxford, R.C.; Lin, W. Hybrid nanomaterials for biomedical applications. *Chem. Commun.* **2010**, *46*, 5832–5849. [CrossRef]

22. Farahani, M.M.; Movassagh, J.; Taghavi, F.; Eghbali, P.; Salimi, F. Magnetite–polyoxometalate hybrid nanomaterials: Synthesis and characterization. *Chem. Eng. J.* **2012**, *184*, 342–346. [CrossRef]

23. Mehryan, S.A.M.; Izadpanahi, E.; Ghalambaz, M.; Chamkha, A.J. Mixed convection flow caused by an oscillating cylinder in a square cavity filled with Cu-Al$_2$O$_3$/water hybrid nanofluid. *J. Therm. Anal. Calorim.* **2019**. [CrossRef]

24. Suresh, S.; Venkitaraj, K.P.; Selvakumar, P.; Chandrasekar, M. Synthesis of Al$_2$O$_3$–Cu/water hybrid nanofluids using two step method and its thermo physical properties. *Colloids Surf. A* **2011**, *388*, 41–48. [CrossRef]

25. Suresh, S.; Venkitaraj, K.P.; Selvakumar, P.; Chandrasekar, M. Effect of Al$_2$O$_3$–Cu/water hybrid nanofluid in heat transfer. *Exp. Fluid Sci.* **2012**, *38*, 54–60. [CrossRef]

26. Rokni, H.B.; Alsaad, D.M.; Valipour, P. Electro hydrodynamic nanofluid flow and heat transfer between two plates. *J. Mol. Liq.* **2016**, *216*, 583–589. [CrossRef]

27. Shehzad, S.A.; Hayat, A.; Alsaedi, A. MHD flow of Jeffrey nanofluid with convective boundary conditions. *Braz. Soc. Mech. Sci. Eng.* **2014**, *3*, 873–883. [CrossRef]

28. Mahmoodi, M.; Kandelousi, S.H. Kerosene∓alumina nanofluid flow and heat transfer for cooling application. *J. Cent. South Univ.* **2016**, *23*, 983–990. [CrossRef]

29. Fakour, M.; Ganji, D.D.; Abbasi, M. Scrutiny of underdeveloped nanofluid MHD flow and heat conduction in a channel with porous walls. *Int. J. Case Stud. Therm. Eng.* **2014**, *4*, 202–214. [CrossRef]

30. Fakour, M.; Rahbari, A.; Khodabandeh, E. Nanofluid thin film flow and heat transfer over an unsteady stretching elastic sheet by LMS. *J. Mech. Sci. Technol.* **2018**, *32*, 177–183. [CrossRef]

31. Hatami, M.; Sheikholeslami, M.; Ganji, D.D. Laminar flow and heat transfer of nanofluid between contracting and rotating disks by least square method. *Power Technol.* **2014**, *253*, 769–779. [CrossRef]

32. Nadeem, S.; Mehmood, R.; Akbar, N.S. Nonorthogonal stagnation point flow of a nano non-Newtonian fluid towards a stretching surface with heat transfer. *Int. J. Heat Mass Transf.* **2013**, *57*, 679–689. [CrossRef]

33. Sheikholeslami, M.; Hatami, M.; Ganji, D.D. Analytical investigation of MHD nanofluid flow in a semi-porous channel. *Powder Technol.* **2013**, *246*, 327–336. [CrossRef]

34. Akbar, N.S.; Tripathi, D.; Khan, Z.H.; Beg, O.A. A numerical study of magneto hydrodynamic transport of nanofluids over a vertical stretching sheet with exponential temperature-dependent viscosity and buoyancy effects. *Chem. Phys. Lett.* **2016**, *661*, 20–30. [CrossRef]

35. Fakour, M.; Vahabzadeh, A.; Ganji, D.D. Scrutiny of mixed convection flow of a nanofluid in a vertical channel. *Int. J. Case Stud. Therm. Eng.* **2014**, *4*, 15–23. [CrossRef]

36. Maskeen, M.M.; Zeeshan, A.; Mehmood, O.U.; Hassan, M. Heat transfer enhancement in hydromagnetic alumina–copper/water hybrid nanofluid flow over a stretching cylinder. *J. Anal. Calorim.* **2019**, *138*, 1127–1136. [CrossRef]

37. Akilu, S.; Sharma, K.V.; Baheta, A.T.; Mamat, R. A review of thermophysical properties of water based composite nanofluids. *Renew. Sustain. Energy Rev.* **2016**, *66*, 654–678. [CrossRef]

38. Hayat, T.; Khan, M.I.; Waqas, M.; Alsaedi, A. Newtonian heating effect in nanofluid flow by a permeable cylinder. *Result Phys.* **2017**, *7*, 256–262. [CrossRef]

39. Jamaludin, A.; Naganthran, K.; Nazar, R.; Pop, I. Thermal radiation and MHD effects in the mixed convection flow of Fe$_3$O$_4$–water ferrofluid towards a nonlinearly moving surface. *Processes* **2020**, *8*, 95. [CrossRef]

40. Zaib, A.; Khan, U.; Khan, I.; Seikh, A.H.; Sherif, E.-S.M. Entropy generation and dual solutions in mixed convection stagnation point flow of micropolar Ti$_6$Al$_4$V nanoparticle along a riga surface. *Processes* **2020**, *8*, 14. [CrossRef]

41. Saeed, A.; Islam, S.; Dawar, A.; Shah, Z.; Kumam, P.; Khan, W. Influence of Cattaneo–Christov heat flux on MHD Jeffrey, Maxwell, and Oldroyd-B nanofluids with homogeneous-heterogeneous reaction. *Symmetry* **2019**, *11*, 439. [CrossRef]

42. Ahmad Farooq, A.; Shah, Z.; Alzahrani, E.O. Heat transfer analysis of a magneto-bio-fluid transport with variable thermal viscosity through a vertical ciliated channel. *Symmetry* **2019**, *11*, 1240. [CrossRef]

43. Jena, P.K.; Brocchi, E.A.; Motta, M.S. In-situ formation of Cu–Al$_2$O$_3$ nano-scale composites by chemical routes and studies on their microstructures. *Mater. Sci. Eng. A* **2001**, *313*, 180–186. [CrossRef]

44. Niihara, K. New design concept of structural ceramics–ceramic nanocomposites. *J. Ceram. Soc. Jpn.* **1991**, *99*, 974–982. [CrossRef]

45. Oh, S.-T.; Sekino, T.; Niihara, K. Effect of particle size distribution and mixing homogeneity on microstructure and strength of alumina/copper composites. *Nanostruct. Mater.* **1998**, *10*, 327–332. [CrossRef]

46. Oh, S.-T.; Sekino, T.; Niihara, K. Fabrication and mechanical properties of 5 vol% copper dispersed alumina nanocomposite. *J. Eur. Ceram. Soc.* **1998**, *18*, 31–37. [CrossRef]

47. Forchheimer, P. *Wasserbewegung Durch Boden*; Zeitschrift des Vereins Deutscher Ingenieure: Düsseldorf, Germany, 1901; Volume 45, pp. 1782–1788.

48. Muskat, M. *The Flow of Homogeneous Fluids through Porous Media*; JW Edwards: Edwards, MI, USA, 1946.

49. Pal, D.; Mondal, H. Hydromagnetic convective diffusion of species in Darcy-Forchheimer porous medium with non-uniform heat source/sink and variable viscosity. *Int. Commun. Heat Mass Transf.* **2012**, *39*, 913–917. [CrossRef]

50. Ganesh, N.V.; Hakeem, A.K.A.; Ganga, B. Darcy-Forchheimer flow of hydromagnetic nanofluid over a stretching/shrinking sheet in a thermally stratified porous medium with second order slip, viscous and Ohmic dissipations effects. *Ain Shams Eng. J.* **2016**, *9*, 939–951. [CrossRef]

51. Hayat, T.; Muhammad, T.; Al-Mezal, S.; Liao, S.J. Darcy-Forchheimer flow with variable thermal conductivity and Cattaneo-Christov heat flux. *Int. J. Numer. Methods Heat Fluid Flow* **2016**, *26*, 2355–2369. [CrossRef]

52. Muhammad, T.; Alsaedi, A.; Shahzad, S.A.; Hayat, T. A revised model for Darcy-Forchheimer flow of Maxwell nanofluid subject to convective boundary condition. *Chin. J. Phys.* **2017**, *55*, 963–976. [CrossRef]

53. Jawad, M.; Shah, Z.; Islam, S.; Bonyah, E.; Khan, A.Z. Darcy-Forchheimer Flow of MHD Nanofluid thin Film Flow with Joule dissipation and Navier's partial slip. *J. Phys. Commun.* **2018**, *2*, 115014. [CrossRef]

54. Uddin, I.; Akhtar, R.; Zhiyu, Z.; Islam, S.; Shoaib, M.; Raja, M.A.Z. Numerical treatment for Darcy-Forchheimer flow of Sisko nanomaterial with nonlinear thermal radiation by lobatto IIIA technique. *Math. Probl. Eng.* **2019**, *2019*. [CrossRef]

55. Mohamed, R.E.; Mahny, K.L.; Dar, A.; Muhammad, T. Numerical study for Carreau nanofluid flow over a convectively heated nonlinear stretching surface with chemically reactive species. *Physica A* **2020**, *540*, 123063.

56. Lahmar, S.; Kezzar, M.; Eid, M.R.; Sari, M.R. Heat transfer of squeezing unsteady nanofluid flow under the effects of an inclined magnetic field and variable thermal conductivity. *Physica A* **2020**, *540*, 123138. [CrossRef]

57. Eid, M.R.; Al-Hossainy, A.F.; Zoromba, M.S. FEM for blood-based SWCNTs flow through a circular cylinder in a porous medium with electromagnetic radiation. *Commun. Theor. Phys.* **2019**, *71*, 1425–1434. [CrossRef]

58. Al-Hossainy, A.F.; Eid, M.R.; Zoromba, M.S. SQLM for external yield stress effect on 3D MHD nanofluid flow in a porous medium. *Phys. Scr.* **2019**, *94*, 105208. [CrossRef]

59. Mohamed, R.E. Chemical reaction effect on MHD boundary-layer flow of two-phase nanofluid model over an exponentially stretching sheet with a heat generation. *J. Mol. Liq.* **2016**, *220*, 718–725.

60. Mohamed, R.E. Time-dependent flow of water-NPs over a stretching sheet in a saturated porous medium in the stagnation-point region in the presence of chemical reaction. *J. Nanofluids* **2017**, *6*, 550–557.

61. Sheikholeslami, M. New computational approach for exergy and entropy analysis of nanofluid under the impact of Lorentz force through a porous media. *Comput. Methods Appl. Mech. Eng.* **2019**, *344*, 319–333. [CrossRef]

62. Awais, M.; Shah, Z.; Parveen, N.; Ali, A.; Kumam, P.; Rehman, H.; Thounthong, P. MHD effects on ciliary-induced peristaltic flow coatings with rheological hybrid nanofluid. *Coatings* **2020**, *10*, 186. [CrossRef]

63. Alsagri, A.S.; Nasir, S.; Gul, T.; Islam, S.; Nisar, K.; Shah, Z.; Khan, I. MHD thin film flow and thermal analysis of blood with CNTs nanofluid. *Coatings* **2019**, *9*, 175. [CrossRef]

Article

Cattaneo-Christov Heat Flux Model for Second Grade Nanofluid Flow with Hall Effect through Entropy Generation over Stretchable Rotating Disk

Muhammad Wakeel Ahmad [1], Luthais B. McCash [2,*], Zahir Shah [3,*] and Rashid Nawaz [1]

[1]   Department of Mathematics, Abdul Wali Khan University, Mardan 23200, KPK, Pakistan;
      wakeel.maths@yahoo.com (M.W.A.); rashid_uop@yahoo.com (R.N.)
[2]   Department of Mathematics, College of Science and Engineering, University of Leicester, University Road,
      Leicester LE1 7RH, UK
[3]   Center of Excellence in Theoretical and Computational Science (TaCS-CoE), SCL 802 Fixed Point Laboratory,
      Science Laboratory Building, King Mongkut's University of Technology Thonburi (KMUTT),
      126 Pracha-Uthit Road, Bang Mod, Thrung Khru, Bangkok 10140, Thailand
*   Correspondence: lm460@leicester.ac.uk (L.B.M.); zahir.sha@kmutt.ac.th (Z.S.)

Received: 1 June 2020; Accepted: 22 June 2020; Published: 28 June 2020

Abstract: The second grade nanofluid flow with Cattaneo-Christov heat flux model by a stretching disk is examined in this paper. The nanofluid flow is characterized with Hall current, Brownian motion and thermophoresis influences. Entropy optimization with nonlinear thermal radiation, Joule heating and heat absorption/generation is also presented. The convergence of an analytical approach (HAM) is shown. Variation in the nanofluid flow profiles (velocities, thermal, concentration, total entropy, Bejan number) via influential parameters and number are also presented. Radial velocity, axial velocity and total entropy are enhanced with the Weissenberg number. Axial velocity, tangential velocity and Bejan number are heightened with the Hall parameter. The total entropy profile is enhanced with the Brinkman number, diffusion parameter, magnetic parameter and temperature difference. The Bejan number profile is heightened with the diffusion parameter and temperature difference. Arithmetical values of physical quantities are illustrated in Tables.

Keywords: entropy; second grade nanofluid; Cattaneo-Christov heat flux model; nanofluid; nonlinear thermal radiation; Joule heating

---

1. Introduction

The enhancement of heat transfer utilizing nanofluids, specifically in solar collectors, has been gaining much attention among researchers. The necessity of heat transfer improvement by ordinary fluids, like ethylene glycol, water, kerosene oil, etc., cannot be achieved. The researchers have conducted many experiments in order to develop the thermal transfer rate. Erosion and blockage are the major disadvantages in the drop of higher pressure and heat transfer rates. To reduce such problems, nanofluids are introduced. The suspension of particles of size 1–100 nm in base fluids can improve the thermal conduction in nanofluids. Using nanoparticles suspension, the thermophysical properties of conventional fluids was first proposed by Choi [1]. The applications of nanofluids are energy storage, heat exchangers, chemical industry, refrigeration process, power production, etc. Choi and Eastman [1] introduced the idea of augmenting fluids thermal conductivity. The radiation influence on nanofluid flow was discussed by Farooq et al. [2]. Sajjid et al. [3] investigated the magnetohydrodynamic (MHD) Fe_3O_4 nanofluid flow with radiation effect. The thermal and mass transmission in a nanofluid flow with chemical reaction and thermal radiation influences was presented by Sreedevi et al. [4]. The flow of silver and copper based nanofluid with radiation impact was determined by Qayyum et al. [5].

Furthermore, the same study with mixed convection and thermal radiation influences was extended by Hayat et al. [6]. The heat transfer analysis in nanofluids multi walled carbon nanotubes was discussed by Goodarzi et al. [7]. The enhancement of thermal transfer in MHD ferrofluid using different geometrical features was investigated by Goshayeshi et al. [8,9]. Other studies are cited in references [10–13]. Different materials have different properties in nature, and those materials are named viscoelastic material. Shampoo, care products, many oils and fuels, ketchup, food stuff are few examples of viscoelastic material. For describing these fluids, Jeffrey, Maxwell, Oldroyd-B, Burgers, generalized Burgers, Williamson, etc., are developed. The joule heating influence on MHD upper convected Maxwell fluid was presented by Zaidi and Mohyud-Din [14]. The MHD Maxwell fluid flow with chemical reaction was investigated by Afify and Elgazery [15]. The MHD flow of Oldroyd-B nanofluid with a heat flux model was analytically proposed by Mustafa [16]. The magnetic field impact on Williamson fluid flow in a channel was discussed by Hayat et al. [17]. The unsteady flow of Williamson fluid with radiation and heat source/sink influences was examined by Khan and Hamid [18].

Furthermore, in 1822, the mechanism of thermal transmission was suggested by Fourier's law [19]. This law leads us to the argument that the medium under consideration is immediately identifying the initial temperature. To resolve this problem, a thermal relaxation time to Fourier's law has been added by Cattaneo [20]. This term explains the required time of the medium to transmit heat to its bordering particles. Further, Christov [21] improved this model. The new model is named the heat flux model of Cattaneo-Christov (C-C). Using the C-C model, Hayat et al. [22] studied the fluid flow with homogeneous-heterogeneous reactions. The thermal transfer in an upper convected Maxwell fluid flow with C-C model was invested by Mustafa [23]. Tibullo et al. [24] presented the C-C model of heat flux, which is applicable to incompressible fluids. Han et al. [25] analyzed the couple flow of viscoelastic fluid with C-C model. Khan et al. [26] investigated the viscoelastic fluid flow over a stretching surface with a C-C model.

The consequence of Hall current on nanofluid flow has not been studied in the above mentioned literature. The modern tendency of research is in the direction of low density and strong magnetic field, due to its frequent applications like nuclear fusion, space flight, refrigeration coils, Hall accelerators, magnetohydrodynamic (MHD) generators, electric transformers, etc. The situation when the magnetic field is very strong with low density leads to conductivity reduction normal to the magnetic field. This refers to an induced current, which is called the Hall current. Under different flow configurations, numerous research works have been found. Raptis and Ram [27] examined the electrically conducting rotatory fluid flow with the Hall current. Unsteady hydrodynamic flow over a porous plate with the Hall current was inspected by Das et al. [28]. The unsteady MHD Couette flow through a rotating system with Hall and ion-slip currents combined influences was probed by Jha and Apere [29]. Aurangzeeb et al. [30] investigated the mixed convection flow with chemical reaction, heat generation and the Hall current. The convective heat transmission flow with Hall and ion-slip currents with slip conditions was determined by Ferdows et al. [31]. The numerical investigation of MHD viscous flow with Hall influence was presented by Beg et al. [32]. The MHD viscoelastic flow of fluid with hall current and convective conditions was analyzed by Kumar and Chand [33].

Entropy is the irreversibility process in a system. The heat transmission is associated with the least possible change of entropy in thermodynamics processes. Entropy generation minimization (EGM) is developed to improve the machines' ability. Spin moment, kinetic energy, vibration and internal-molecular friction are some applications of EGM. Such types of energy loss cannot be recovered deprived of additional work. That is why the measure of irreversibility process through mass and heat transfers is called entropy. The EGM process is used by investigators in many systems, like gas turbines, cooling by evaporation, natural convection, fuel cells, etc. Li and Faghri [34] investigated the EGM on high concentration direct methanol fuel cell. Hayat et al. [35] observed the EGM for peristaltic flow in a rotating frame. Nouri et al. [36] analyzed the EGM in a nanofluid flow inside a channel with a heat source. Dalir et al. [37] presented the EGM in MHD Jeffrey nanofluid over a stretching sheet.

Khan et al. [38] investigated the EGM in MHD flow of nanomaterial with binary chemical reaction and Arrhenius activation energy. The EGM in a nanomaterial mixed convective flow with slip condition was presented by Khan et al. [39]. Hayat et al. [40] presented the EGM in a second grade fluid with thermal radiation influence.

Previous studies of fluid flow over stretchable rotating disk utilizing nanofluids have not been able to consider the impacts of Cattaneo-Christov heat flux, Brownian motion and thermophoresis distribution. This paper reports on a study which considers the implications of Hall effect, Brownian motion and thermophoresis distribution on the second grade nanofluid flow with the Cattaneo-Christov heat flux model, with entropy optimization over a stretching disk. The aim of this paper is to explore a relationship between nanofluid heat transfer rate, entropy, Bejan number, Brownian motion, thermophoresis distribution and Hall effect.

2. Problem Modeling

The second grade nanofluid flow by stretchable rotating disk is assumed. The heat model of C-C is also taken in the nanofluid flow. The Hall current influence is considered in this nanofluid flow. Furthermore, EGM is considered with heat generation/absorption, Joule heating and non-linear thermal radiation. At $z = 0$ the disk rotates with angular velocity α_1. The ambient and disk temperatures are T_∞ and T_w respectively. Similarly, the ambient and surface concentrations are C_∞ and C_w. Geometry of the fluid is displayed in Figure 1.

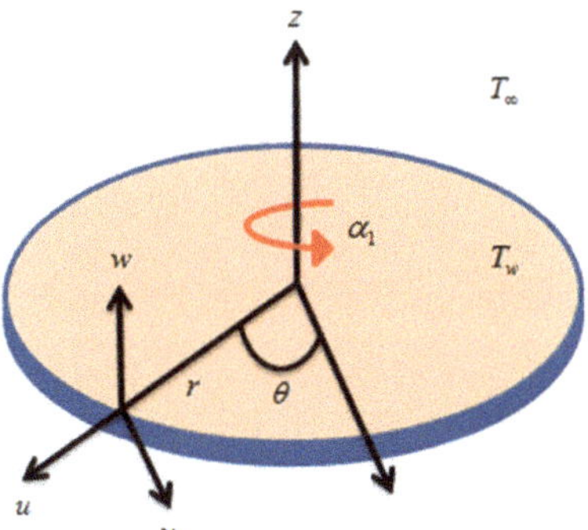

Figure 1. Fluid flow geometry.

The continuity, momentum, energy and concentration equations are taken as [40]:

$$\frac{\partial u}{\partial r} + \frac{u}{r} + \frac{\partial w}{\partial z} = 0 \tag{1}$$

$$u\frac{\partial u}{\partial r} - \frac{v^2}{r} + w\frac{\partial u}{\partial z} = v_f\frac{\partial^2 u}{\partial z^2} + \frac{\beta_1}{\rho_f}\left[u\frac{\partial^3 u}{\partial r\partial z^2} - \frac{1}{r}\left(\frac{\partial u}{\partial z}\right)^2 + 2w\frac{\partial u}{\partial r}\frac{\partial^2 u}{\partial z^2}\frac{\partial^3 u}{\partial z^3} + \right.$$
$$\left.\frac{\partial v}{\partial r}\frac{\partial^2 v}{\partial z^2} + \frac{\partial w}{\partial z}\frac{\partial^2 u}{\partial z^2} + \frac{\partial v}{\partial z}\frac{\partial^2 v}{\partial r\partial z} + 3\frac{\partial u}{\partial r}\frac{\partial^2 u}{\partial r\partial z} - \frac{\partial^2 v}{\partial z^2}\frac{v}{r}\right] - \frac{\sigma_f B_0^2}{\rho_f(1+m^2)}(u + mv) \tag{2}$$

$$u\frac{\partial v}{\partial r} + \frac{uv}{r} + w\frac{\partial v}{\partial z} = v_f\frac{\partial^2 v}{\partial z^2} + \frac{\beta_1}{\rho_f}\left[\frac{u}{r}\frac{\partial^2 u}{\partial z^2} - 2u\frac{\partial v}{\partial z}\frac{\partial^2 u}{\partial r\partial z}\frac{\partial^3 v}{\partial r\partial z^2}\right.$$
$$\left. + w\frac{\partial^3 v}{\partial z^3} - \frac{1}{r}\frac{\partial v}{\partial z}\frac{\partial u}{\partial z}\right] + \frac{\sigma_f B_0^2}{\rho_f(1+m^2)}(mu - v) \tag{3}$$

$$\left(\rho c_p\right)_f\left(u\frac{\partial T}{\partial r} + w\frac{\partial T}{\partial z}\right) + \gamma\left[u^2\frac{\partial^2 T}{\partial r^2} + w^2\frac{\partial^2 T}{\partial z^2} + 2uw\frac{\partial^2 T}{\partial r\partial z} + \frac{\partial T}{\partial r}\left(u\frac{\partial u}{\partial r} + w\frac{\partial u}{\partial z}\right)\right.$$

$$\left. + \frac{\partial T}{\partial z}\left(u\frac{\partial w}{\partial r} + w\frac{\partial w}{\partial z}\right)\right] = k_f\frac{\partial^2 T}{\partial z^2} + Q(T - T_\infty) + \frac{\sigma_f B_0^2}{(1+m^2)}\left(u^2 + v^2\right) - \tag{4}$$

$$\frac{16\sigma^*}{3k^*}\left[T^3\frac{\partial^2 T}{\partial z^2} + 3T^2\left(\frac{\partial T}{\partial z}\right)^2\right] + \left(\rho c_p\right)_s\left[\frac{D_T}{T_\infty}\left(\frac{\partial T}{\partial z}\right)^2 + D_B\left(\frac{\partial T}{\partial z}\frac{\partial C}{\partial z}\right)\right]$$

$$u\frac{\partial C}{\partial r} + w\frac{\partial C}{\partial z} = D_B\frac{\partial^2 C}{\partial z^2} + \frac{D_T}{T_\infty}\frac{\partial^2 T}{\partial z^2} \tag{5}$$

with

$$u = ra, \ v = ra_1, \ w = 0, \ T = T_w, \ C = C_w, \ \text{at } z = 0,$$
$$u = v = 0, \ T \to T_\infty, \ C \to C_\infty \text{ when } z \to \infty \tag{6}$$

where u, v and w are the components of velocity in r, θ and z directions, respectively, v_f is the kinematic viscosity, β_1 is the material parameter, k_f is the thermal conductivity, ϱ_f is the density, cp is the specific heat, Q is the heat absorption/generation, σ_f is the electrical conductivity, $(\varrho cp)_f$ is the heat capacitance, D_T is the thermophoretic diffusion coefficient and D_B is the Brownian diffusion coefficient.

Similarity transformations are defined as [40]

$$u = ra_1 f'(\zeta), \ v = ra_1 g(\zeta), \ w = -2ha_1 f(\zeta), \ \theta = \frac{T - T_\infty}{T_w - T_\infty}, \ \phi = \frac{C - C_\infty}{C_w - C_\infty}, \ \zeta = \frac{z}{h} \tag{7}$$

The dimensionless forms of the leading equations are

$$f''' + ReWe\left(2f''^2 + g'^2 - 2ff'''' + f'f'''\right) - Re\left(f'^2 - 2ff'' - g^2\right) - \frac{M}{(1+m^2)}(f' + mg) = 0 \tag{8}$$

$$g'' + ReWe(2f'g'' - 2fg''' - 3f''g') - Re(2f'g - 2fg') + \frac{M}{(1+m^2)}(mf' - g) = 0 \tag{9}$$

$$\theta'' + 2RePrf\theta' + RePrq\theta + NbPr\theta'\phi' + NtPr\theta'^2 - 4\lambda PrRe\left(f^2\theta'' + ff'\theta'\right)$$

$$+ Rd\left[3(\theta_w - 1)\left(\theta'^2 + \theta^2\theta'^2(\theta_w - 1)^2 + 2\theta\theta'^2(\theta_w - 1)\right) + \theta'' + \theta^3\theta''(\theta_w - 1)^3\right.$$ \tag{10}

$$\left. + 3(\theta_w - 1)\theta\theta'' + 3(\theta_w - 1)^2\theta^2\theta''\right] + \frac{MPrEc}{(1+m^2)}\left(f'^2 + g^2\right) = 0$$

$$\phi'' + 2ReScf\phi' + \frac{Nt}{Nb}\theta'' = 0 \tag{11}$$

with

$$f(0) = 0, \ f'(0) = A, \ f'(\infty) = 0, \ g(0) = 1,$$
$$g(\infty) = 0, \ \theta(0) = 1, \ \theta(\infty) = 0, \ \phi(0) = 1, \ \phi(\infty) = 0 \tag{12}$$

where $We = \beta_1/\varrho h^2$ is the Weissenberg number, $Re = \alpha_1 h^2/v_f$ is the Reynolds number, $A = a/\alpha_1$ is the stretching parameter, $Pr = (\varrho cp)v_f/k_f$ is the Prandtl number, $q = Q/\varrho cp\alpha_1$ is the heat absorption/generation parameter, $Nb = \tau D_B(C_w - C_\infty)/v$ Brownian motion parameter, $Nt = \tau D_B(T_w - T_\infty)/T_\infty v$ thermophoresis parameter, $M = \sigma B_0^2/\alpha_1\varrho$ magnetic parameter, $Sc = v/D_B$ Schmidt number, $\theta_w = T_w/T_\infty$ temperature difference, $Ec = (ra_1)^2/cp(T_w - T_\infty)$ is the Eckert number, $Rd = 16\sigma^*T_\infty^3/3k_f k^*$ radiation parameter and $\lambda = ra_1$ is the thermal relaxation parameter.

2.1. Skin Friction and Nusselt Number

Skin frictions along radial and tangential directions are

$$C_{fr} = \frac{\tau_{z\theta}}{\varrho(ra_1)^2}, \ C_{f\theta} = \frac{\tau_{zr}}{\varrho(ra_1)^2} \tag{13}$$

in which τ_{zr} and $\tau_{z\theta}$ are called shear stresses in radial and tangential directions respectively, and are defined as

$$
\begin{aligned}
\tau_{zr} = {}& \mu\left(\frac{\partial w}{\partial r} + \frac{\partial u}{\partial z}\right) + \beta_1\left[2\frac{\partial v}{\partial z}\left(\frac{\partial v}{\partial r} - \frac{v}{r}\right) + \left(\frac{\partial w}{\partial r} + \frac{\partial u}{\partial z}\right)\left(\frac{\partial u}{\partial r} + \frac{\partial w}{\partial z}\right) + \frac{\partial w}{\partial z}\frac{\partial u}{\partial z} + \right. \\
& \frac{\partial u}{\partial r}\frac{\partial w}{\partial r} + 3\left(\frac{\partial w}{\partial r}\frac{\partial w}{\partial z} + \frac{\partial u}{\partial z}\frac{\partial u}{\partial r}\right)\Big] - \beta_1\left[\frac{\partial v}{\partial z}\left(\frac{\partial v}{\partial r} - \frac{v}{r}\right) + \left(\frac{\partial u}{\partial r} + \frac{\partial w}{\partial z}\right)\left(\frac{\partial w}{\partial r} + \frac{\partial u}{\partial z}\right)\right]
\end{aligned}
\tag{14}
$$

$$
\begin{aligned}
\tau_{z\theta} = {}& \mu\frac{\partial v}{\partial z} + \beta_1\left[w\frac{\partial^2 v}{\partial z^2} - \frac{v}{r}\frac{\partial u}{\partial z} + \frac{\partial v}{\partial z}\frac{\partial w}{\partial z} + u\frac{\partial^2 v}{\partial r\partial z} + \frac{\partial u}{\partial z}\frac{\partial v}{\partial r} + 3\frac{u}{r}\frac{\partial v}{\partial z}\right] \\
& -\beta_1\left[\frac{\partial u}{\partial z}\frac{\partial v}{\partial r} + 2\frac{u}{r}\frac{\partial v}{\partial z} + 2\frac{\partial v}{\partial z}\frac{\partial w}{\partial z} - \frac{v}{r}\frac{\partial w}{\partial r} - \frac{v}{r}\frac{\partial u}{\partial z} + \frac{\partial w}{\partial r}\frac{\partial v}{\partial r}\right]
\end{aligned}
\tag{15}
$$

Dimensionless forms are

$$
Re_r C_{fr} = f''(\zeta) + ReWe[3f'(\zeta)f''(\zeta) - 2f(\zeta)f'''(\zeta)]\big|_{\zeta=0}
\tag{16}
$$

$$
Re_r C_{f\theta} = g'(\zeta) + ReWe[4f'(\zeta)g'(\zeta) - 2f(\zeta)g''(\zeta)]\big|_{\zeta=0}
\tag{17}
$$

Heat transfer rate is defined as

$$
Nu_x = \frac{hq_w}{k(T_w - T_\infty)}
\tag{18}
$$

where q_w is the wall heat flux, which is defined as

$$
q_w\big|_{z=0} = -\left(k + \frac{16\sigma^* T^3}{3k^*}\right)\frac{\partial T}{\partial z}\bigg|_{z=0}
\tag{19}
$$

The dimensionless form is

$$
Nu_x = -\left(1 + Rd\theta_w^3\right)\theta'(\zeta)\big|_{\zeta=0}
\tag{20}
$$

2.2. Entropy Generation

$$
\begin{aligned}
S_G = {}& \frac{k_f}{T_\infty^2}\left[1 + \frac{16\sigma^* T^3}{3k^* k_f}\right]\left(\frac{\partial T}{\partial z}\right)^2 + \frac{1}{T_\infty}\left[\mu_f\left(\left(\frac{\partial u}{\partial z}\right)^2 + \left(\frac{\partial v}{\partial z}\right)^2\right) + \beta_1\left[2\frac{\partial u}{\partial r}\left(\frac{\partial u}{\partial z}\right)^2\right.\right. \\
& -\left(\frac{\partial u}{\partial z}\frac{\partial v}{\partial z}\frac{v}{r} + \frac{\partial v}{\partial r}\frac{\partial u}{\partial z}\frac{\partial v}{\partial z}\right) + u\frac{\partial u}{\partial z}\frac{\partial^2 u}{\partial r\partial z} + w\frac{\partial v}{\partial z}\frac{\partial^2 v}{\partial z^2} + \frac{\partial w}{\partial z}\left(\frac{\partial u}{\partial z}\right)^2 - \frac{\partial v}{\partial r}\left(\frac{\partial u}{\partial z}\right)^2 \\
& \left.\left. + \frac{v}{r}\left(\frac{\partial u}{\partial z}\right)^2\right] + \frac{\sigma_f B_0^2}{T_\infty(1+m^2)}\left(u^2 + v^2\right) + \frac{R_D}{C_\infty}\left(\frac{\partial C}{\partial z}\right)^2 + \frac{R_D}{T_\infty}\left(\frac{\partial C}{\partial z}\frac{\partial T}{\partial z}\right)\right.
\end{aligned}
\tag{21}
$$

Equation (21) is reduced as

$$
\begin{aligned}
N_G = {}& \gamma_1\left[1 + Rd(1 + \theta(\theta_w - 1))^3\right]\theta'^2 + Br\left(f''^2 + g'^2\right) + L\theta'\phi' + \\
& WeBrRe\left(f'f''^2 - 2f''gg' - 2fg'g''\right) + \frac{BrM}{(1+m^2)}\left(f'^2 + g^2\right) + L\frac{\gamma_2}{\gamma_1}\phi'^2
\end{aligned}
\tag{22}
$$

The Bejan number is defined as

$$
Be = \frac{\gamma_1\left[1 + Rd(1 + \theta(\theta_w - 1))^3\right]\theta'^2 + L\theta'\phi' + L\frac{\gamma_2}{\gamma_1}\phi'^2}{\left[\begin{array}{l} \gamma_1\left[1 + Rd(1 + \theta(\theta_w - 1))^3\right]\theta'^2 + Br\left(f''^2 + g'^2\right) + L\theta'\phi' + \\[4pt] WeBrRe\left(\begin{array}{l} f'f''^2 - 2f''gg' \\ -2fg'g'' \end{array}\right) + \frac{BrM}{(1+m^2)}\left(f'^2 + g^2\right) + L\frac{\gamma_2}{\gamma_1}\phi'^2 \end{array}\right]}
\tag{23}
$$

where $N_G = T_\infty h^2 S_G/k_f(T_w - T_\infty)$ is the entropy generation, $Br = \mu_f(r\alpha_1)^2/k_f(T_w - T_\infty)$ is the Brinkman number, $L = R_D(C_w - C_\infty)/k_f$ is the diffusion parameter, $\gamma_1 = (T_w - T_\infty)/T_\infty$ and $\gamma_2 = (C_w - C_\infty)/C_\infty$ are the temperature and concentration ratios, respectively.

3. HAM Solution

Linear operators are

$$L_f[f] = f''' - f', \; L_g[g] = g'' - g, \; L_\theta[\theta] = \theta'' - \theta, \; L_\phi[\phi] = \phi'' - \phi \tag{24}$$

Initial guesses are taken as

$$f_0[\zeta] = A\left[1 - e^{-\zeta}\right], \; g_0[\zeta] = \theta_0[\zeta] = \phi_0[\zeta] = e^{-\zeta} \tag{25}$$

with

$$L_f[f]\left[p_1 + p_2 e^{-\zeta} + p_3 e^\zeta\right] = 0, \; L_g[g]\left[p_4 e^{-\zeta} + p_5 e^\zeta\right] = 0,$$
$$L_\theta[\theta]\left[p_6 e^{-\zeta} + p_7 e^\zeta\right] = 0, \; L_\phi[\phi]\left[p_8 e^{-\zeta} + p_9 e^\zeta\right] = 0 \tag{26}$$

where $P_1, P_2, P_3 \ldots, P_9$ are called constants.

Convergence Investigation

By using the auxiliary parameters $\hbar_f$, $\hbar_g$, $\hbar_\theta$ and $\hbar_\phi$, we analyzed the convergence regions for $f'(\zeta), g(\zeta)$ and $\theta(\zeta)$ of the modeled system of equations. At the 25th deformation order, the $\hbar-$ curves are presented here (see Figure 2). Convergence regions for $f''(0), g'(0), \theta'(0)$ and $\varphi'(0)$ are $-0.06 \leq \hbar_f \leq 0.01$, $-0.08 \leq \hbar_g \leq 0.03$, $-0.11 \leq \hbar_\theta \leq 0.05$ and $-0.11 \leq \hbar_\phi \leq 0.05$, respectively.

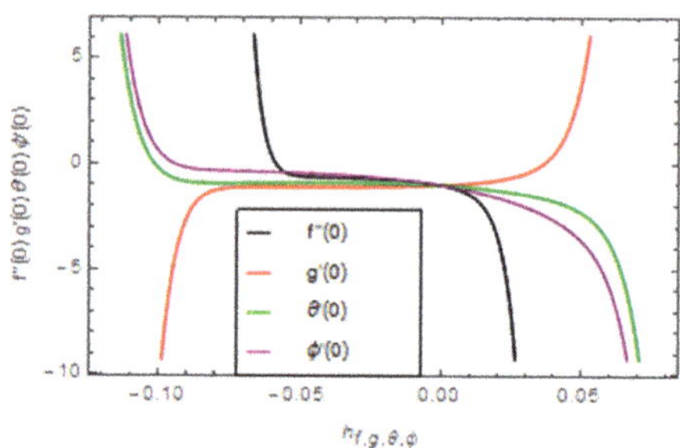

Figure 2. $\hbar-$curves for $f''(0), g'(0), \theta'(0)$ and $\varphi'(0)$.

4. Results and Discussion

Variation in the second grade nanofluid flow due to influential variables on $f'(\zeta), g(\zeta), \theta(\zeta), \varphi(\zeta),$ $N_G(\zeta)$ and $Be(\zeta)$ are illustrated in Figures 3–29. The parameters are taken fixed as Re = 0.9, We = 0.3, $Nb = Nt = 0.3$, $Ec = 0.4$, $q = 0.7$, $\theta_w = 0.2$, $Rd = 0.5$, $M = 0.5$, $m = 0.5$, $Sc = 1.0$ and $\lambda = 0.1$.

4.1. Velocities, Temperature and Concentration Fields

Figures 3–5 exemplify the variation in velocity profiles due to the Reynolds number. Greater Reynolds number reduces the velocity profiles in axial, radial and tangential directions. The Reynolds number is associated with the inertial forces of the fluid flow. The greater the Reynolds number, the stronger the inertial forces that reduce the motion of the fluid flow. Figures 6–8 depict the change in $f'(\zeta), g(\zeta)$ and $\theta(\zeta)$ for greater m. The last terms in the principle equations of the velocity function leads us to a smaller conductivity of the fluid flow via increasing m. So, due to less conductivity, the damping force in fluid flow produces which intensifies the velocity components $f'(\zeta)$ and $g(\zeta)$. These phenomena reduce the thermal field of the fluid flow. Figures 9 and 10 illustrate the change in velocity components via the Weissenberg number. Greater values of the Weissenberg number heighten the velocity components. Increasing the Weissenberg number reduces the fluid flow viscosity. As the viscosity of the fluid is reduced, the motion of the fluid particles increase. Thus, the velocity components escalate with a greater Weissenberg number. Figures 11–13 illustrate the change in $f'(\zeta)$, $g(\zeta)$ and $\theta(\zeta)$ via the magnetic field parameter. A higher magnetic parameter reduces the velocity components, while a reverse impact of the magnetic parameter on the thermal field is observed. The heightening magnetic field produces higher resistive force to the flow of fluid, which drops the motion of the fluid flow. Thus, the velocity components decline. On the other hand, the higher resistive force increases the electrons collision, which produces more heat to fluid flow. Therefore, the thermal field rises with the higher values of magnetic parameter. Figures 14 and 15 exhibit the change in velocity components $(f(\zeta), g(\zeta))$ via a stretching parameter. The momentum boundary layer escalates with higher values of stretching parameter and, consequently, the velocity profile $f(\zeta)$ heightens. On the other hand, the velocity profile $g(\zeta)$ declines with a higher stretching parameter. This influence is due to the fact that the higher values of stretching parameter reduce the angular velocity of the fluid flow. Figure 16 displays the change in thermal field via Brownian motion and thermophoresis parameters. The rising values of Brownian motion and thermophoresis parameters intensify the thermal field. The rising thermophoretic force pushes the fluid particles to move form heated to cold regions and, consequently, the temperature field increases. A similar impact is also depicted against the Brownian motion parameter. Figure 17 shows the change in thermal profile via thermal relaxation parameter. Higher values of relaxation parameter decline the temperature profile. With higher values of thermal relaxation parameter, the material particles need more potential to transmit energy to its surrounding particles. Additionally, this behavior is less for the C-C model as compared to Fourier's law. Figure 18 illustrates the change in thermal field via temperature difference parameter. An escalating conduct is detected in thermal field by heightening the temperature difference parameter. By increasing the temperature difference parameter, the temperature at the wall increases, and then the ambient temperature also increases. Consequently, the temperature field heightens. Figure 19 clarifies the change in thermal field via heat generation/absorption parameter. Clearly, the increasing heat parameter increases the temperature profile. The heat generation/absorption parameter acts like a heat generator. The increasing generation/absorption increases the thermal field of the fluid flow. Figure 20 indicates the change in thermal profile via the Prandtl number. A declining impact is detected via increasing Prandtl number. Figure 21 indicates the variation in concentration field via thermophoresis parameter. As the increasing thermophoresis parameter increases the thermal field (see Figure 16), consequently, the concentration of the fluid flow also increases. The opposite impact of Brownian motion parameter is depicted against the concentration profile (see Figure 22). The change in concentration field via the Schmidt number is displayed in Figure 23. A declining impact is observed here. The concentration distribution in inversely related with the Schmidt number. The intensifying estimations of the Schmidt number reduce the thickness of the boundary layer flow. The concentration distribution therefore declines.

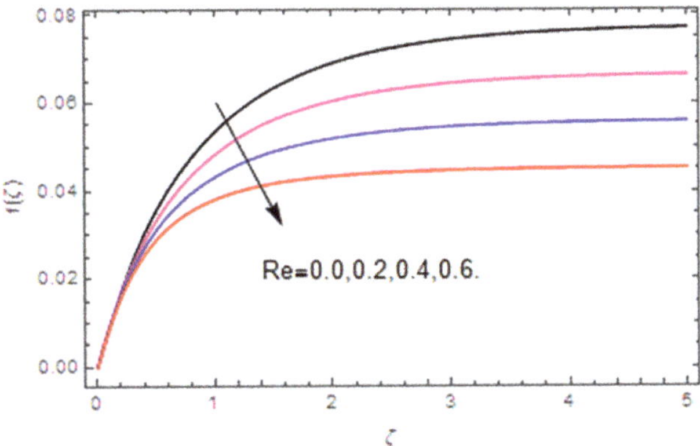

Figure 3. Re on $f(\zeta)$.

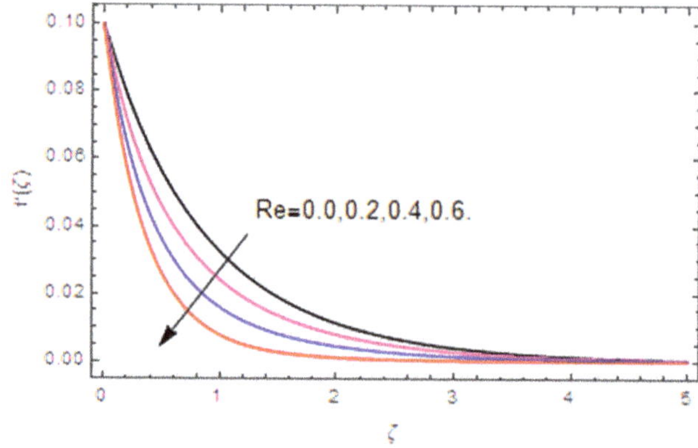

Figure 4. Re on $f'(\zeta)$.

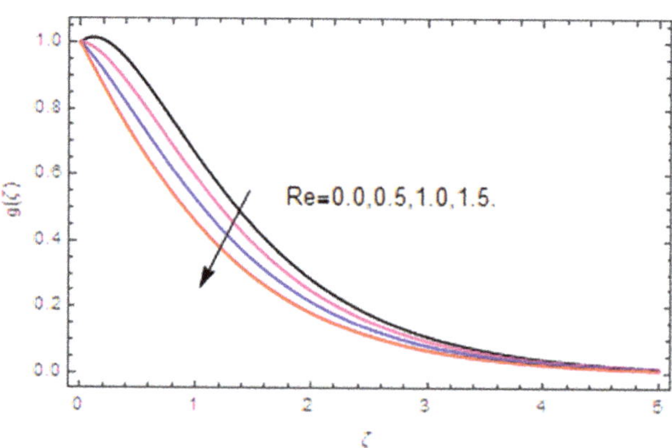

Figure 5. Re on $g(\zeta)$.

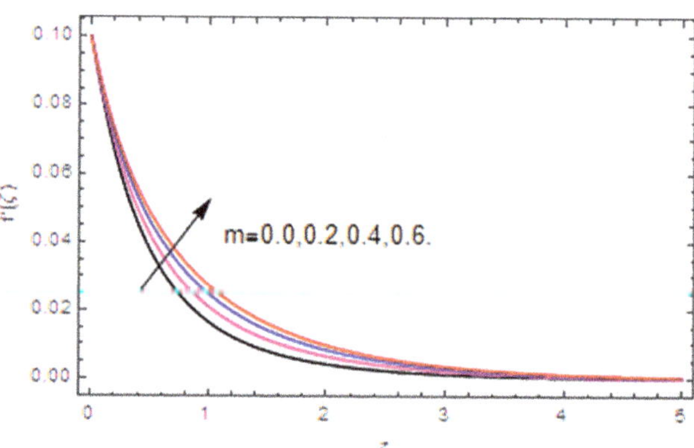

Figure 6. m on $f'(\zeta)$.

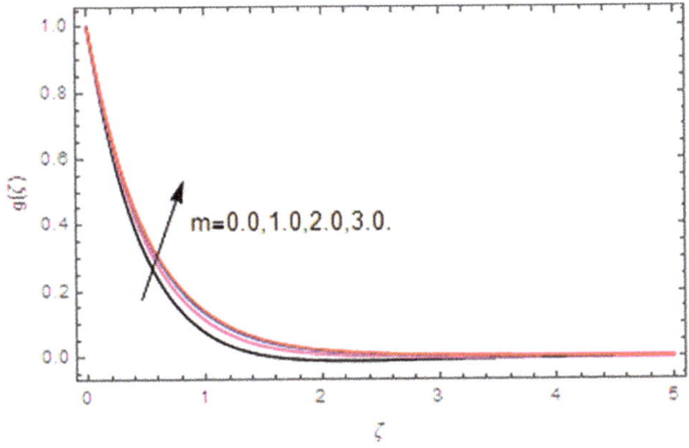

Figure 7. *m* on $g(\zeta)$.

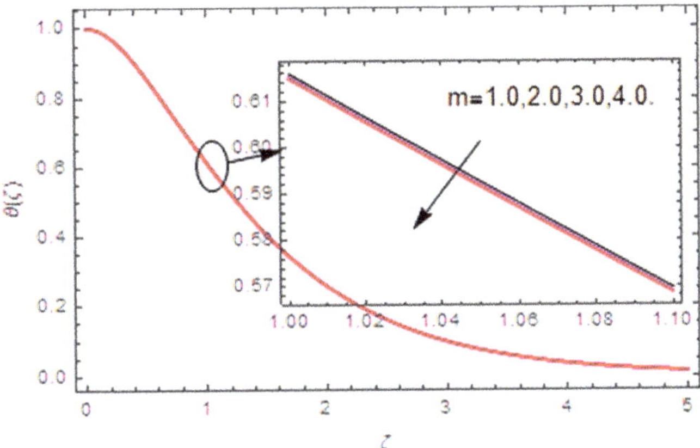

Figure 8. *m* on $\theta(\zeta)$.

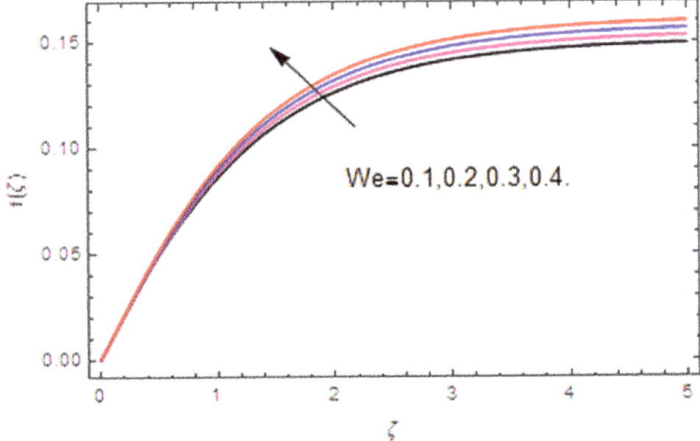

Figure 9. *We* on $f(\zeta)$.

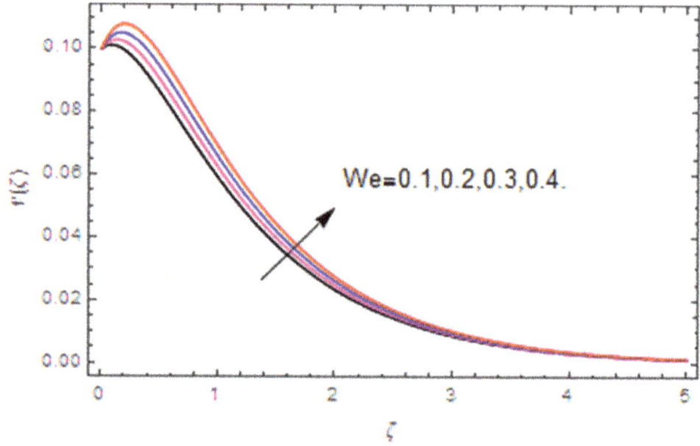

Figure 10. *We* on $f'(\zeta)$.

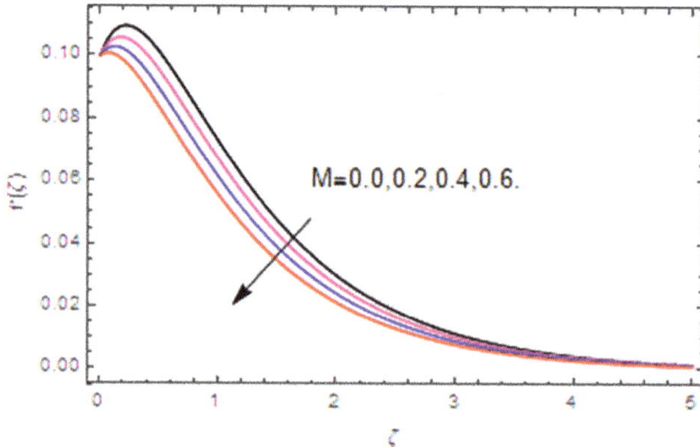

Figure 11. *M* on $f'(\zeta)$.

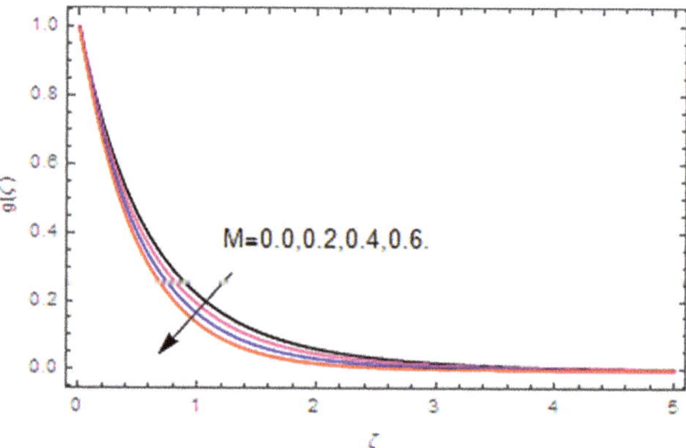

Figure 12. *M* on $g(\zeta)$.

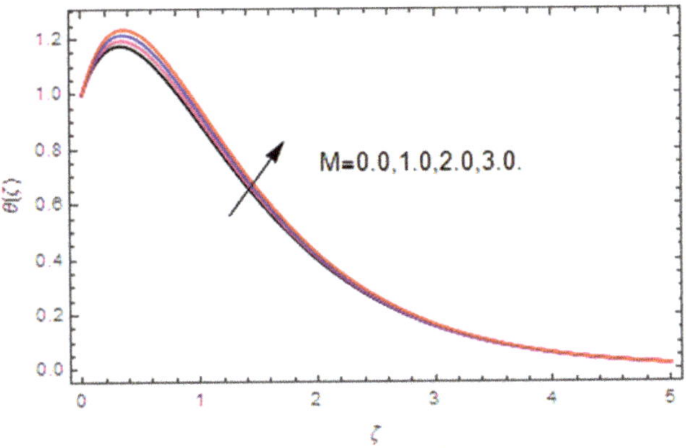

Figure 13. *M* on $\theta(\zeta)$.

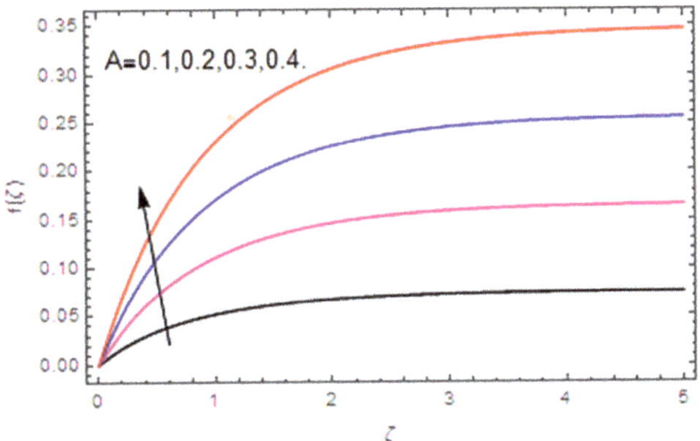

Figure 14. *A* on $f(\zeta)$.

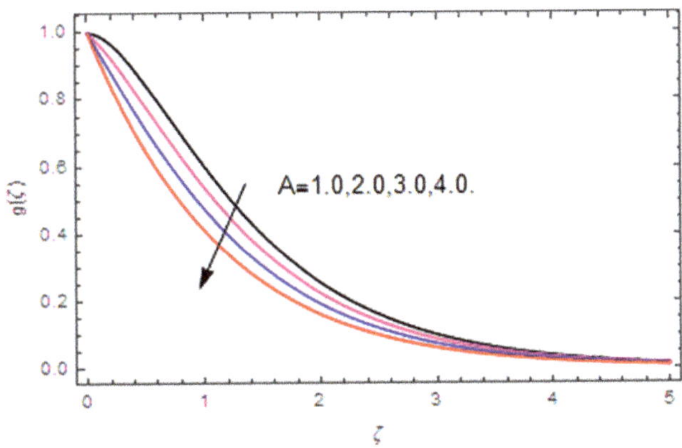

Figure 15. *A* on $g(\zeta)$.

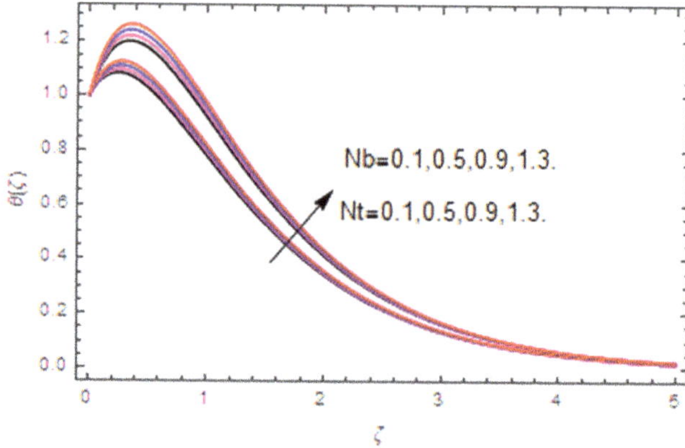

Figure 16. *Nb* and *Nt* on $\theta(\zeta)$.

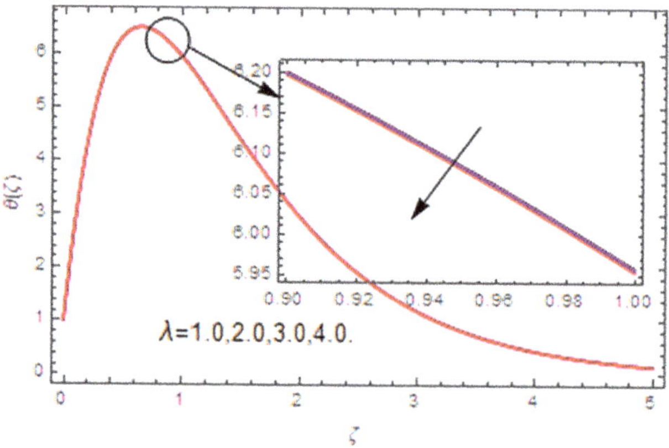

Figure 17. λ on $\theta(\zeta)$.

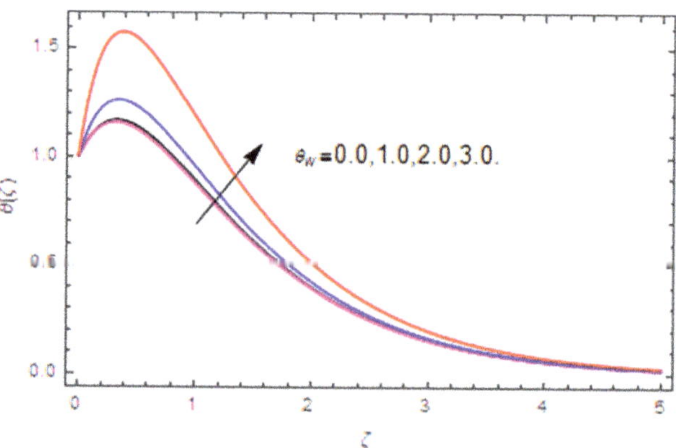

Figure 18. θ_w on $\theta(\zeta)$.

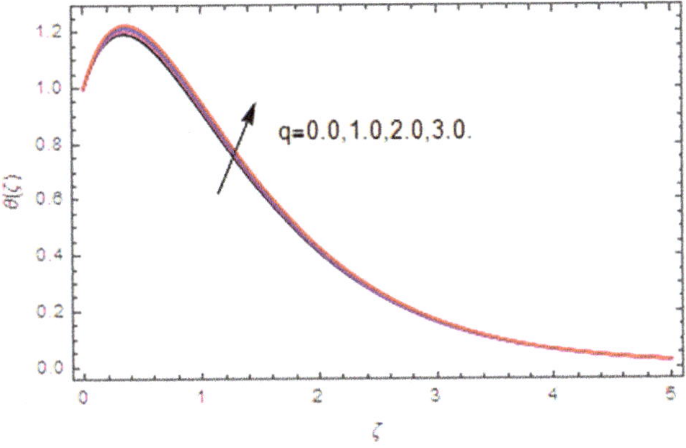

Figure 19. q on $\theta(\zeta)$.

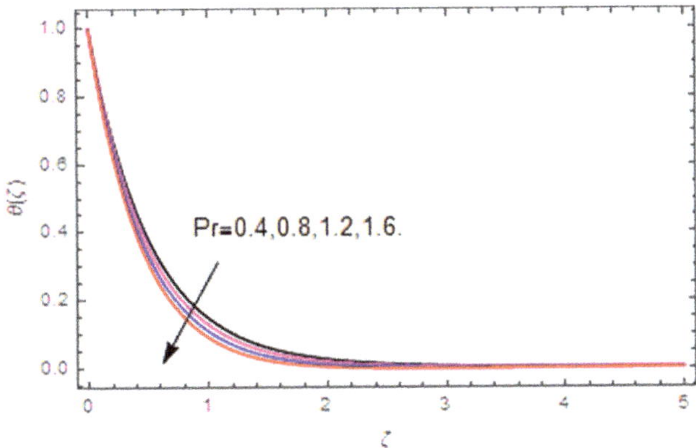

Figure 20. Pr on $\theta(\zeta)$.

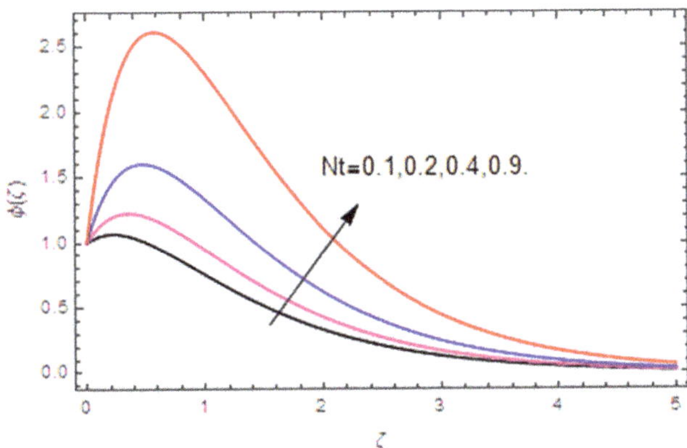

Figure 21. Nt on $\varphi(\zeta)$.

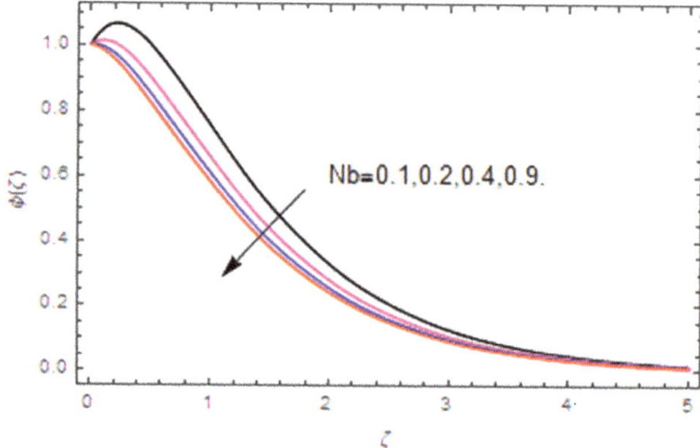

Figure 22. *Nb* on $\varphi(\zeta)$.

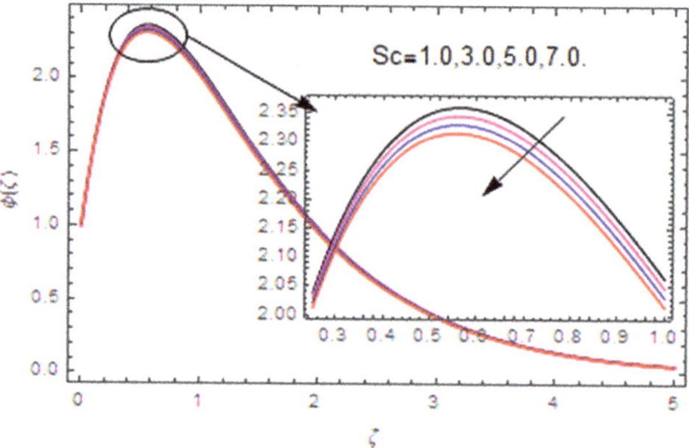

Figure 23. *Sc* on $\varphi(\zeta)$.

4.2. Entropy Optimization and Bejan Number

Figures 24 and 25 depict the impact of Brinkman number on the total entropy profile and the Bejan number profile, respectively. The total entropy profile escalates with a higher Brinkman number, while the opposite trend is observed on the Bejan number profile. The heat rises in the fluid moving in related region heightens with greater Brinkman number. Therefore, the total entropy profile rises with the Brinkman number, and the Bejan number declines (see Figure 25). Figures 26 and 27 illustrate the change in the total entropy profile and the Bejan number profile, via greater values of the diffusion parameter. Both total entropy and the Bejan number profiles escalate with larger values of diffusion parameter. The rising diffusion parameter increases the nanoparticle's diffusion rate. Thus, both profiles escalate with a greater diffusion parameter. Figures 28 and 29 display the change in total entropy and Bejan number profiles via the magnetic parameter. The total entropy profile escalates, while the Bejan number profile deescalates with a higher magnetic parameter. According to the Lorentz force, the total entropy of the system heightens, while the Bejan number reduces with higher magnetic parameter. So, the total entropy profile escalates, while the Bejan number profile deescalates with a higher magnetic parameter. Figures 30 and 31 illustrate the change in total entropy and the Bejan number profiles via temperature difference parameters. Both profiles escalate with higher temperature difference parameters. Figures 32 and 33 show the change in total entropy and Bejan number profiles via the Weissenberg number. Increasing the Weissenberg number reduces the fluid flow viscosity.

This behavior escalates both total entropy and the Bejan number profiles. Figures 34 and 35 indicate the variation the total entropy and Bejan number profiles via the Hall parameter. The total entropy profile declines, while the Bejan number profile escalates with the higher Hall parameter. The Hall parameter has a direct effect on the Lorentz force term and current density of the nanofluid. Thus, the electrical conductivity of the nanofluid heightens with the higher Hall parameter and, consequently, the total entropy profile is reduced. The opposite impact of Hall parameter on the Bejan number is displayed in Figure 35.

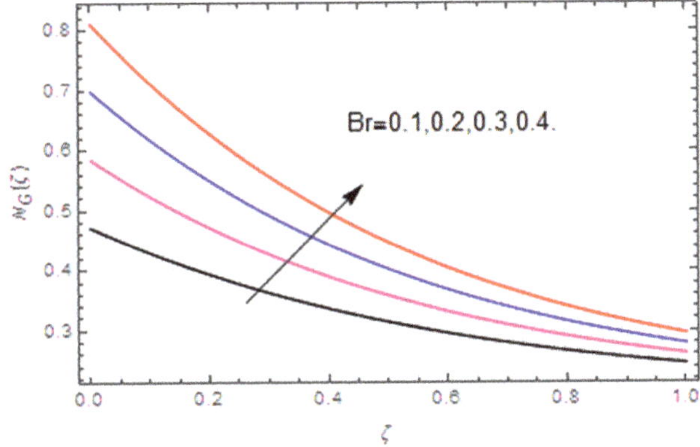

Figure 24. *Br* on $N_G(\zeta)$.

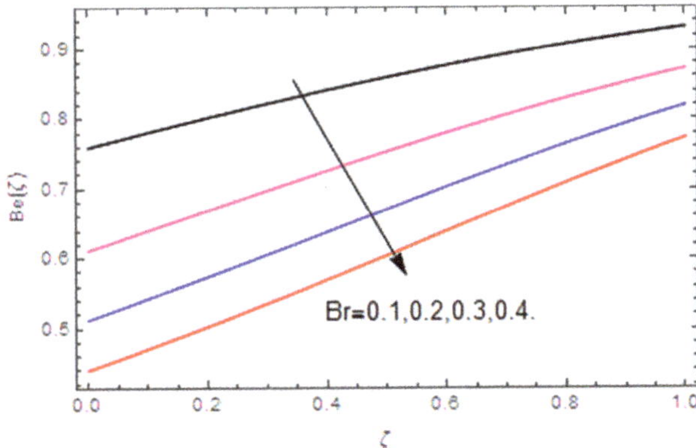

Figure 25. *Br* on $Be(\zeta)$.

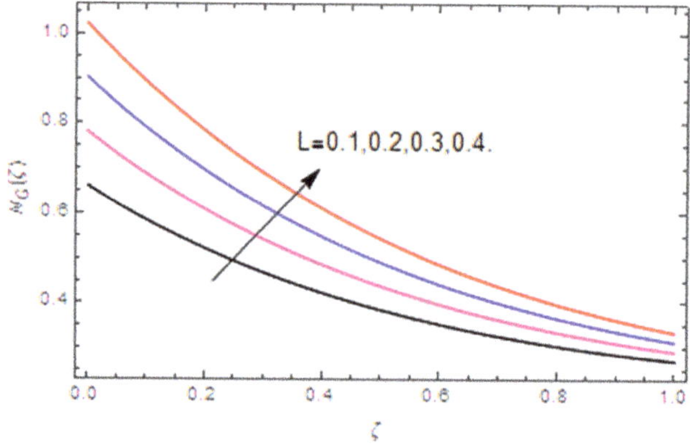

Figure 26. L on $N_G(\zeta)$.

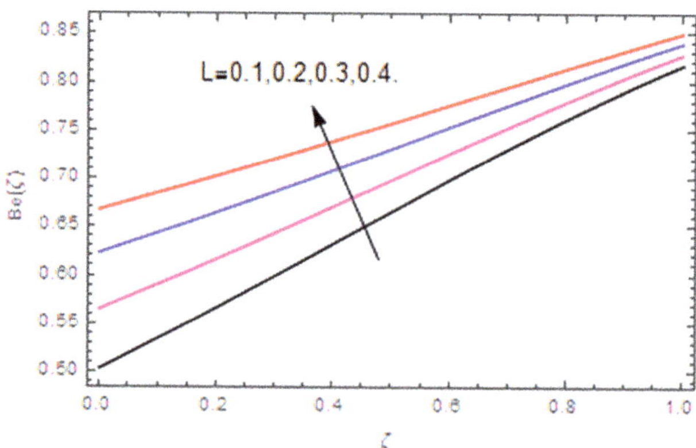

Figure 27. L on $Be(\zeta)$.

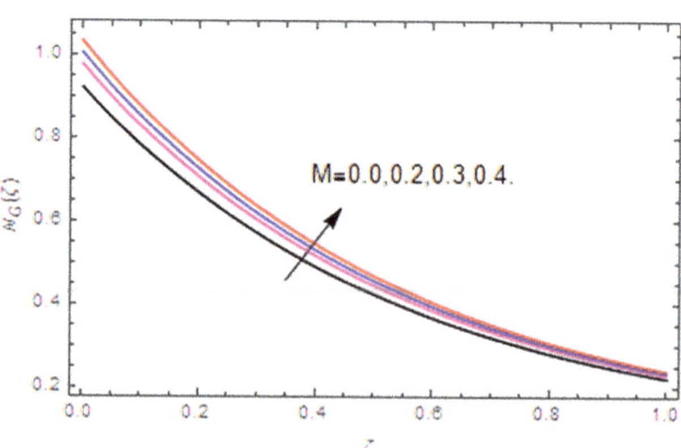

Figure 28. M on $N_G(\zeta)$.

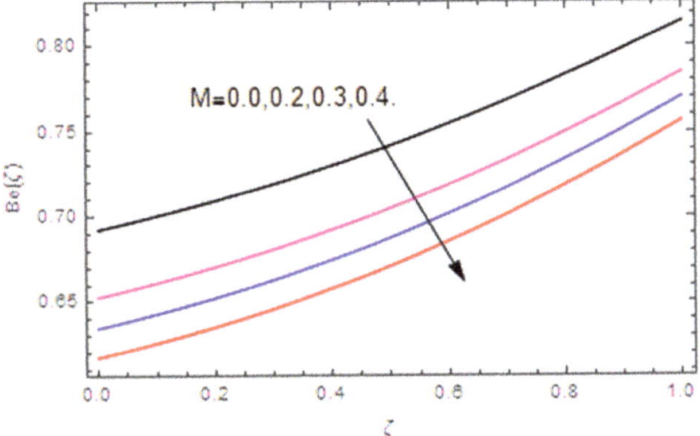

Figure 29. *M* on *Be*(ζ).

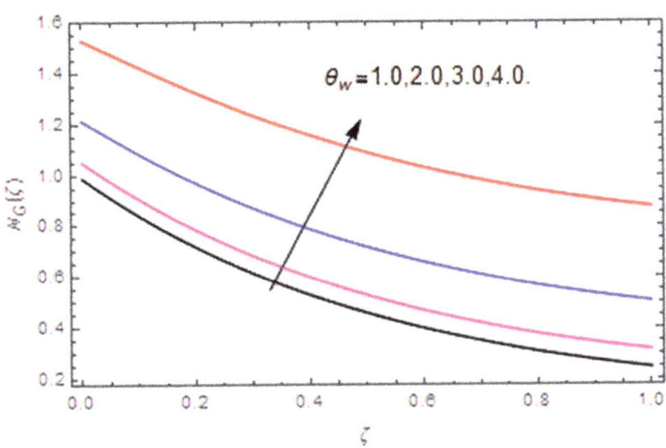

Figure 30. θ$_w$ on N_G(ζ).

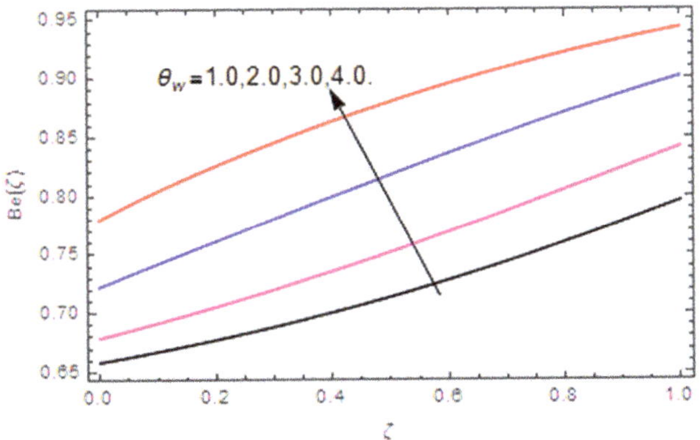

Figure 31. θ$_w$ on *Be*(ζ).

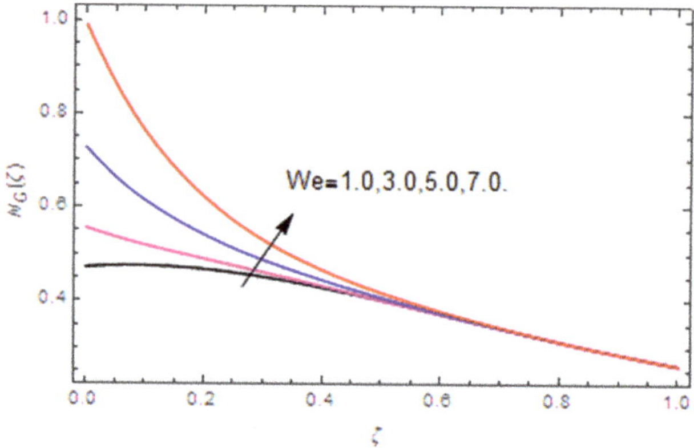

Figure 32. *We* on $N_G(\zeta)$.

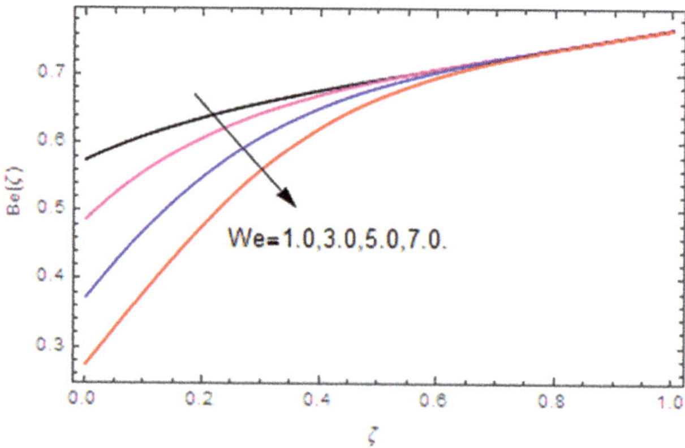

Figure 33. *We* on $Be(\zeta)$.

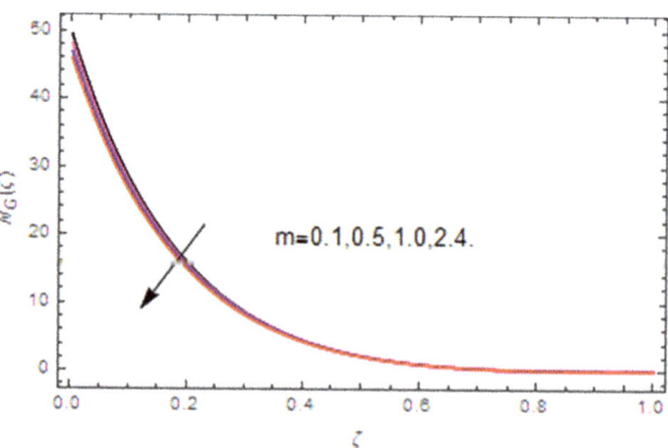

Figure 34. *m* on $N_G(\zeta)$.

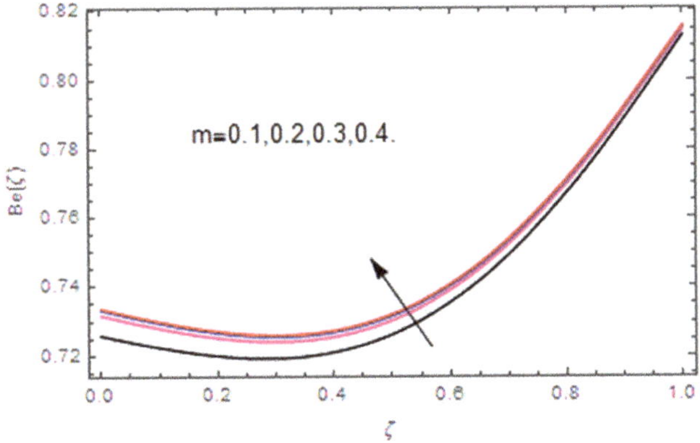

Figure 35. m on $Be(\zeta)$.

4.3. Physical Quantities

The values of skin friction along radial and tangential directions are presented in Table 1. The higher Weissenberg and Reynolds numbers increase the skin friction, while the magnetic, Hall and stretching parameters have a reducing influence on the skin friction along the radial direction. The heightening magnetic parameter escalates the skin friction while opposite trend is observed via the Weissenberg number, Reynods number, Hall parameter and stretching parameter along a tangential direction. Table 2 represents the values of Nu_x, via influential parameters and numbers. The rising values of λ, heightens Nu_x, while increasing Re, Pr, q, Nb, Nt, Ec and θ_w reducing Nu_x.

Table 1. Numerical values of skin friction along radial (Re_r, C_{fr}) and tangential (Re_θ, $C_{f\theta}$) directions.

We	Re	M	m	A	$\mathrm{Re}_r C_{fr}$	$\mathrm{Re}_\theta C_{f\theta}$
0.2	0.5	0.3	0.2	0.1	0.076871	−0.578141
0.3	–	–	–	–	0.109318	−0.601784
0.4	–	–	–	–	0.126191	−0.613634
–	0.0	–	–	–	−0.073910	−0.529679
–	1.0	–	–	–	0.349447	−0.703740
–	1.5	–	–	–	0.595858	−0.800002
–	–	0.0	–	–	0.608339	−0.928553
–	–	0.2	–	–	0.600019	−0.842866
–	–	0.5	–	–	0.587538	−0.714335
–	–	–	0.3	–	0.588645	−0.726248
–	–	–	0.5	–	0.539113	−0.755780
–	–	–	0.7	–	0.521225	−0.786661
–	–	–	–	0.2	0.536944	−1.122600
–	–	–	–	0.4	0.531917	−1.973220
–	–	–	–	0.5	0.514140	−2.487880

Table 2. Numerical values of Nusselt number (Nu_x).

Re	Pr	q	Nb	Nt	λ	Ec	θ_w	Nu_x
0.0	0.72	0.3	0.4	0.4	0.1	0.2	0.3	0.436148
1.0	–	–	–	–	–	–	–	0.425559
1.5	–	–	–	–	–	–	–	0.420265
–	0.4	–	–	–	–	–	–	0.502356
–	0.8	–	–	–	–	–	–	0.399743
–	1.2	–	–	–	–	–	–	0.297129
–	–	0.4	–	–	–	–	–	0.230951
–	–	0.5	–	–	–	–	–	0.164773
–	–	0.6	–	–	–	–	–	0.098594
–	–	–	0.5	–	–	–	–	0.065505
–	–	–	0.6	–	–	–	–	0.032416
–	–	–	0.7	–	–	–	–	−0.000672
–	–	–	–	0.5	–	–	–	−0.033761
–	–	–	–	0.6	–	–	–	−0.066850
–	–	–	–	0.7	–	–	–	−0.099939
–	–	–	–	–	0.3	–	–	0.111830
–	–	–	–	—	0.5	–	–	0.125066
–	–	–	–	–	0.7	–	–	0.138302
–	–	–	–	–	–	0.3	–	0.129974
–	–	–	–	–	–	0.4	–	0.121646
–	–	–	–	–	–	0.5	–	0.113318
–	–	–	–	–	–	–	1.0	0.116974
–	–	–	–	–	–	–	1.5	0.041694
–	–	–	–	–	–	–	2.0	−0.296988

5. Conclusions

The second grade nanofluid flow by a stretching disk is examined here. The nanofluid flow is characterized with Hall current, Brownian motion and thermophoresis influences. Entropy optimization with nonlinear thermal radiation, Joule heating and heat absorption/generation is also presented. Concluding remarks are mentioned below:

- Velocity along the radial direction is enhanced with the Weissenberg number and stretching parameter, however, the conflicting influence is observed via the Reynolds number.
- Velocity along the axial direction is heightened with the Hall parameter and Weissenberg number, while it is decreased with the magnetic parameter and the Reynolds number.
- Velocity along the tangential direction is escalated with the Hall parameter, while it decays with the magnetic parameter, Reynolds number and stretching parameter.
- The temperature profile is increased with the magnetic parameter, Brownian motion parameter, thermophoresis parameter, temperature difference and heat generation/absorption parameter, whereas it is reduced with the Hall parameter, thermal relaxation parameter and Prandtl number.
- The concentration profile is increased with the thermophoresis parameter, while it is reduced with the Brownian motion parameter and the Schmidt number.
- The total entropy profile is enhanced with the Brinkman number, diffusion parameter, magnetic parameter, temperature difference and Weissenberg number, while a declining influence is detected via the Hall parameter.
- The Bejan number profile is heightened with diffusion parameter, temperature difference and Hall parameter, while it is diminished with the Brinkman number, magnetic parameter and Weissenberg number.

Author Contributions: Conceptualization, Z.S. and L.B.M.; methodology, M.W.A.; software, Z.S.; validation, R.N. and Z.S.Y.; formal analysis, L.B.M.; investigation, Z.S.; resources, L.B.M.; data curation, M.W.A.; writing—original draft preparation, M.W.A and L.B.M.; writing—review and editing, Z.S. and R.N.; visualization, L.B.M.; supervision, Z.S. and R.N.; project administration, L.B.M.; funding acquisition, L.B.M. All authors have read and agreed to the published version of the manuscript.

Funding: This research received no external funding.

Acknowledgments: This research is supported by Postdoctoral Fellowship from King Mongkut's University of Technology Thonburi (KMUTT), Thailand.

Conflicts of Interest: The authors declare no conflict of interest.

References

1. Choi, S.U.; Eastman, J.A. *Enhancing Thermal Conductivity of Fluids with Nanoparticles*; No. ANL/MSD/CP-84938; CONF-951135-29; Argonne National Lab.: Cass Ave, IL, USA, 1995.
2. Farooq, M.; Khan, M.I.; Waqas, M.; Hayat, T.; Alsaedi, A.; Khan, M.I. MHD stagnation point flow of viscoelastic nanofluid with non-linear radiation effects. *J. Molecul. Liq.* **2016**, *221*, 1097–1103. [CrossRef]
3. Sajid, M.; Iqbal, S.A.; Naveed, M.; Abbas, Z. Effect of homogeneous-heterogeneous reactions and magnetohydrodynamics on Fe_3O_4 nanofluid for the Blasius flow with thermal radiations. *J. Molecul. Liq.* **2017**, *233*, 115–121. [CrossRef]
4. Sreedevi, P.; Reddy, P.S.; Chamkha, A.J. Heat and mass transfer analysis of nanofluid over linear and non-linear stretching surfaces with thermal radiation and chemical reaction. *Powder Technol.* **2017**, *315*, 194–204. [CrossRef]
5. Qayyum, S.; Khan, M.I.; Hayat, T.; Alsaedi, A. A framework for nonlinear thermal radiation and homogeneous-heterogeneous reactions flow based on silver-water and copper-water nanoparticles: A numerical model for probable error. *Results Phys.* **2017**, *7*, 1907–1914. [CrossRef]
6. Hayat, T.; Qayyum, S.; Imtiaz, M.; Alsaedi, A. Comparative study of silver and copper water nanofluids with mixed convection and nonlinear thermal radiation. *Int. J. Heat Mass Transf.* **2016**, *102*, 723–732. [CrossRef]
7. Goodarzi, M.; Amiri, A.; Goodarzi, M.S.; Safaei, M.R.; Karimipour, A.; Languri, E.M.; Dahari, M. Investigation of heat transfer and pressure drop of a counter flow corrugated plate heat exchanger using MWCNT based nanofluids. *Int. Commun. Heat Mass Transf.* **2015**, *66*, 172–179. [CrossRef]
8. Goshayeshi, H.R.; Goodarzi, M.; Safaei, M.R.; Dahari, M. Experimental study on the effect of inclination angle on heat transfer enhancement of a ferrofluid in a closed loop oscillating heat pipe under magnetic field. *Exp. Therm. Fluid Sci.* **2016**, *74*, 265–270. [CrossRef]
9. Goshayeshi, H.R.; Safaei, M.R.; Goodarzi, M.; Dahari, M. Particle size and type effects on heat transfer enhancement of Ferro-nanofluids in a pulsating heat pipe. *Powder Technol.* **2016**, *301*, 1218–1226. [CrossRef]
10. Hosseini, S.M.; Safaei, M.R.; Goodarzi, M.; Alrashed, A.A.; Nguyen, T.K. New temperature, interfacial shell dependent dimensionless model for thermal conductivity of nanofluids. *Int. J. Heat Mass Transf.* **2017**, *114*, 207–210. [CrossRef]
11. Peng, Y.; Zahedidastjerdi, A.; Abdollahi, A.; Amindoust, A.; Bahrami, M.; Karimipour, A.; Goodarzi, M. Investigation of energy performance in a U-shaped evacuated solar tube collector using oxide added nanoparticles through the emitter, absorber and transmittal environments via discrete ordinates radiation method. *J. Therm. Anal. Calorim.* **2020**, *139*, 2623–2631. [CrossRef]
12. Shamshirband, S.; Malvandi, A.; Karimipour, A.; Goodarzi, M.; Afrand, M.; Petković, D.; Dahari, M.; Mahmoodian, N. Performance investigation of micro-and nano-sized particle erosion in a 90 elbow using an ANFIS model. *Powder Technol.* **2015**, *284*, 336–343. [CrossRef]
13. Tian, Z.; Etedali, S.; Afrand, M.; Abdollahi, A.; Goodarzi, M. Experimental study of the effect of various surfactants on surface sediment and pool boiling heat transfer coefficient of silica/DI water nano-fluid. *Powder Technol.* **2019**, *356*, 391–402. [CrossRef]
14. Zaidi, Z.A.; Mohyud-Din, S.T. Effect of joule heating and MHD in the presence of convective boundary condition for upper convected Maxwell fluid through wall jet. *J. Mol. Liq.* **2017**, *230*, 230–234. [CrossRef]

15. Afify, A.A.; Elgazery, N.S. Effect of a chemical reaction on magnetohydrodynamic boundary layer flow of a Maxwell fluid over a stretching sheet with nanoparticles. *Particuology* **2016**, *29*, 154–161. [CrossRef]
16. Mustafa, M. An analytical treatment for MHD mixed convection boundary layer flow of Oldroyd-B fluid utilizing non-Fourier heat flux model. *Int. J. Heat Mass Transf.* **2017**, *113*, 1012–1020. [CrossRef]
17. Hayat, T.; Nawaz, S.; Alsaedi, A.; Rafiq, M. Influence of radial magnetic field on the peristaltic flow of Williamson fluid in a curved complaint walls channel. *Results phys.* **2017**, *7*, 982–990. [CrossRef]
18. Khan, M.; Hamid, A. Influence of non-linear thermal radiation on 2D unsteady flow of a Williamson fluid with heat source/sink. *Results phys.* **2017**, *7*, 3968–3975. [CrossRef]
19. Baron Fourier, J.B.J. Théorie Analytique de la Chaleur. In *Landmark Writings in Western Mathematics 1640–1940*; Elsevier: Amsterdam, The Netherlands, 2005.
20. Cattaneo, C. Sulla conduzione del calore. *Atti Sem. Mat. Fis. Univ. Modena* **1948**, *3*, 83–101.
21. Christov, C.I. On frame indifferent formulation of the Maxwell–Cattaneo model of finite-speed heat conduction. *Mech. Res. Commun.* **2009**, *36*, 481–486. [CrossRef]
22. Hayat, T.; Qayyum, S.; Imtiaz, M.; Alsaedi, A. Impact of Cattaneo-Christov heat flux in Jeffrey fluid flow with homogeneous-heterogeneous reactions. *PLoS ONE* **2016**, *11*, e0148662. [CrossRef]
23. Mustafa, M. Cattaneo-Christov heat flux model for rotating flow and heat transfer of upper-convected Maxwell fluid. *AIP Adv.* **2015**, *5*, 047109. [CrossRef]
24. Tibullo, V.; Zampoli, V. A uniqueness result for the Cattaneo-Christov heat conduction model applied to incompressible fluids. *Mech. Res. Commun.* **2011**, *38*, 77–99. [CrossRef]
25. Han, S.; Zheng, L.; Li, C.; Zhang, X. Coupled flow and heat transfer in viscoelastic fluid with Cattaneo-Christov heat flux model. *Appl. Math. Lett.* **2014**, *38*, 87–93. [CrossRef]
26. Ahmad Khan, J.; Mustafa, M.; Hayat, T.; Alsaedi, A. Numerical study of Cattaneo-Christov heat flux model for viscoelastic flow due to an exponentially stretching surface. *PLoS ONE* **2015**, *10*, e0137363. [CrossRef]
27. Raptis, A.; Ram, P.C. Effects of hall current and rotation. *Astrophys. Space Sci.* **1984**, *106*, 257–264. [CrossRef]
28. Das, S.; Maji, S.L.; Jana, R.N. Hall effects on unsteady hydromagnetic flow induced by a porous plate. *Int. J. Comput. Appli.* **2012**, *57*, 37–44.
29. Jha, B.K.; Apere, C.A. Combined effect of hall and ion-slip currents on unsteady MHD couette flows in a rotating system. *J. Phys. Soc. Jpn.* **2010**, *79*, 1–9.
30. Aurangzeeb, S.; Shafie, S. Effects of Soret and Dufour on unsteady MHD flow by mixed convection over a vertical surface in porous media with internal heat generation, chemical reaction and Hall current. *Can. J. Sci. Eng. Math.* **2011**, *2*, 153–162.
31. Ferdows, M.; Jahan, E.; Hamad, M.M.; Masahiro, O.T.A. Effects of Hall and ion-slip currents on free convective heat transfer flow past a vertical plate considering slip conditions. *Can. J. Sci. Eng. Math.* **2011**, *2*, 70–76.
32. Anwar Bég, O.; Sim, L.; Zueco, J.; Bhargava, R. Numerical study of magnetohydro-dynamic viscous plasma flow in rotating porous media with Hall currents and inclined magnetic field influence. *Commun. Nonlinear Sci. Numer. Simul.* **2010**, *15*, 345–359.
33. Kumar, R.; Chand, K. Effect of slip conditions and hall current on unsteady MHD flow of a viscoelastic fluid past an infinite vertical porous plate through porous medium. *Int. J. Eng. Sci. Technol.* **2011**, *3*, 3124–3133.
34. Li, X.; Faghri, A. Local entropy generation analysis on passive high concentration DMFCs (direct methanol fuel cell) with different cell structures. *Energy* **2011**, *36*, 403–414. [CrossRef]
35. Hayat, T.; Rafiq, M.; Ahmad, B.; Asghar, S. Entropy generation analysis for peristaltic flow of nanoparticles in a rotating frame. *Int. J. Heat Mass Transf.* **2017**, *108*, 1775–1786. [CrossRef]
36. Nouri, D.; Pasandideh-Fard, M.; Oboodi, M.J.; Mahian, O.; Sahin, A.Z. Entropy generation analysis of nanofluid flow over a spherical heat source inside a channel with sudden expansion and contraction. *Int. J. Heat Mass Transf.* **2018**, *116*, 1036–1043. [CrossRef]
37. Dalir, N.; Dehsara, M.; Nourazar, S.S. Entropy analysis for magnetohydrodynamic flow and heat transfer of a Jeffrey nanofluid over a stretching sheet. *Energy* **2015**, *79*, 351–362. [CrossRef]
38. Khan, M.I.; Qayyum, S.; Hayat, T.; Waqas, M.; Khan, M.I.; Alsaedi, A. Entropy generation minimization and binary chemical reaction with Arrhenius activation energy in MHD radiative flow of nanomaterial. *J. Mol. Liq.* **2018**, *259*, 274–283. [CrossRef]

39. Khan, M.I.; Hayat, T.; Waqas, M.; Khan, M.I.; Alsaedi, A. Entropy generation minimization (EGM) in nonlinear mixed convective flow of nanomaterial with Joule heating and slip condition. *J. Mol. Liq.* **2018**, *256*, 108–120. [CrossRef]

40. Hayat, T.; Kanwal, M.; Qayyum, S.; Khan, M.I.; Alsaedi, A. Entropy generation optimisation in the nanofluid flow of a second grade fluid with nonlinear thermal radiation. *Pramana* **2019**, *93*, 54. [CrossRef]

Article

Fluid Morphologies Governed by the Competition of Viscous Dissipation and Phase Separation in a Radial Hele-Shaw Flow

Ryuta X. Suzuki [1], Risa Takeda [1], Yuichiro Nagatsu [1], Manoranjan Mishra [2] and Takahiko Ban [3,*]

[1] Department of Chemical Engineering, Tokyo University of Agriculture and Technology, Nakacho 2-24-16, Koganei, Tokyo 184-8588, Japan; ryuta.x.suzuki@gmail.com (R.X.S.); risaadekat@gmail.com (R.T.); nagatsu@cc.tuat.ac.jp (Y.N.)

[2] Department of Mathematics, Indian Institute of Technology Roper, Rupnagar 140001, India; manoranjan.mishra@gmail.com

[3] Division of Chemical Engineering, Department of Materials Engineering Science, Graduate School of Engineering Science, Osaka University, Machikaneyamacho 1-3, Toyonaka, Osaka 560-8531, Japan

* Correspondence: ban@cheng.es.osaka-u.ac.jp

Received: 16 September 2020; Accepted: 1 October 2020; Published: 6 October 2020

Abstract: The displacement of a less viscous fluid by a more viscous fluid in a radial Hele-Shaw cell makes a circular pattern because the interface is hydrodynamically stable in this condition. Very recently, it has been experimentally reported that the hydrodynamically stable displacement in a partially miscible system induces fingering patterns while stable circular patterns are made at fully miscible and immiscible systems. The fingering instability in the partially miscible system results from complex and entangled elements involving viscous dissipation, molecular diffusion, and phase separation. The analyzing mechanism requires a quantitative relationship between the hydrodynamic interfacial fingering patterns and underlying physicochemical properties. Here, we experimentally investigated the change in fluid patterns formed by the progression of phase separation in the partially miscible systems and categorized them into three patterns: finger-like pattern, annular-like pattern, and circular pattern. Moreover, we propose the mechanism of the pattern formation by an interfacial tension measurement and evaluate the patterns by modified capillary number and newly defined body force ratio, B_f. Our analysis revealed that the deformation index of the pattern can be expressed as a function of B_f on a single curve regardless of the miscibility.

Keywords: fluid displacement; inverse Saffman–Taylor instability; partially miscible; Korteweg force

---

1. Introduction

The displacement of one fluid by another in porous media is ubiquitous in several processes such as chromatographic separation [1], transportation of digestive juice [2], CO_2 sequestration [3], frontal polymerization [4], and secondary and tertiary oil recovery [5]. It is generally accepted that when the displacing fluid is more viscous, no interfacial instability occurs, while a finger-like pattern forms at the interface in the reverse situation (Figure 1). The latter case, where a less viscous fluid displaces a more viscous one in porous media, is called Saffman–Taylor instability [6] or viscous fingering (VF) [7]. Saffman–Taylor instability, based on Darcy's law [8,9], can be explained by Figure 2 and Equation (1):

$$u_A - u_B = \left[\frac{u_B(\mu_1 - \mu_2)}{\mu_2 + (\mu_1 - \mu_2)\frac{z}{L}} \right]\frac{\delta z}{L},$$

$$(1)$$

where u is flow velocity, μ is viscosity, δz is an initial disturbance, L is a length scale of porous media, and indexes A and B represent location. u_A and u_B represent the velocity at the different locations in the interface between fluid 1 and fluid 2 (Figure 2a). We assume that a velocity fluctuation occurs at the interface, and the velocity at location B becomes faster than the remaining interface at the initial state. Therefore, the interface at location B protrudes from the initial flat interface by the displacement δz. We have two different combinations of viscosity of the fluids. When displacing fluid 1 is more viscous than the other ($\mu_1 > \mu_2$), the right-hand side of Equation (1) becomes positive, and then $u_A > u_B$ according to Equation (1). In this case, the initial disturbance δz becomes small. As a result, the disturbance is suppressed, and hydrodynamically stable displacement proceeds. However, when the displacing fluid is less viscous than the other ($\mu_1 < \mu_2$), the right-hand side of Equation (1) becomes negative, and then $u_A < u_B$, and thus the initial disturbance δz is enhanced. Therefore, the interface at location B protrudes further, and a finger-like pattern (VF) appears.

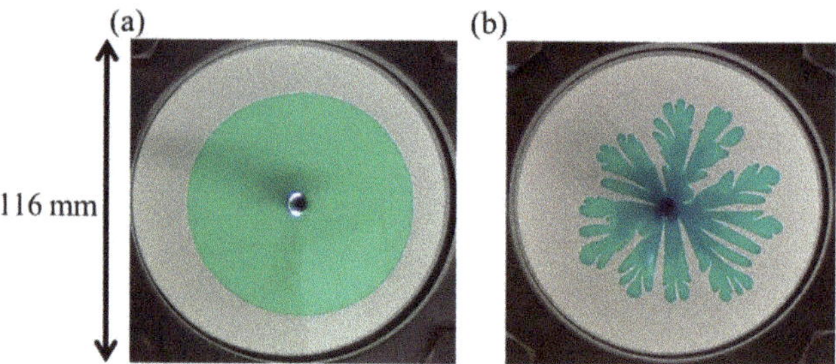

Figure 1. Fluid displacements for (**a**) a hydrodynamically stable condition of the interface, (**b**) a hydrodynamically unstable condition of the interface. The blue solution displaces the surrounding white solution. A more viscous solution is displacing a less viscous one in (**a**) while a less viscous solution is displacing a more viscous one in (**b**). The patterns like fingering in (**b**) are called "Saffman–Taylor instability" or "viscous fingering" (VF).

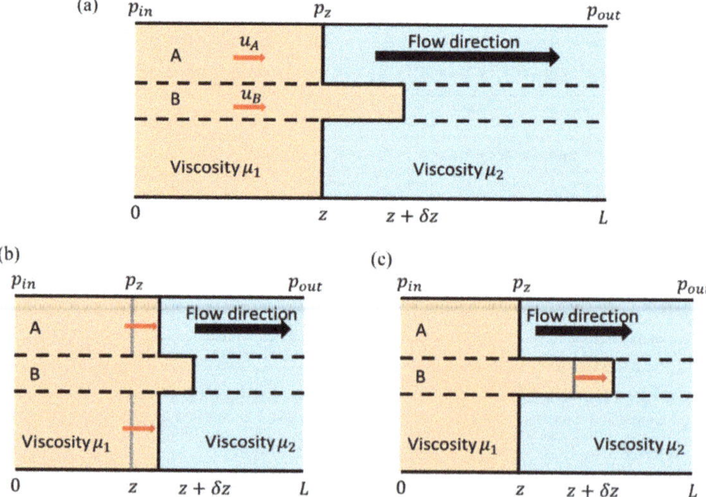

Figure 2. The explanation for Saffman–Taylor instability. The solution flows from left to right in the figures. (**a**) The setup situation and the results under (**b**) a hydrodynamically stable condition ($\mu_1 > \mu_2$) and (**c**) a hydrodynamically unstable condition ($\mu_1 < \mu_2$), i.e., Saffman–Taylor instability. p represents pressure at the position z.

The displacement fluid pair can be categorized into three types: fully miscible, partially miscible, and immiscible. The fully miscible system has infinite mutual solubility, like glycerol–water, the partially miscible system has finite mutual solubility, and the immiscible system has zero mutual solubility, like oil–water. The hydrodynamically stable displacement in fully miscible and immiscible systems makes the pattern circular for a radial geometry or flat for a rectangular geometry. Any fluctuation in the interface should always decay as fluid displacement proceeds. However, the hydrodynamically stable displacement in the partially miscible system shows not a circular but a finger-like pattern [10]. There are many studies on the displacement in the porous media or Hele-Shaw cells under hydrodynamically unstable and stable conditions of the interface with and without chemical reactions [7,11–14]. Although the unstable interfaces on the fully miscible and immiscible systems in hydrodynamically stable displacement have been reported by many papers, research on specialized conditions such as prewetting condition [15–17], including particles [18–21], viscosity-change reaction at the interface [22–25], precipitation reaction [26], and double diffusivity effect [27] is limited. Studies on the partially miscible system have recently received a lot of attention because of its application in enhanced oil recovery [5]. However, past work was unable to acquire an appropriate partially miscible system because changes in the composition and temperature lead to changes in multiple hydrodynamic properties of the fluid systems. These undesirable changes, which occur simultaneously, represent a significant obstacle in elucidating the exact properties of the miscibility that play a crucial role in fingering pattern formation. Therefore, hydrodynamically unstable displacement (Saffman–Taylor instability or VF) in the partially miscible systems has mostly been studied through numerical simulations [28–30]. One simulation [30] shows that the solubility in a partially miscible system greatly influences the degree of fingerings and also demonstrates that fluid dissolution or exsolution due to partial miscibility can hinder or enhance viscous fingering, respectively. However, partial miscibility is not mutual solubility but one-direction solubility, where one species dissolves into the other one. Very recently, Suzuki et al. have experimentally shown that a partially miscible system affects displacement such as a fingering pattern occurring in a hydrodynamically stable displacement [10] and a droplet formation in Saffman–Taylor instability or viscous fingering [31]. They have proven that the morphologies are driven by spontaneous convection induced by Korteweg force due to chemical potential gradient during spinodal decomposition-type phase separation. The force, first proposed by Korteweg in 1901 [32], is thermodynamically defined as the functional derivative of free energy [33] and is characterized as a body force. Korteweg force tends to minimize the free energy stored at the interface and induce spontaneous convection. The free energy is closely involved with an interfacial tension because the interfacial tension is defined as free energy per unit area. Therefore, measuring the interfacial tension is important for considering a partially miscible system or Korteweg force. We note that in Rayleigh–Taylor instability, where a more dense fluid is on a less dense fluid in gravitational field, the experimental studies and theoretical model for a partially miscible system have been reported [34,35].

Detailed experimental studies without any hydrodynamic interfacial instabilities are needed to thoroughly understand the influence of the partially miscible system. Here, we investigated the effect of different progressions of phase separation in a partially miscible system on the patterns in a hydrodynamically stable condition of the interface, like inverse Saffman–Taylor instability. An aqueous two-phase system was employed by following the previous reports [10,31], where the system consisted of polyethylene glycol (PEG; weight-average molecular weight $M_w = 8000$), Na_2SO_4, and water, allowing quantitative and qualitative control over the thermodynamic stability of the fluid system as well as the hydrodynamic interfacial stability. We changed the progression of the phase separation by changing the concentrations of PEG and Na_2SO_4 and investigated the effect on the fluid patterns to make a phase diagram of patterns. In addition, we measured the dynamic interfacial tension between the displacing and the displaced fluids to clarify the mechanism of pattern formation and to evaluate the patterns quantitatively.

2. Experiment

2.1. Solutions

We used an aqueous two-phase system (ATPS), the same as the one used in [10,31], where the more viscous fluid is PEG solution and the less viscous fluid is Na_2SO_4 solution. Figure 3 shows the phase diagram of the ATPS [36]. The partially miscible zone is the green Region II, where the system can be separated into two phases because of thermodynamic instability. Region II is called the spinodal region, where concentration fluctuations spontaneously grow due to thermodynamic instability as the system reduces the total free energy, but the interfacial free energy increases because the interfaces of domains gradually grow. After the domain growth, the domains coalesce with each other, and phase separation is eventually completed. The progression of phase separation becomes faster with the increase in the concentration of Na_2SO_4, according to the previous paper [10,31], as this leads the system far from equilibrium. Region I is a one-phase region where the system is fully miscible because of the thermodynamically stable region. Here, we consider that the progression of phase separation represents the extent of phase separation. The details are described in [10,31]. The solutions used here are shown in Table 1; seven more viscous solutions and 11 less viscous solutions were used for the fully and partially miscible systems and a combination of Phases L and H was used for the immiscible system. Since Phases L and H are the solution after the phase separation, they are thermodynamically stable. The composition of Phases L and H is 36.5 wt % PEG solution with 3.2 wt % Na_2SO_4 and 1.4 wt % PEG solution with 16.0 wt % Na_2SO_4 [36]. To visualize the displacement process, the more viscous solutions were dyed blue by 0.1 wt % indigo carmine, which cannot dissolve into Na_2SO_4 solution because of the salting-out effect.

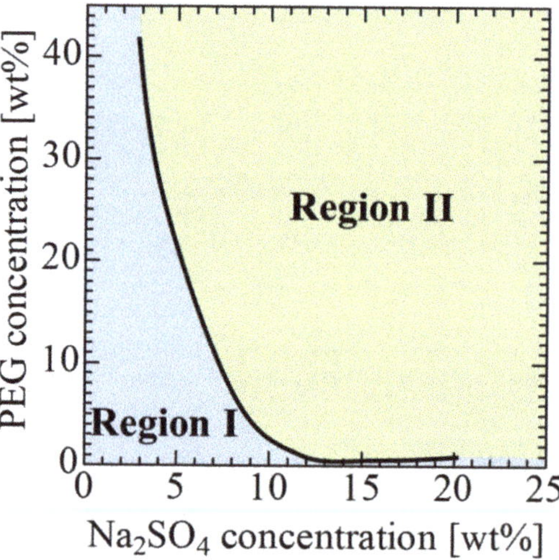

Figure 3. Phase diagram of polyethylene glycol (PEG)–Na_2SO_4–water system.

The composition of the system is important to decide whether the solution system is in Region I or II. Here, we assumed that the concentration at the interface, where the displacing and displaced fluids are brought into contact with each other, can be estimated to be half of the initial concentrations. For example, as the displacing and displaced fluids are 40 wt % PEG solution and 10 wt % Na_2SO_4 solution, respectively, the concentration at the interface was considered to be 20 wt % PEG–5 wt % Na_2SO_4–75 wt % water. This assumption is very simple, but we think this is appropriate. In previous studies [10,31], the same assumption was taken into account, and it can be explained by the assumption that the result of the fluid displacement, calculation of free energy, and the result of the interfacial tension measurement all have good agreement.

Table 1. The solution system used.

Displacing More Viscous Liquid	Displaced Less Viscous Liquid
	0 wt % Na_2SO_4 solution
	2 wt % Na_2SO_4 solution
10 wt % PEG solution	4 wt % Na_2SO_4 solution
15 wt % PEG solution	6 wt % Na_2SO_4 solution
20 wt % PEG solution	8 wt % Na_2SO_4 solution
25 wt % PEG solution	10 wt % Na_2SO_4 solution
30 wt % PEG solution	12 wt % Na_2SO_4 solution
35 wt % PEG solution	14 wt % Na_2SO_4 solution
40 wt % PEG solution	16 wt % Na_2SO_4 solution
	18 wt % Na_2SO_4 solution
	20 wt % Na_2SO_4 solution
Phase L	Phase H

2.2. Displacement Experiment

We performed the fluid displacement under hydrodynamically stable conditions, i.e., a more viscous fluid displaced a less viscous one. The schematic for the fluid displacement is shown in Figure 4. Hele-Shaw cells were used instead of porous media because the Hele-Shaw cell consists of a thin gap between two parallel glass plates and is the simplest two-dimensional model of porous media. The gap between the Hele-Shaw cells was fixed at 0.3 mm. First, we filled up the cell with the less viscous Na_2SO_4 solution and then injected the more viscous PEG solution into the cell using a syringe pump. The PEG solutions were dyed blue to visualize the experimental process. The displacement experiments were recorded by a video camera from the bottom. The injection flow rate of the more viscous liquid was fixed as 3.35 mL/h. All experiments were performed at room temperature (25 ± 1 °C) and atmospheric pressure (approx. 1 atm).

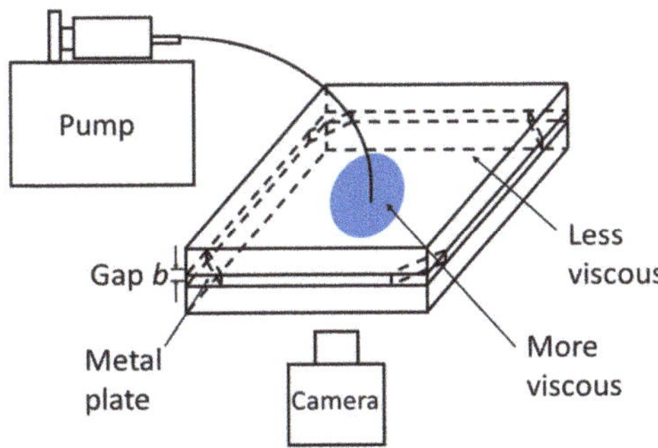

Figure 4. The schematic for the hydrodynamically stable displacement.

2.3. Physical Property Measurements

The density of the solutions was measured using a pycnometer with an electric balance. However, the density effect can be ignored when using a horizontal Hele-Shaw cell and when the gap between the cells is small enough.

Another important property is interfacial tension between the displacing and displaced solutions. The interfacial tension was measured by a spinning drop tensiometer, SITE100 from KRÜSS, Germany. The interfacial tension measurement was carried out in a rotating horizontal capillary filled with a high-density liquid. A drop of low-density fluid was placed inside the liquid. Because the rotation of the horizontal capillary creates a centrifugal force, which is against the interfacial tension of the fluids, the drop elongates, making the drop radius smaller, until it reaches a certain radius where the interfacial tension and centrifugal force are balanced. This device calculates the interfacial tension of the system using the following equation:

$$\gamma = \frac{r^3 \omega^2 \Delta\rho}{4},\tag{2}$$

where r (m) is the drop radius, ω (rad/s) is the angular frequency of rotation, and $\Delta\rho$ (kg/m^3) is the density difference between the two fluids. Here, ω was fixed at 6000 rpm (628 rad/s).

The other important physical property is the viscosity of the solutions. The viscosity was measured using an AR-G2 rheometer from TA Instruments, New Castle, DE, USA. We measured shear viscosity to investigate the viscosity itself and confirmed whether our solutions were Newtonian using a cone–plate-type sensor.

3. Results and Discussion

3.1. Physical Properties

The measured physical properties are shown in Figure 5. Figure 5a shows that the viscosities of PEG solutions are constant against shear rates, which means the PEG solutions are Newtonian. Figure 5b depicts the dependence of viscosity on the solute concentrations, and the inset shows the results of Na_2SO_4 for easy observation. The viscosities of PEG solutions depend greatly on the concentration of PEG, while the viscosities of Na_2SO_4 are very low compared to those of PEG solutions. Figure 5c shows the densities, which have a linear relationship with the concentration of the solution. In the horizontal Hele-Shaw cells, however, the density effect (gravitational effect) can be ignored when the gap between the cells is small enough.

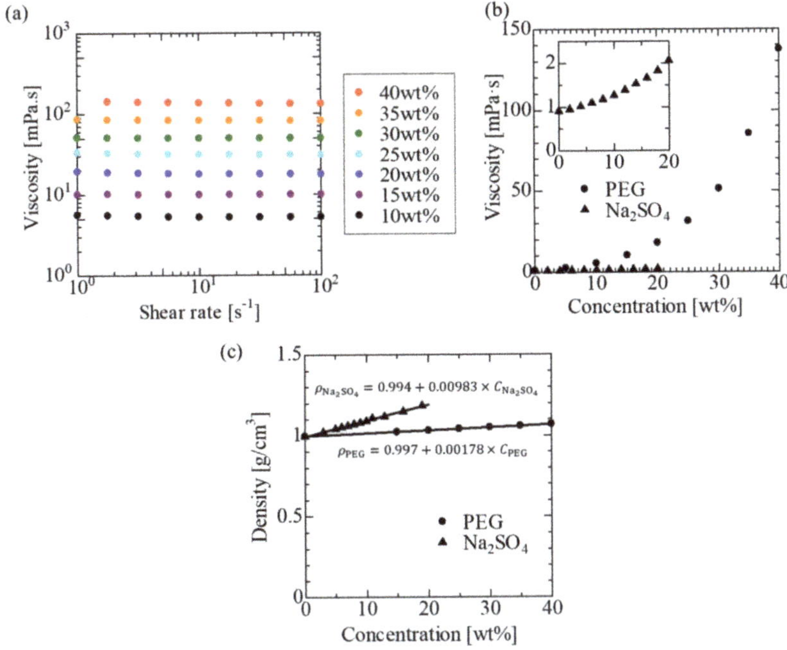

Figure 5. (**a**) Viscosity of PEG solutions against shear rate, (**b**) viscosity of PEG and Na$_2$SO$_4$ solutions against concentration, (inset) viscosity of Na$_2$SO$_4$ solutions against concentrations, (**c**) density of PEG and Na$_2$SO$_4$ solutions against concentrations.

3.2. Fluid Displacements

We first investigated the pattern formation of three fluid systems during fluid displacement under different thermodynamic conditions and the same hydrodynamic conditions or different hydrodynamic conditions and similar thermodynamic conditions. System I is at 20 wt % PEG and 0 wt % Na$_2$SO$_4$, System II is at 20 wt % PEG and 20 wt % Na$_2$SO$_4$, and System III is at 40 wt % PEG and 20 wt % Na$_2$SO$_4$. System I lies in Region I, whereas Systems II and III lie in Region II in Figure 3. Figure 6a shows the time evolution of the pattern formation of System I. A circular pattern expands as time proceeds because System I is absent from the thermodynamic instability and hydrodynamic interfacial instability. Similarly, the immiscible system, which is thermodynamically and hydrodynamically stable, approaches a perfectly circular pattern because of the high interfacial tension. The displacement patterns in the immiscible system have already been reported in [10].

Figure 6b shows the time evolution of the pattern of System II. At the beginning of the formation, the color of the outer region of the displacing fluid changes from indigo blue to light blue at time $t = 30$ s. The color change means that water and Na$_2$SO$_4$ molecules diffuse from the interface into the displacing fluid to reduce the PEG concentration inside the diffusion region. At $t = 50$ s, many indigo blue domains newly form in the outermost part of the displacing fluid, and the light blue region is left inside the outer indigo domains to produce a light blue circular ring. The formation of the outer indigo blue domains and the light blue circular ring indicates the spontaneous formation of PEG-rich and Na$_2$SO$_4$-rich phases, respectively. The interface becomes distorted, and the annular-like pattern expands with distortion as time proceeds. System I has a similar viscous ratio to System II because the viscous ratio depends on the PEG concentration, as shown in Figure 5b. Thus, thermodynamic instability, i.e., phase separation, is estimated to contribute to the spontaneous formation of the outer indigo domains and to the distorted interfacial pattern.

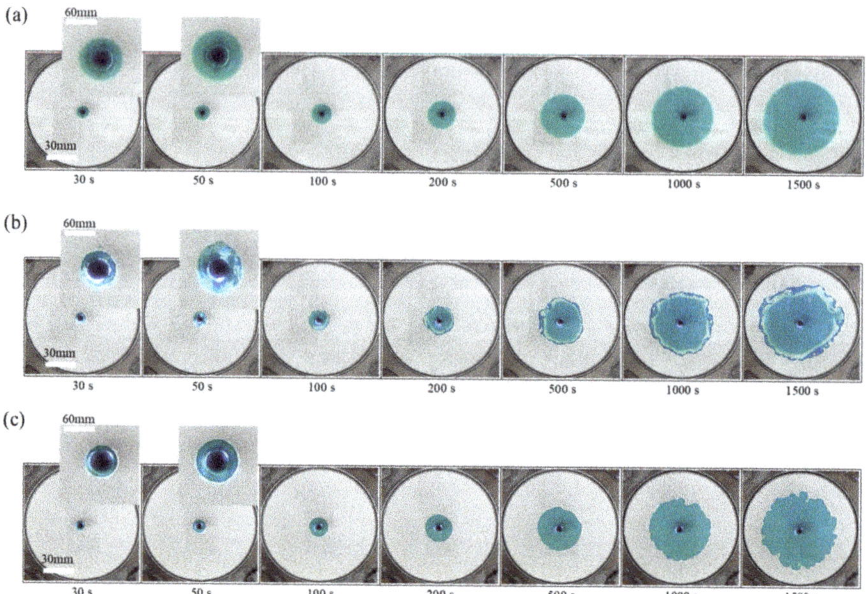

Figure 6. Time evolution of the pattern formation at (**a**) System I, where 20 wt % PEG solution displaces 0 wt % Na$_2$SO$_4$ solution; (**b**) System II, where 20 wt % PEG solution displaces 20 wt % Na$_2$SO$_4$ solution; and (**c**) System III, where 35 wt % PEG solution displaces 20 wt % Na$_2$SO$_4$ solution.

The pattern formation of System III is shown in Figure 6c. The outer region of the displacing fluid becomes distorted without a color change, and the distorted interface grows continuously. System III is comparable in the extent of thermodynamic instability to System II because both fluid systems lie in Region II at the same Na$_2$SO$_4$ concentration, but they are different in terms of their viscous ratio. A high viscous ratio restricts the diffusion region to the periphery of the displacing fluid. Thus, phase separation occurs at the outmost interface of the displacing fluid, inducing the interfacial distortion. This fingering-like pattern is already reported by Suzuki et al. [10]. The distorted interfacial pattern is created by the Korteweg force exerted in a direction toward the higher region of PEG concentration [10]. Thus, the pattern of System II is a phase separation-dominated displacement in comparison with System III, where the process of viscous dissipation weakens the phase separation effect.

Figure 7 shows the results of the hydrodynamically stable displacement using various composition combinations. A circular, stable pattern forms as PEG concentration increases and Na$_2$SO$_4$ concentration decreases, i.e., the composition goes to the upper left in Figure 7. In contrast, the finger-like pattern forms as PEG concentration decreases and Na$_2$SO$_4$ concentration increases, i.e., the composition moves to the lower right in Figure 7. For PEG concentrations higher than 30 wt %, the light blue circular ring disappears, and the outer indigo blue interface becomes distorted. As the Na$_2$SO$_4$ concentration increases, the interface becomes sharp because the interfacial tension increases with the increase in Na$_2$SO$_4$ concentration [31].

It is reported that the Korteweg force becomes stronger as the concentration of Na$_2$SO$_4$ increases [10,31]. As shown in Figure 5b, the viscosity decreases as the concentration of PEG solution decreases, which means that the solutions with less viscosity easily move, and phase separation is thought to easily occur.

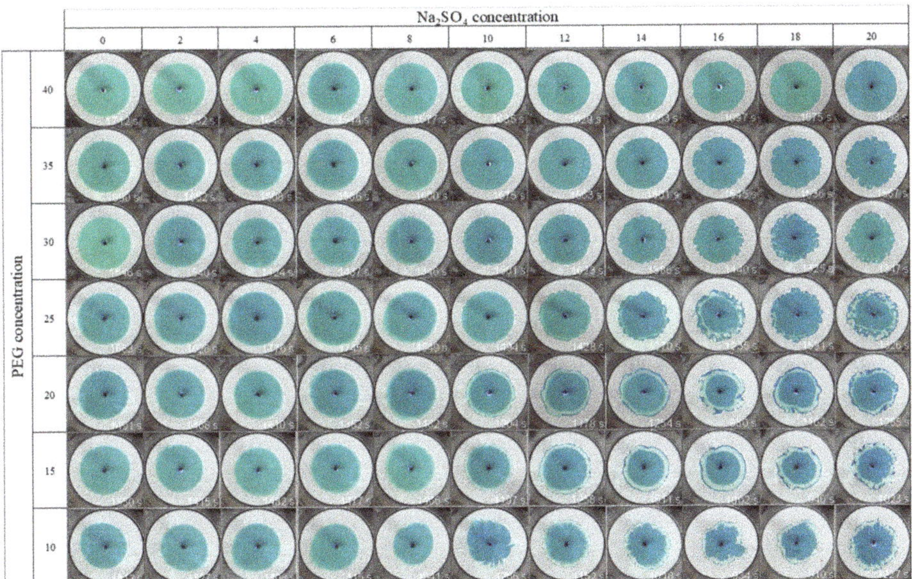

Figure 7. The results of the hydrodynamically stable displacement. The time shown in the right bottom corner is the time when the longest radius reached 42 mm.

Figure 8 shows the fluid displacement patterns with the phase diagram shown in Figure 3. The curve in the figure is from [36] and indicates the boundary between fully miscible and partially miscible zones. The patterns are categorized into three types: the circular pattern (•), the finger-like pattern (♦), and the annular-like pattern (▲) in Figure 8. If we pay close attention to the boundary between the finger-like pattern (♦) and the annular-like pattern (▲), both PEG and Na_2SO_4 concentrations are related to the pattern formation, which means that the morphologies are affected by the complexity of the hydrodynamic effect, such as the viscosity and thermodynamic effect such as phase separation. On the other hand, the patterns are circular (•) when the Na_2SO_4 concentration is less than 5 wt % in Figure 8, regardless of the PEG and Na_2SO_4 concentrations, because the displacement patterns are determined only by the hydrodynamic effect of viscosity.

The extent of the phase separation can be expressed by the progression of the phase separation. The progression of the phase separation can be considered from the growth rate of the interfacial tension (IFT) because the formation of an interface due to phase separation induces the increase in interfacial free energy. Figure 9a shows the time evolution of IFT between 20 wt % Na_2SO_4 solution and PEG solutions with several concentrations. IFT, at all concentration ranges, increases with time. The steady value of the IFT increases with the decrease in PEG concentration. The IFT of the fully miscible systems decreases with time because the width of the interface becomes wider due to molecular diffusion [10]. In contrast, the IFT of partially miscible systems increases with time because the interface becomes sharp due to phase separation. Figure 9b depicts the rate constant, k, defined as $\gamma = (\gamma_0 - \gamma_\infty)e^{-kt} + \gamma_\infty$, where γ_0 and γ_∞ are initial and steady values of IFT in Figure 9a, respectively. The relaxation process of IFT corresponds to that of phase separation. Therefore, the rate constant, k, is thought to represent the progress of phase separation. The rate is higher with the decrease in PEG concentration because the formation rate of the interface decreases with the increase in the viscosity of the displacing fluid. Therefore, the progress of phase separation is important for forming patterns, for example, the annular-like pattern (Figure 6b) for high rates and the finger-like patterns (Figure 6c) for low rates. Moreover, the progression is affected by the mobility-like viscosity contrast, and the viscosity is affected by the concentrations of the components (here, PEG). Thus, we compared the patterns using a dimensionless number considering those effects, such as modified capillary number, Ca′, including the viscosity, interfacial tension, and flow rate.

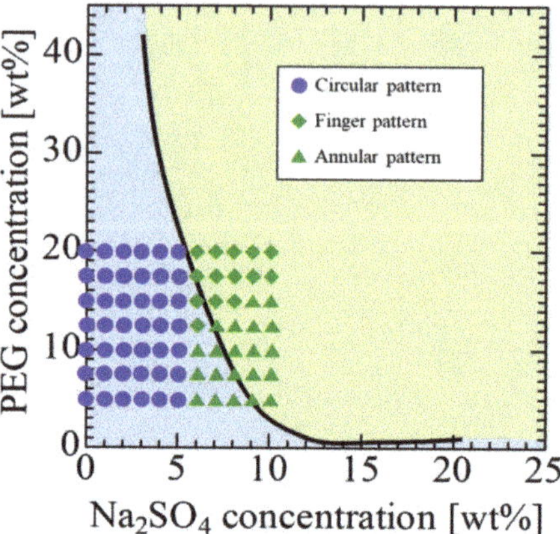

Figure 8. Phase diagram of the displacement patterns.

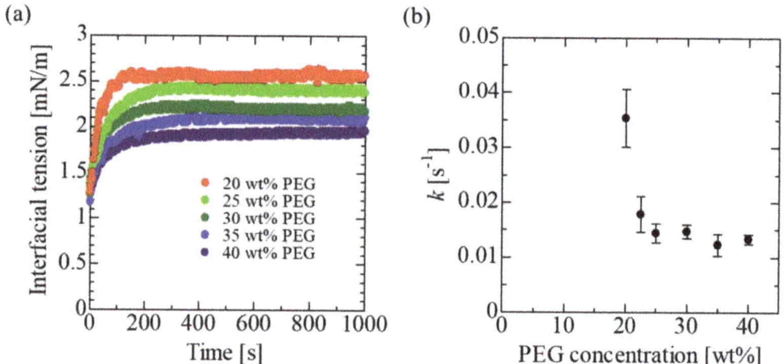

Figure 9. (**a**) Time evolution of the interfacial tension between PEG solutions and 20 wt % Na$_2$SO$_4$ solution; (**b**) the phase separation rate constant, k, of (**a**).

In order to explore the factors on which the fluid displacement patterns depend, we evaluated the effects of physicochemical and hydrodynamic properties on the patterns using two dimensionless numbers: the well-known capillary number, Ca′, and the newly defined body force number, B$_f$. The modified capillary number, Ca′, is defined as Ca′ $= \frac{\mu q}{b^2 \gamma}$, where μ (Pa·s) is viscosity, q (m^3/s) is the flow rate, b (m) is the gap between the cells, and γ (N/m) is an interfacial tension between displacing and displaced liquids [37–41]. It is noted that Ca′ at fully miscible systems cannot be defined because IFT in the fully miscible systems is almost zero. A dimensionless number, B$_f$, represents the relative effect of the body force driven by thermodynamic instability versus the pressure gradient related to Darcy's law, which was introduced in [10]:

$$B_f = \frac{\frac{\Delta\gamma}{2\pi r_m b}}{\frac{\mu V}{\kappa}} = \frac{\Delta\gamma b^2}{\mu q}, \tag{3}$$

where $\Delta\gamma = \gamma^* - \gamma_0$, γ_0 is the initial value of the measured interfacial tension, γ^* is interfacial tension at a time when the longest radius, r_m, reaches 42 mm, κ $(= b^2/12)$ is the permeability of the Hele-Shaw cell, μ is the viscosity of the displacing fluid, V $(= q/2\pi r_m b)$ is linear velocity, and q is the injected flow rate. We neglected the numerical coefficient on the right-hand side for simplicity. We introduced a deformation index as a quantitative evaluation index for the extent of interfacial deformation. The index, ρ_{di}, is defined as the white area involving areas of Na_2SO_4-rich regions generated by phase separation and created by deformation, divided by the area of a circle with maximum radius of 42 mm as shown in Figure 10a. Here, we measured the patterns of the immiscible system, the conditions of which are described in [10], to better understand the mechanism of the pattern formation of all cases. The ρ_{di} of the immiscible system is almost zero because the patterns are perfectly circular, as mentioned in the Introduction.

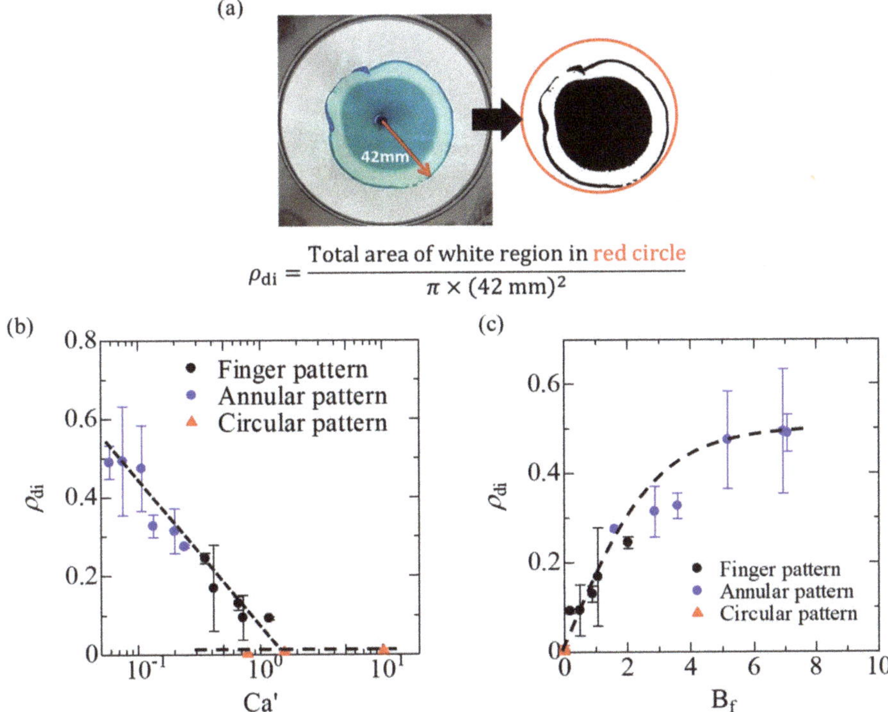

$$\rho_{di} = \frac{\text{Total area of white region in red circle}}{\pi \times (42 \text{ mm})^2}$$

Figure 10. (a) The definition of ρ_{di}. (b) The relationship between the modified capillary number, Ca', and the ρ_{di} for the patterns. (c) The pattern evaluation with B_f. The fitted curves are for a better visualization of the effects of Ca' and B_f.

Figure 10b shows the relationship between the deformation index and Ca' in the partially miscible and the immiscible systems. The value of ρ_{di} decreases with increasing Ca', but the immiscible system indicates that the value of ρ_{di} remains at zero regardless of Ca' because the displacement of the immiscible system is hydrodynamically and thermodynamically stable. Thus, the well-known Ca' is not appropriate for a comprehensive analysis of the fluid displacement using various types of fluid systems. Therefore, we investigated the relationship between ρ_{di} and the newly defined B_f. For the immiscible system, IFT cannot change with time because the system is under thermodynamic equilibrium, and the value of B_f is zero. Figure 10c shows the relationship between ρ_{di} and B_f. The value of ρ_{di} increases with increasing B_f, and the patterns change from circular to finger-like to annular-like patterns. The data of the patterns seem to collapse well onto a single curve, even with the immiscible patterns with the variation in B_f, which is not with the case for the effects on Ca'. This relation indicates that, as the thermodynamic instability is dominant over the viscous dissipation, the degree of deformation becomes larger. The increase in IFT due to phase separation enhances the interface deformation because the increase in IFT corresponds to the increase in Korteweg force, which is a driving force for spontaneous convection. The pattern formation is attributed to the combination of the hydrodynamic effect (viscosity), molecular diffusion, and phase separation (thermodynamic effect). Because B_f includes these effects, the ρ_{di} can be expressed as a function of B_f on a single curve, regardless of the miscibility.

4. Conclusions

We investigated the deformation index of various patterns such as circular, finger-like, and annular-like patterns during fluid displacement, while changing the progression of phase separation using the newly defined dimensionless number, B_f. The fully miscible cases under the hydrodynamically stable displacement condition, where the more viscous PEG solution displaces the less viscous Na_2SO_4 solution, show constant circular patterns. For the partially miscible systems, phase separation occurs at the interface between displacing and displaced solutions, creating a separated region with domain growth for the annular patterns and creating a deformed interface for the finger pattern. Moreover, the deformation index of the patterns can be scaled with B_f, which involves viscous dissipation, molecular diffusion, and phase separation. Therefore, the patterns in the partially miscible systems are proved to be formed by the competition of viscous dissipation and phase separation. B_f is a crucial factor to describe the complex morphologies induced by the fluid displacement in partially miscible systems. The investigated morphologies and B_f will directly contribute to predicting and/or controlling CO_2-enhanced oil recovery, where more viscous water displaces less viscous CO_2 under the ground, the conditions of which are partially miscible.

Author Contributions: Experiments, R.T. and R.X.S.; discussion and review, R.X.S., R.T., Y.N., M.M., and T.B.; writing, R.X.S. and T.B. All authors have read and agreed to the published version of the manuscript.

Funding: This is study was supported by JSPS KAKENHI Grant No. 19J12553.

Conflicts of Interest: The authors declare no conflict of interest.

References

1. Broyles, B.S.; Shalliker, R.A.; Cherrak, D.E.; Guiochon, G. Visualization of viscous fingering in chromatographic columns. *J. Chromatogr. A* **1998**, *822*, 173–187. [CrossRef]
2. Bhaskar, K.R.; Garik, P.; Turner, B.S.; Bradley, J.D.; Bansil, R.; Stanley, H.E.; LaMont, J.T. Viscous fingering of HCl through gastric mucin. *Nature* **1992**, *360*, 458–461. [CrossRef] [PubMed]
3. Nordbotten, J.M.; Celia, M.A.; Bachu, S. Injection and storage of CO2 in deep saline aquifers: Analytical solution for CO2 plume evolution during injection. *Transp. Porous Media* **2005**, *58*, 339–360. [CrossRef]
4. Pojman, J.A.; Gunn, G.; Patterson, C.; Owens, J.; Simmons, C. Frontal dispersion polymerization. *J. Phys. Chem. B* **1998**, *102*, 3927–3929. [CrossRef]
5. Lake, L.W.; Johns, R.T.; Rossen, W.R.; Pope, G.A. *Fundamentals of Enhanced Oil Recovery*; Society of Petroleum Engineers: Richardson, TX, USA, 2014; ISBN 978-1-61399-328-6.
6. Saffman, P.G.; Taylor, G. The penetration of a fluid into a porous medium or Hele-Shaw cell containing a more viscous liquid. *Proc. R. Soc. A* **1958**, *245*, 312–329. [CrossRef]
7. Homsy, G.M. Viscous fingering in porous media. *Annu. Rev. Fluid Mech.* **1987**, *19*, 271–311. [CrossRef]
8. Darcy, H.P.G. *Les Fontaines Publiques de la Ville de Dijon*; Victor Dalmont: Paris, France, 1856.
9. Wyckoff, R.D.; Botset, H.G.; Muskat, M.; Reed, D.W. The measurement of the permeability of porous media for homogeneous fluids. *Rev. Sci. Instrum.* **1933**, *4*, 394–405. [CrossRef]
10. Suzuki, R.X.; Nagatsu, Y.; Mishra, M.; Ban, T. Fingering pattern induced by spinodal decomposition in hydrodynamically stable displacement in a partially miscible system. *Phys. Rev. Fluids* **2019**, *4*, 104005. [CrossRef]

11. McCloud, K.V.; Maher, J.V. Experimental perturbations to Saffman-Taylor flow. *Phys. Rep.* **1995**, *260*, 139–185. [CrossRef]

12. Nagatsu, Y. Viscous fingering phenomena with chemical reactions. *Curr. Phys. Chem.* **2015**, *5*, 52–63. [CrossRef]

13. De Wit, A. Chemo-hydrodynamic patterns in porous media. *Philos. Trans. R. Soc. A Math. Phys. Eng. Sci.* **2016**, *374*, 20150419. [CrossRef] [PubMed]

14. De Wit, A. Chemo-hydrodynamic patterns and instabilities. *Annu. Rev. Fluid Mech.* **2020**, *52*, 531–555. [CrossRef]

15. Chan, C.K.; Liang, N.Y. Observations of surfactant driven instability in a Hele-Shaw cell. *Phys. Rev. Lett.* **1997**, *79*, 4381–4384. [CrossRef]

16. Krechetnikov, R.; Homsy, G.M. On a new surfactant-driven fingering phenomenon in a Hele-Shaw cell. *J. Fluid Mech.* **2004**, *509*, 103–124. [CrossRef]

17. Fernandez, J.; Krechetnikov, R.; Homsy, G.M. Experimental study of a surfactant-driven fingering phenomenon in a Hele-Shaw cell. *J. Fluid Mech.* **2005**, *527*, 197–216. [CrossRef]

18. Tang, H.; Grivas, W.; Homentcovschi, D.; Geer, J.; Singler, T. Stability considerations associated with the meniscoid particle band at advancing interfaces in Hele-Shaw suspension flows. *Phys. Rev. Lett.* **2000**, *85*, 2112–2115. [CrossRef]

19. Bihi, I.; Baudoin, M.; Butler, J.E.; Faille, C.; Zoueshtiagh, F. Inverse Saffman-Taylor experiments with particles lead to capillarity driven fingering instabilities. *Phys. Rev. Lett.* **2016**, *117*, 034501. [CrossRef]

20. Xu, F.; Kim, J.; Lee, S. Particle-induced viscous fingering. *J. Nonnewton. Fluid Mech.* **2016**, *238*, 92–99. [CrossRef]

21. Kim, J.; Xu, F.; Lee, S. Formation and destabilization of the particle band on the fluid-fluid interface. *Phys. Rev. Lett.* **2017**, *118*, 074501. [CrossRef]

22. Podgorski, T.; Sostarecz, M.C.; Zorman, S.; Belmonte, A. Fingering instabilities of a reactive micellar interface. *Phys. Rev. E* **2007**, *76*, 016202. [CrossRef]

23. Gérard, T.; De Wit, A. Miscible viscous fingering induced by a simple A+B→C chemical reaction. *Phys. Rev. E* **2009**, *79*, 016308. [CrossRef] [PubMed]

24. Nagatsu, Y.; De Wit, A. Viscous fingering of a miscible reactive A + B → C interface for an infinitely fast chemical reaction: Nonlinear simulations. *Phys. Fluids* **2011**, *23*, 043103. [CrossRef]

25. Riolfo, L.A.; Nagatsu, Y.; Iwata, S.; Maes, R.; Trevelyan, P.M.J.; De Wit, A. Experimental evidence of reaction-driven miscible viscous fingering. *Phys. Rev. E* **2012**, *85*, 015304. [CrossRef] [PubMed]

26. Nagatsu, Y.; Ishii, Y.; Tada, Y.; De Wit, A. Hydrodynamic fingering instability induced by a precipitation reaction. *Phys. Rev. Lett.* **2014**, *113*, 024502. [CrossRef]

27. Mishra, M.; Trevelyan, P.M.J.; Almarcha, C.; De Wit, A. Influence of double diffusive effects on miscible viscous fingering. *Phys. Rev. Lett.* **2010**, *105*, 204501. [CrossRef]

28. Fu, X.; Cueto-Felgueroso, L.; Juanes, R. Thermodynamic coarsening arrested by viscous fingering in partially miscible binary mixtures. *Phys. Rev. E* **2016**, *94*, 033111. [CrossRef]

29. Amooie, M.A.; Soltanian, M.R.; Moortgat, J. Hydrothermodynamic mixing of fluids across phases in porous media. *Geophys. Res. Lett.* **2017**, *44*, 3624–3634. [CrossRef]

30. Fu, X.; Cueto-Felgueroso, L.; Juanes, R. Viscous fingering with partially miscible fluids. *Phys. Rev. Fluids* **2017**, *2*, 104001. [CrossRef]

31. Suzuki, R.X.; Nagatsu, Y.; Mishra, M.; Ban, T. Phase separation effects on a partially miscible viscous fingering dynamics. *J. Fluid Mech.* **2020**, *898*, A11. [CrossRef]

32. Korteweg, D.J. Sur la forme que prennent les équations du mouvement des fluides si l'on tient compte des forces capillaires causées par des variations de densité. *Arch. Néerl. Sci. Exactes Nat.* **1901**, *6*, 1–24.

33. Molin, D.; Mauri, R. Enhanced heat transport during phase separation of liquid binary mixtures. *Phys. Fluids* **2007**, *19*, 074102. [CrossRef]

34. Budroni, M.A.; Riolfo, L.A.; Lemaigre, L.; Rossi, F.; Rustici, M.; De Wit, A. Chemical control of Hydrodynamic instabilities in partially miscible two-layer systems. *J. Phys. Chem. Lett.* **2014**, *5*, 875–881. [CrossRef] [PubMed]

35. Budroni, M.A.; Thomas, C.; De Wit, A. Chemical control of dissolution-driven convection in partially miscible systems: Nonlinear simulations and experiments. *Phys. Chem. Chem. Phys.* **2017**, *19*, 7936–7946. [CrossRef] [PubMed]

36. Snyder, S.M.; Cole, K.D.; Sziag, D.C. Phase compositions, viscosities, and densities for aqueous two-phase systems composed of polyethylene glycol and various salts at 25 °C. *J. Chem. Eng. Data* **1992**, *37*, 268–274. [CrossRef]
37. Rauseo, S.N.; Barnes, P.D.; Maher, J.V. Development of radial fingering patterns. *Phys. Rev. A* **1987**, *35*, 1245–1251. [CrossRef]
38. Chen, J.D. Growth of radial viscous fingers in a Hele-Shaw cell. *J. Fluid Mech.* **1989**, *201*, 223–242. [CrossRef]
39. Fernandez, J.; Homsy, G.M. Viscous fingering with chemical reaction: Effect of in-situ production of surfactants. *J. Fluid Mech.* **2003**, *480*, 267–281. [CrossRef]
40. Tsuzuki, R.; Ban, T.; Fujimura, M.; Nagatsu, Y. Dual role of surfactant-producing reaction in immiscible viscous fingering evolution. *Phys. Fluids* **2019**, *31*, 022102. [CrossRef]
41. Tsuzuki, R.; Tanaka, R.; Ban, T.; Nagatsu, Y. Deviation from capillary number scaling of nonlinear viscous fingering formed by the injection of Newtonian surfactant solution. *Phys. Fluids* **2019**, *31*, 042108. [CrossRef]

Article

Microstructure and Inertial Characteristics of MHD Suspended SWCNTs and MWCNTs Based Maxwell Nanofluid Flow with Bio-Convection and Entropy Generation Past a Permeable Vertical Cone

Zahir Shah [1],*, Ebraheem Alzahrani [2], Muhammad Jawad [3] and Umair Khan [4]

1   Department of Mathematics, University of Lakki Marwat, Lakki Marwat 28420, Pakistan
2   Department of Mathematics, Faculty of Science, King Abdulaziz University, P. O. Box 80203, Jeddah 21589, Saudi Arabia; eoalzahrani@kau.edu.sa
3   Department of Mathematics, Abdul Wali Khan University, Mardan 23200, Pakistan; muhammadjawad175@yahoo.com
4   Department of Mathematics and Social Sciences, Sukkur IBA University, Sukkur 65200, Pakistan; umairkhan@iba-suk.edu.pk
*   Correspondence: zahir@ulm.edu.pk

Received: 28 August 2020; Accepted: 16 October 2020; Published: 19 October 2020

Abstract: In this research work, our goal is to scrutinize the case, where water-based nanofluids having single-wall and multi-wall carbon nanotubes (CNTs) flow through a vertical cone. The second law of thermodynamic is taken for the aim of scheming effective heat storage units. The body package is layered in convective heat and diluted permeable medium. The effects of Joule heating, rotary microorganisms, heat generation/absorption, chemical reactions, and heat radiation increase the novelty of the established model. By using a local similarity transformation technique, the partial differential equations (PDEs) change into a coupled differential equation. By using the numerical technique, bvp4c, to get the solution of the conservation equations and their relevant boundary conditions. The parameters appearing in the distribution analysis of the alliance are scrutinized in detail, and the consequences are depicted graphically. It can be perceived that in the situation of composed nanotubes, the velocity of fluid decreases as the magnetic field is increased.

Keywords: gyrotactic microorganisms; micropolar magnetohydrodynamics (MHD); Maxwell nanofluid; single wall carbon nanotubes (SWCNTs) and multi wall carbon nanotubes (MWCNTs); thermal radiation; chemical reaction; mixed convection; permeability

---

1. Introduction

In the present era, nanofluid is getting importance from the researchers due to its diverse application in the industrial field. For instance, they are available in polymer manufacturing, gas turbines, power generators, glass fabric, paper production, wire drawing, and many more. Nanofluid is a sort of heat transport medium containing nanoparticles under 100 nm, which are reliably and consistently scattered in a base fluid like water, oil, and ethylene glycol. These scattered nanoparticles, for the most part, a metal or metal oxide massively improve the thermal conductivity of the nanofluid, upgrades conduction, and convection coefficients, mulling over more heat transport. Enhancing the thermal specifications of liquids leads to a greater level of connective flow in thermal units. For heat transfer enhancement, adding additives to the operant liquids for modifying their thermal features are very attractive method. For this, a way has been represented by enhancement in nanotechnology. The meaning of 'nanofluid' has been expressed by Choi [1] in 1995 for increasing heat transfer specifications of convectional fluids. A total report on the productivity of temperature transformation in the sunlight-based authority with the nanoliquid was done by Chen et al. [2]. They originate that the presentation of photo thermal change in the gatherer expanded by 96.93% and 52% at 30 and 75 °C separately, an entirely reasonable liquid for sun powered authorities which have the shortest fascination in low-temperatures. Oudina [3,4] examined nanomaterial conduct esoteric an annulus with different designs of temperature foundations. They used an arithmetical methodology and introduced soundness investigation. Mesoscopic line for investigating nanomaterial course through permeable area was introduced by Sheikholeslami [5]. They utilized Lorentz force to switch the stream style inside an opening. Chougule et al. [6] initiate that because of the low association of nano-powders, a pressure drop is critical in the curved cylinder when they utilized carbon nanotubes (CNTs) instead of unadulterated water. Besides, a curved cylinder with loop supplements improves better Nusselt number as a result of the rate of energy altercation increment because of this reality nano-powders have arbitrary and unpredictable movement in the liquids. Numerically, the plan assessment of a whirling stream microchannel for incredible warmth transition uses had been examined by Hartmann-Priesnitz et al. [7] who displayed the operant liquids as Cu-H_2O nanomaterial in the laminar stream. Ding et al. [8] portrayed nanofluids that depend on CNTs, which depend upon the heat of the base liquid, and set up that when the liquid heat is 25 °C, the warm conductivity can be expanded by up to 30%, but can be increased by 79% observed at 40 °C. Pop and Watanabe [9] carried out a theoretical analysis with the main aim to discuss the influence of injection/suction on fluid flowing over a cone with free convection and heat flux. The authors used different differential methods to solve the existing equation describing the flow. Xu [10] recently studied time-dependent hybrid nanofluid with mixed convection in rotating disks multiple kinds of nanoparticles are taken here. A numerical approach is used for the solution. Flaccid devices are used by investigators in an earlier investigation to improve the convective coefficient [11]. Many researchers used a combination of both concepts to enhance convective coefficient by use of insert with nanofluids. Heat transfer and friction element characteristics on warped tape with Al_2O_3/water nanofluid are analyzed by Sharma et al. [12]. Zhang [13] has investigated adapted computational approaches in which a 2-D effective heat capacity model is used for forecasting the fleeting heat transmission procedure of the building envelopes equipped with phase change materials (PCM). The deviations against test data made by the principal procedures were viewed as especially bigger than the changed methodologies adjusted techniques. Sun et al. [14] have experimentally deliberated the heat transmission rate augmentation produced by natural convection of PCMs experiencing melting. The thermophysical properties of nanofluids have been discussed by Phuoc et al. [15]. The increase in the transfer of heat of SWCNTs-glycol-based nanofluids was examined by Harish et al. [16]. It was found that while 0.2% by volume of SNTs was added to ethylene glycol, the thermal conductivity increased by 14.8%. The magnetohydrodynamic (MHD) 3-D Maxwell nanofluid boundary layer flow with convective boundary conditions on a biaxially stretched sheet was explored by Hayat et al. [17].

Nowadays the main concerns of scientists are to make the strategies that control the ingesting of skilled vitality. In the field of thermal structuring, the key objective is to achieve the best viability of contraptions and with the base loss of warmth, scouring, and spread during the mechanical procedures. All the heat gadgets take a shot at the guideline of thermodynamics and produce. Thermodynamic second laws used to look at the irreversibility in terms of the entropy age rate. Entropy growth is abused to elucidate the exhibition of various settings in present-day and structure solicitations. Entropy is imitative from the Greek word Entropia, which suggests "moving toward" or "alteration". Entropy figuring is basic as it orders the factors for energy forfeiture. Bejan [18] offered the clue of an entropy generation problem. Ellahi et al. [19] studied the influence of entropy optimization on natural convective nanoliquid stream. A multiple turbulator has been engaged by Sheikholeslami et al. [20] to enlarge the involvement of nanomaterial inside a tube. They verified that thermal irreversibility improves with the upsurge of subordinate flow. The related research work in a similar filed can be seen in [21,22].

The attractive properties of carbon nanotubes (CNTs) include mechanical and chemical stability, excellent thermal and electrical conductivity, lightweight and physicochemical reliability, making them a desirable material in the manufacture of electrochemical devices. Considering this exciting feature of carbon nanotubes, in this research work, our goal was to scrutinize the case where water-based nanofluids having single-wall and multi-wall CNTs flow through a vertical cone. The body package is layered in convective heat and diluted permeable medium. The effects of Joule heating, rotary microorganisms, and heat generation/absorption, chemical reactions, and heat radiation increase the novelty of the established model. By using a local similarity transformation technique, the PDE is changed into a coupled differential equation. By using the Homotopy analysis method to get the solution of the conservation equations and their relevant Boundary conditions. The parameters appearing in the distribution analysis of the alliance are scrutinized, and the consequences are depicted graphically. It can be perceived that on account of the two nanotubes, the velocity of fluid decreases as the magnetic is increased. Moreover, the thickness of moving microorganisms is decreased compared to more estimation of biological convection constants.

2. Mathematical Analysis

We suppose the flow of magnetohydrodynamic (MHD) mixed convective viscous water-based micropolar nanofluidic of CNTs on a vertical cone in a penetrable medium. The coordinate system for the flow phenomena is chosen is such a way that x-coordinate is parallel with the direction of the fluid. Energy expression with thermal radiation dissipation, thermal flux, and Joule heating is measured. Irreversibility investigation with chemical reaction is investigated. Entropy rate is determined. Slip impact is likewise talked about. A magnetic field of constant strength β_0 is applied vertically. The temperature (T) and concentration ponder (C). Speeding up because gravity acts downwards. Problem geometry is featured in Figure 1.

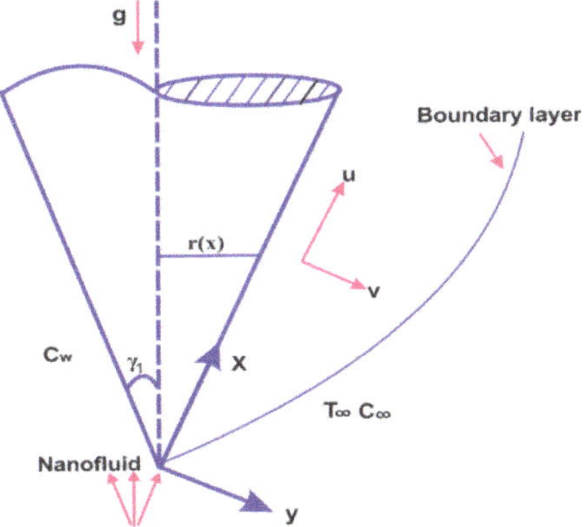

Figure 1. Schematic diagram for the flow direction.

From the above suppositions, the resulting modeled equations are:

$$(ru)_x + (rv)_y = 0 \tag{1}$$

$$uu_x + vu_x + \lambda\left(u^2 u_{xx} + v^2 u_{yy} + 2uvu_{xy}\right) = \frac{\mu_{nf}}{\rho_{nf}}u_{yy} - \frac{\mu_{nf}}{\rho_{nf}}\frac{u}{K} +$$
$$g[\beta(T-T_\infty) - \beta*(C-C_\infty) - \beta*\gamma(n-n_\infty)]\cos\gamma_1 - \frac{\sigma_{nf}\beta_0^2}{\rho_{nf}}u + KN_y \tag{2}$$

$$uT_x + vT_y = \alpha_{nf}T_{yy} - \frac{1}{\left(\rho_{cp}\right)_{nf}}(q_r)_y + \frac{Q_0}{\left(\rho_{cp}\right)_{nf}}(T-T_\infty) + \frac{\sigma\beta_0^2}{\left(\rho_{cp}\right)_{nf}}u^2 \tag{3}$$

$$uN_x + vN_y = \frac{\gamma^*_{nf}}{\rho_{nf}}N_{yy} - \frac{k}{\rho_{nf}}\left(2N + u_y\right) \tag{4}$$

$$uC_x + vC_y = D_m C_{yy} - k_r(C - C_\infty) \tag{5}$$

$$un_x + vn_y + \frac{bW}{(C_w - C_0)}\left(nC_y\right)_y = D_n n_{yy} \tag{6}$$

with the corresponding boundary conditions

$$
\begin{aligned}
&u = 0, v = V_1, N = 0, T_y = \frac{h_f(T_f - T)}{-k_{nf}}, \quad C_w = C = dx + C_0, n = n_w, \quad At \quad y = a\\
&u \to 0, N \to 0, C \to C_\infty = C_0 + ex, T \to T_\infty, n \to n_w \quad At \quad y \to \infty
\end{aligned}
\tag{7}
$$

where, (β, β^*), (μ_{nf}, μ_f), (ρ_{CHT}, ρ_f), β_0, α_{nf}, V_0, h_f, (d, e), Q_0, $((\rho_{cp})_f, (\rho_{cp})_{nf})$, (k_f, k_{nf}, k), D_n, k_r, W_c, D_m, q_r, γ_1 indicate coefficients of thermal and solutal expansion, dynamic viscosities, densities, magnetic strength, thermal modified diffusivity, suction/injection parameter, convective parameter, dimensionless constants of concentration, heat generation/absorption parameter, heat capacities, thermal conductivity, diffusivity of microorganisms, chemical reaction rate, coefficient, extreme cell swimming motion, Brownian diffusion, radiation coefficient, and cone half-angle, respectively. In Equation (7) the term V_1 characterizes the mass transmission and defined as $V_1 = -\left(\frac{3}{4}ax^{-1}R(a_x)^{\frac{1}{4}}\right)V_0$. In case of $V_1 < 0$, the mass transfer is for injection and $V_1 > 0$ shows suction.

$$
\begin{aligned}
&\alpha_{nf} = \frac{k_{nf}}{\rho_{nf}(c_p)_{nf}}, \; \frac{k_{nf}}{k_f} = \frac{(1-\phi)+2\phi\frac{k_{CNT}}{k_{CNT}-k_f}\ell n\left(\frac{k_{CNT}+k_f}{2k_f}\right)}{(1-\phi)+2\phi\frac{k_{nf}}{k_{CNT}-k_f}\ell n\left(\frac{k_{CNT}+k_f}{2k_f}\right)}\\
&v_{nf} = \frac{\mu_{nf}}{\rho_{nf}}, \gamma^*_{nf} = \left(\mu_{nf} + 0.5k\right)j, \rho_{nf} = (1-\phi)\rho_f + \phi\rho_{CNT}, \mu_{nf} = \mu(1-\phi)^{-2.5}\\
&\frac{\sigma_{nf}}{\sigma_f} = 1 - \frac{3\left(\frac{\sigma_s}{\sigma_f}-1\right)\phi}{\left(\frac{\sigma_s}{\sigma_f}-1\right)\phi-\left(\frac{\sigma_s}{\sigma_f}+2\right)}
\end{aligned}
\tag{8}
$$

$$
\begin{aligned}
&\psi = \alpha Ra_x^{1/4} f(\eta), g(\eta) = \frac{C-C_\infty}{C_w-C_0}, \eta = \frac{y}{x}Ra_x^{1/4}\\
&\theta(\eta) = \frac{T-T_\infty}{T_w-T_\infty}, h(\eta) = \frac{n-n_\infty}{n_w-n_\infty}, S(\eta) = \frac{\rho_f x^2 NRa_x^{-3/4}}{\mu_f}
\end{aligned}
\tag{9}
$$

Using Equations (8) and (9), Equation (1) is satisfied and Equations (2)–(6) are written as

$$
\begin{aligned}
&f''' + -k_1 f' + (1-\phi)^{2.5}\left(1-\phi+\phi\frac{\rho_{CNT}}{\rho_f}\right)[\theta - N_r g - R_b h] - (1-\phi)^{2.5}Mf' +\\
&K(1-\phi)^{2.5}\left(1-\phi+\phi\frac{\rho_{CNT}}{\rho_f}\right)S' + \frac{1}{2Pr}(1-\phi)^{2.5}\left(1-\phi+\phi\frac{\rho_{CNT}}{\rho_f}\right)\\
&\left\{f'^2 + \frac{1}{2}ff''\right\} - \frac{\beta}{4Pr}(1-\phi)^{2.5}\left(1-\phi+\phi\frac{\rho_{CNT}}{\rho_f}\right)\left(3f'^3 + \frac{1}{4}f''f^2 - \frac{5}{2}f''f'f\right) = 0
\end{aligned}
\tag{10}
$$

$$\frac{k_{nf}}{k_f}(1+Rd)\theta'' + \frac{3}{4}\left(1-\phi+\phi\frac{\rho_{CNT}}{\rho_f}\right)f\theta' + \gamma\theta + PrEcMf'^2 = 0 \tag{11}$$

$$S'' - \gamma*(2S + \alpha f'') + \frac{1}{4Pr}\frac{(1-\phi)^{2.5}\left(1-\phi+\phi\frac{\rho_{CNT}}{\rho_f}\right)}{(1+0.5K)}(5Sf' + S'f) = 0 \tag{12}$$

$$g'' + \frac{3}{4}Scfg' - Scnf' - G_r g = 0 \tag{13}$$

$$h'' + \frac{3}{4}L_b fh' - P_e(h'g' + (h+\delta)g'') = 0 \tag{14}$$

$$f'(0) = 0, f(0) = V_0, \frac{k_{nf}}{k_f}\theta'(0) = -B_1(1 - \theta(0)), S(0) = 0, h(0) = 1, g(0) = 1 - n,$$
$$f'(\infty) \to 0, S(\infty) \to 0, \theta(\infty) \to 0, h(\infty) \to 0, g(\infty) \to 0 \tag{15}$$

Non-dimensional form of parameters is specified and defined as below in Abbreviations.

3. Entropy Generation Modeling

To include the irreversibility sources, below equations can be used:

$$S'''_{gen} = \frac{k_{nf}}{k_f}\left(1 + \frac{16T_\infty^3 \sigma*}{3k*k_{nf}}\right)\left(T_{yy}\right) + \frac{\mu_{nf}}{T_\infty}\left(u_y\right)^2 + \frac{\sigma}{T_\infty}\beta_0^2 u^2 + \frac{\mu_{nf}}{T_\infty k}u^2 +$$
$$\frac{RD}{C_\infty}\left(C_y\right)^2 + \frac{RD}{T_\infty}\left(T_y\right)\left(C_y\right) \tag{16}$$

where

$$N_G = \frac{S'''_{gen}}{S'''_0} \tag{17}$$

$\left(S'''_{gen}\right)$ is irreversibility optimization rate and (S''') the characteristic irreversibility optimization rate signified by:

$$N_G = \frac{k_{nf}}{k_f}(1 + R)Ra_x\theta'^2 + \frac{1}{(1-\phi)^{2.5}}\frac{BrRa_x}{\alpha}\left(f''^2 + k_1 f'^2\right) +$$
$$\frac{BrRa_x M}{\alpha}f'^2 + \lambda\left(\frac{\xi}{\alpha}\right)Ra_x g'^2 + \frac{\xi}{\alpha}Ra_x\lambda\theta' g' \tag{18}$$

where $Br\left(= \frac{\mu_f U_w}{k_f \Delta T}\right)$ Brinkman number, $\alpha\left(= \frac{\Delta T}{T_\infty}\right)$ diffusion parameter, $\xi\left(= \frac{\Delta C}{C_\infty}\right)$ concentration ratio parameter, and $\lambda\left(= \frac{RDC_\infty}{k_f}\right)$ temperature difference parameter, respectively.

4. Engineering Quantities

4.1. Skin Friction Coefficients C_{Fx}

Mathematically, it is defined as

$$C_{Fx} = \frac{2\varphi_w}{\rho u_\infty^2} \tag{19}$$

The dimensionless form is

$$Re_x^{1/2}C_{Fx} = (1 + K)\frac{1}{(1 - \phi)^{2.5}}f''(0) \tag{20}$$

In which $Re_x^{1/2}$ designates Reynold number.

4.2. Heat Transfer Rate

Nu_x is

$$Nu_x = \frac{xQ_w}{k(T_w - T_\infty)} \tag{21}$$

where the heat flux Q_w is

$$Q_w = -\frac{k_{nf}}{\kappa}\left(\frac{4\sigma_e}{3\kappa\kappa_R}T_\infty^3 + 1\right)T_y\bigg|_{y=0} \tag{22}$$

$$Re_x^{-1/2}Nu_x = -\frac{k_{nf}}{k_f}(1 + Rd)\theta'(0) \tag{23}$$

4.3. Mass Transfer Rate

Sherwood number Sh_x is defined as

$$Sh_x = \frac{xh_w}{D_B(C_w - C_\infty)} \tag{24}$$

$$h_w = -D_B C_y\big|_{y=0} \tag{25}$$

$$Re_x^{-1/2} Sh_x = -g'(0) \tag{26}$$

4.4. Local Density of Motile Microorganisms

Sherwood number N_{nx} are stated as

$$N_{nx} = \frac{xQ_n}{D_n(n_w - n_\infty)} \tag{27}$$

$$Q_n = -D_n n_y\big|_{y=0} \tag{28}$$

After simplification

$$Re_x^{-1/2} N_{nx} = -h'(0) \tag{29}$$

5. Solution Technique

In this present work, the constitutive equations for the given problem are mathematically modeled under the following assumptions, heat generation/absorption, permeable medium, the geometry is taken to be a vertical cone, SWCNTs-water, and MWCNTs-water, convective boundary circumstances. The highly non-linear ODEs were obtained from PDEs by using the transformation technique. Hence, we employed a numerical system on the highly non-linear transformed differential equations.

Equations (12)–(16), along with boundary value problem (17) and (18), called a bvp4c just a name built-in Matlab function which is based on finite difference method and found the solutions computationally which is captured both in tables. Equations (1)–(6) and as well as in graphs. For this purpose, we can transform our differential equations into an arrangement of first order differential equations by letting the new factors

$$f = y_1, f' = y_2, f'' = y_3, \theta = y_4, \theta' = y_5, S = y_6, S' = y_7, g = y_8, g' = y_9, h = y_{10}, h' = y_{11} \tag{30}$$

Exercising the above new variables in the Equations (12)–(16) then the following first order differential equations is achieved:

$$\begin{pmatrix} y'_1 \\ y'_2 \\ y'_3 \\ y'_4 \\ y'_5 \\ y'_6 \\ y'_7 \\ y'_8 \\ y'_9 \\ y'_{10} \\ y'_{11} \end{pmatrix} = \begin{pmatrix} y_2 \\ y_3 \\ \begin{array}{l} k_1 y_2 - (1-\phi)^{2.5}\left(1-\phi+\phi\frac{\rho_{CNT}}{\rho_f}\right)[y_4 - N_r y_8 - R_b y_{10}] + (1-\phi)^{2.5} M y_2 - K(1-\phi)^{2.5}\left(1-\phi+\phi\frac{\rho_{CNT}}{\rho_f}\right)y_7 \\ \quad -\frac{1}{2D_r}(1-\phi)^{2.5}\left(1-\phi+\phi\frac{\rho_{CNT}}{\rho_f}\right)\left\{(y_7)^2 + \frac{1}{8}y_1 y_2\right\}+ \\ \quad \frac{\beta}{4\text{Pr}}(1-\phi)^{2.5}\left(1-\phi+\phi\frac{\rho_{CNT}}{\rho_f}\right)\left(3(y_2)^3 + \frac{1}{4}(y_1)^2 y_3 - \frac{5}{2} y_3 y_1 y_2\right) \end{array} \\ y_5 \\ \dfrac{\frac{-3}{4}\left(1-\phi+\phi\frac{\rho_{CNT}}{\rho_f}\right)y_1 y_5 - \gamma y_4 - \text{Pr}EcM(y_2)^2}{\frac{k_{nf}}{k_f}(1+R_d)} \\ y_7 \\ \gamma * (2y_6 + \alpha y_3) - \frac{1}{4\text{Pr}(1+0.5K)}(1-\phi)^{2.5}\left(1-\phi+\phi\frac{\rho_{CNT}}{\rho_f}\right)(5y_6 y_2 + y_7 y_1) \\ y_9 \\ -\frac{3}{4} Sc y_1 y_9 + Scn y_2 + G_r y_8 \\ y_{11} \\ \frac{-3}{4} L_b y_1 y_{11} + P_e(y_{11} y_9 + (y_{10} + \delta) y y_4) \end{pmatrix} \tag{31}$$

With initial conditions

$$
\begin{pmatrix}
y_1(0) \\
y_2(0) \\
y_2(\infty) \\
y'_5(0) \\
y_4(\infty) \\
y_6(0) \\
y_6(\infty) \\
y_8(0) \\
y_8(\infty) \\
y_{10}(0) \\
y_{10}(\infty)
\end{pmatrix}
=
\begin{pmatrix}
V_0 \\
0 \\
0 \\
-\dfrac{k_f}{k_{nf}}B_1(1-y_4(0)) \\
0 \\
0 \\
0 \\
1-n \\
0 \\
1 \\
0
\end{pmatrix}
\tag{32}
$$

In this procedure, we can fix the pertinent parameters and then the solution starts with the initial guess supplied at the step size and changes the step size values to get the specified accuracy. The final number of the mesh length is attained by the function of Matlab called bvp4c throughout in the study to get the solutions. The region for the numerical solution should be finite, and the value is taken to be approximately 10 using $\eta = \eta_\infty$. Since, in the current problem we can find the solution for the SWCNTs-water and MWCNTs-water for a single guess of η_∞. Here we have taken the value of $\eta_\infty = 10$ for both SWCNT and MWCNT and found the profiles to reach the far field of the boundary layer conditions asymptotically.

6. Validation of the Results

Comparison between present and previous outcomes for justification is given in this section. Table 1 delineated to certify the accuracy of our current outcome with an available outcome in literature. The comparison of concentration gradient $Sh_x Re_x^{-1/2}$ with variation of Sc while the rest of the parameters are associated with Ref. [21] and shown in Table 1. Obviously the outcome is in acceptable understanding.

Table 1. Numerical variation of mass transfer rate Sh_x and their comparison with Ref. [21] via various value of Sc at $G_r = n = 0.1$, $N_r = 0.5$.

	Ramzan et al. [21]		Present Results	
Sc	$-g'(0)$ SWCNT	$-g'(0)$ MWCNT	$-g'(0)$ SWCNT	$-g'(0)$ MWCNT
0.1	0.31891	0.31882	0.3189450	0.3188567
0.5	0.50221	0.50155	0.5022674	0.5015768
0.9	0.74207	0.74087	0.7420467	0.7408564

7. Discussion

This area is dedicated to the conversation and expectation of the effects of numerous parameters modeled from Equation (2) on $f'(\eta)$. The impact of M, ϕ, N_r, R_b, k_1, β on velocity profile is discussed. Figure 2 presents the effect of the solid volume fraction (ϕ) of Maxwell micropolar nanofluid on velocity function. The motion of the nanoparticle increases for enlarging values of (ϕ). It is noted that $f'(\eta)$ increases quickly for SWCNT in comparison to MWCNT. This augmentation in a motion of the micropolar nanoparticle is noted faster for single-wall carbon nanotube as compared to the multiwall carbon nanotube. Figure 3 presents the influence of M on $f'(\eta)$. The converse disparity is seen amongst M and $f'(\eta)$. The impact of the magnetic force is perpendicular to the Maxwell micropolar nanofluid flow direction executes augment to a resistive force. For a larger value of magnetic parameter (M), the Lorentz forces enhance which raises the forces of resistance of the Maxwell micropolar motion which in turn reduces velocity $f'(\eta)$. Figure 4 presents the impact of a suction parameter V_0 on $f'(\eta)$.

Here it is obvious from Figure 4, that enlarged estimation of the V_0 reduces the SWCNTs and MWCNTs Maxwell nanofluid motion. The impression of the buoyancy proportion parameter N_r for SWCNTs and MWCNTs on $f'(\eta)$ is presented in Figure 5. It is the ratio of nanofluid concentration and temperature difference amongst the layers as well as the intended operative resistance ratio at diverse values of β. From mathematical relation of N_r it is clear that increasing concentration difference $(C_W - C_0)$ augmented N_r, while increasing temperature difference $(T_f - T_0)$ enhances N_r Therefore, the augmented N_r reduced the fluid motion. The impact of bio-convection Rayleigh number R_b in Figure 6 as R_b is a dimensionless number related to the buoyancy-driven of Maxwell micropolar nanofluid flow. From Figure 6 it is cleared that augmented value Rayleigh number R_b reduced the Maxwell micropolar nanofluid motion. It is also found that velocity for MWCNT declines more quickly. In Figure 7, the result of the permeable parameter k_1 on $f'(\eta)$ is drawn. As it is obvious that the permeable medium creates resistance to the fluid motion. From the figure it is perceived that $f'(\eta)$ is decreases with higher permeability k_1. Additionally, the momentum boundary layer reduces with enhances value of k_1.

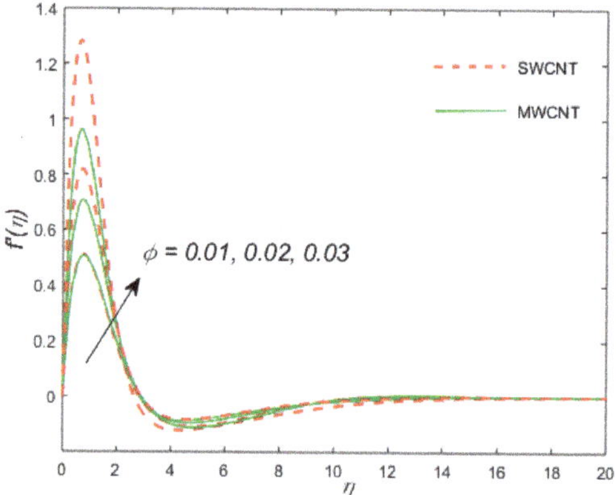

Figure 2. The variation of the velocity distribution profile $f'(\eta)$ for the case of SWCNT and MWCNT versus the similarity variable for the distinct values of the nanoparticle volume fraction (ϕ).

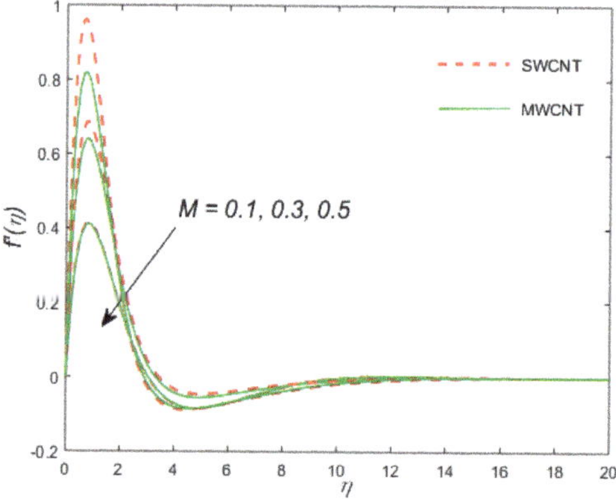

Figure 3. The variation of the velocity distribution profile $f'(\eta)$ for the case of SWCNT and MWCNT versus the similarity variable for the distinct values of the magnetic parameter (M).

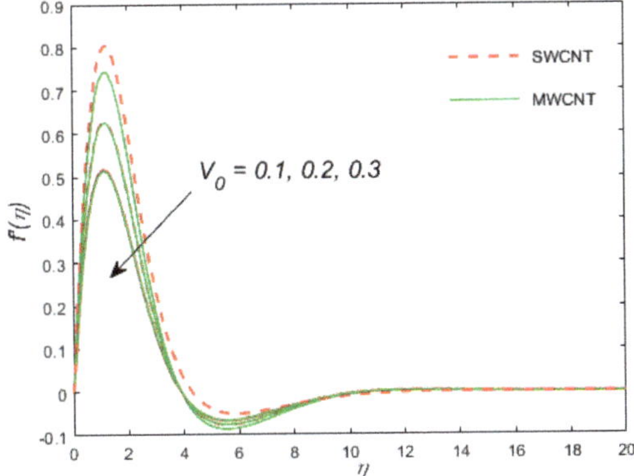

Figure 4. The variation of the velocity distribution profile $f'(\eta)$ for the case of SWCNT and MWCNT versus the similarity variable for the distinct values of the suction parameter (V_0).

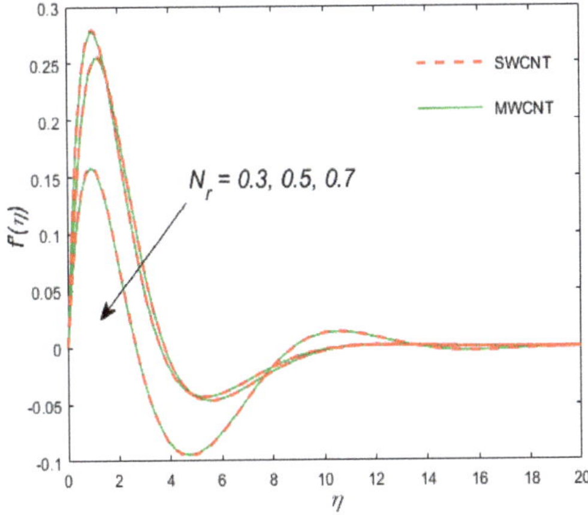

Figure 5. The variation of the velocity distribution profile $f'(\eta)$ for the case of SWCNT and MWCNT versus the similarity variable for the distinct values of the buoyancy ratio parameter (N_r).

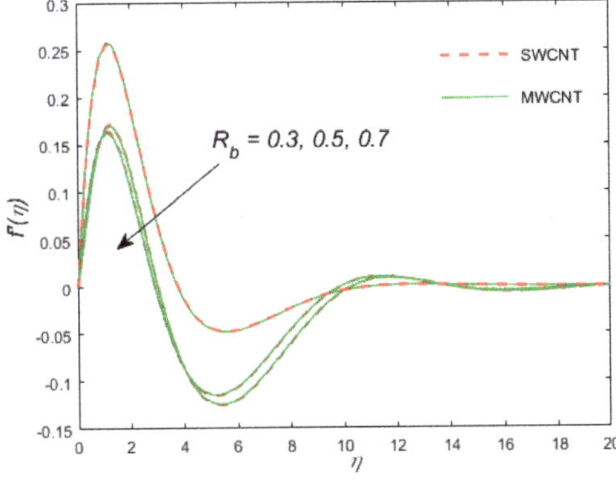

Figure 6. The variation of the velocity distribution profile $f'(\eta)$ for the case of SWCNT and MWCNT versus the similarity variable for the distinct values of the bio-convection Rayleigh number (R_b).

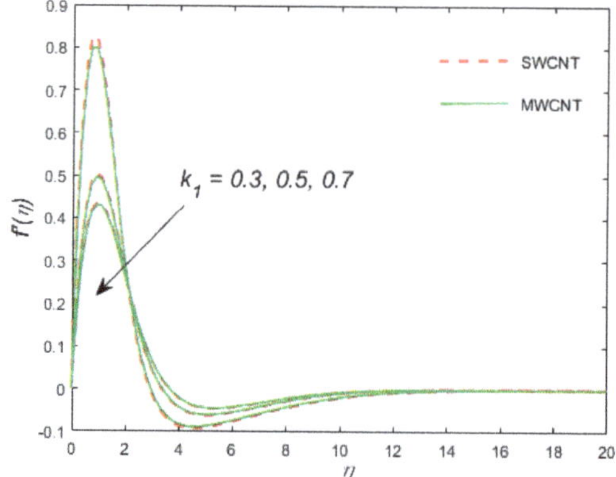

Figure 7. The variation of the velocity distribution profile $f'(\eta)$ for the case of SWCNT and MWCNT versus the similarity variable for the distinct values of the porous parameter (k_1).

7.1. Temperature

The substantial impacts of numerous factors displayed from temperature Equation (3) similar (B_1), (M), (R_d), and (Ec) on temperature distribution profile are shown in Figures 8–11. The impression of Biot number B_1 on temperature distribution function $\theta(\eta)$ is presented in Figure 8. It is seen for the higher value of B_1 the temperature function $\theta(\eta)$ augmented for CNTs Maxwell micropolar nanofluid. Actually, increasing B_1 enhances the heat transmission from the surface becomes equivalent to that added from the exposed field which, in turn, conquers the temperature upsurge at the surface. Figure 9 labeled the impression of radiative parameter R_d on the temperature distribution field $\theta(\eta)$ Augmentation in the $\theta(\eta)$ with enhancement radiation parameter R_d is observed. Actually, intensification radiation causes additional heat which in turn escalates the CNTS Maxwell micropolar nanofluid temperature. Relation between Eckert number Ec and temperature distribution $\theta(\eta)$ is illustrated in Figure 10. Higher value of Ec amplified the kinetic energy of CNTs Maxwell micropolar nanofluid molecules which thus, enhanced the warmth transmission rate. Figure 11 shows the influence of magnetic induction M on temperature distribution $\theta(\eta)$. For higher (M) the strength of Lorentz forces become stronger which enhances the contrasting forces to the Maxwell micropolar nanofluid and results in the temperature distribution being $\theta(\eta)$ enhanced.

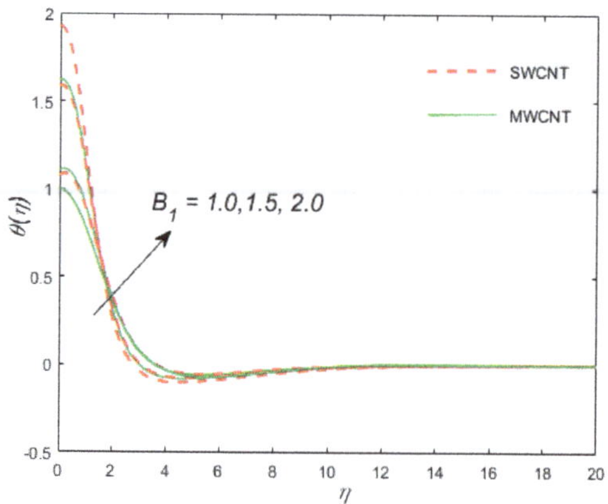

Figure 8. The impact of the temperature distribution profile $\theta(\eta)$ for the case of SWCNT and MWCNT versus the similarity variable for the distinct values of the Boit number (B_1).

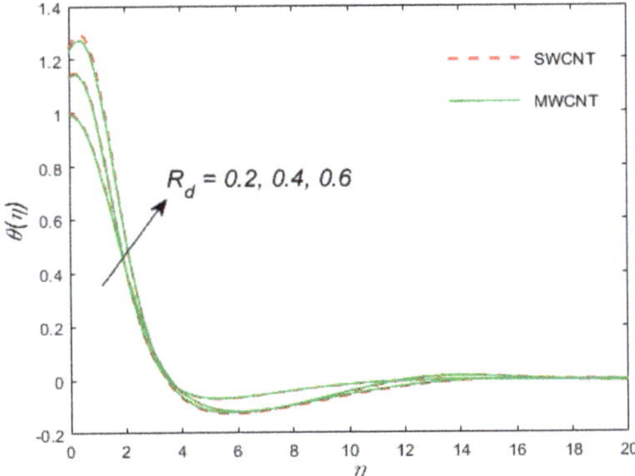

Figure 9. The impact of the temperature distribution profile $\theta(\eta)$ for the case of SWCNT and MWCNT versus the similarity variable for the distinct values of the radiation parameter R_d.

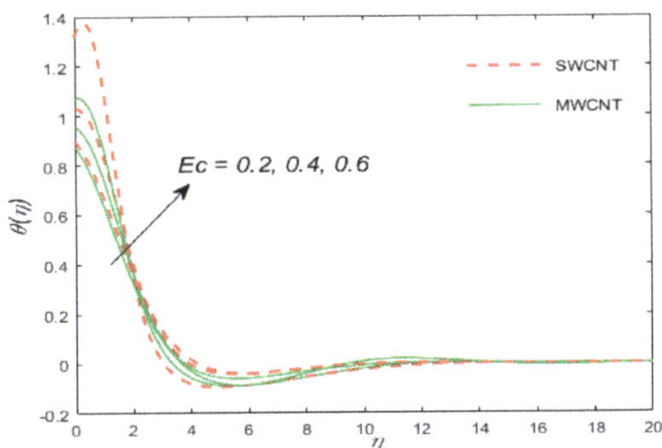

Figure 10. The impact of the temperature distribution profile $\theta(\eta)$ for the case of SWCNT and MWCNT versus the similarity variable for the distinct values of the Eckert number Ec.

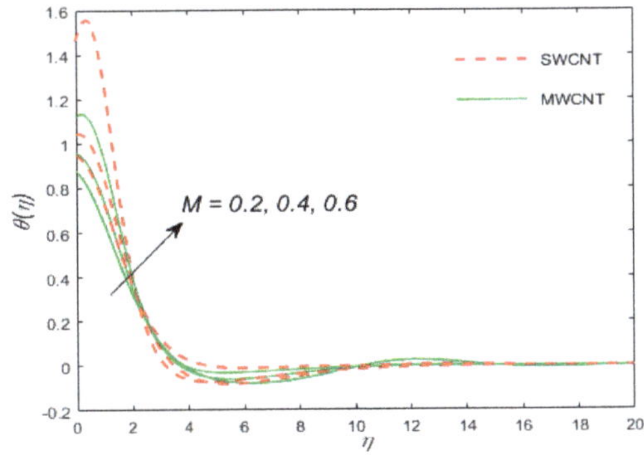

Figure 11. The impact of the temperature distribution profile $\theta(\eta)$ for the case of SWCNT and MWCNT versus the similarity variable for the distinct values of the magnetic parameter M.

7.2. Micro Rotation Profile

This region is stanch to the argument and anticipation of the effects of numerous parameters modeled from Equation (2) on $f'(\eta)$. The impact of K, γ^* and α on micro rotation velocity $S(\eta)$ is discussed, respectively. It is clear from the Figures 12–14, that micro rotation velocity $S(\eta)$ augmented with higher value of K, γ^*, and α.

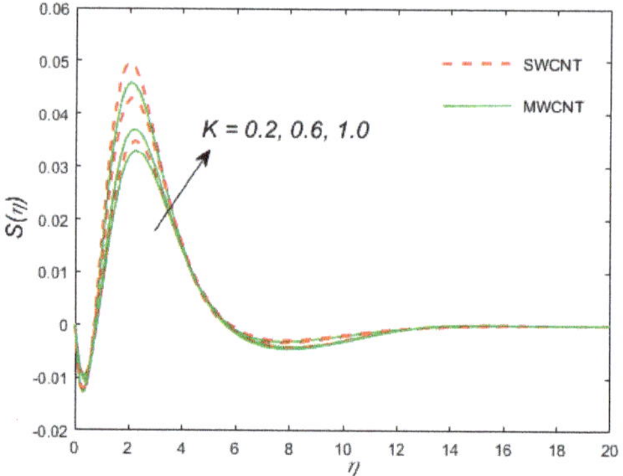

Figure 12. The variation of the micro rotation profile $S(\eta)$ for the case of SWCNT and MWCNT versus the similarity variable for the distinct values of the material parameter K.

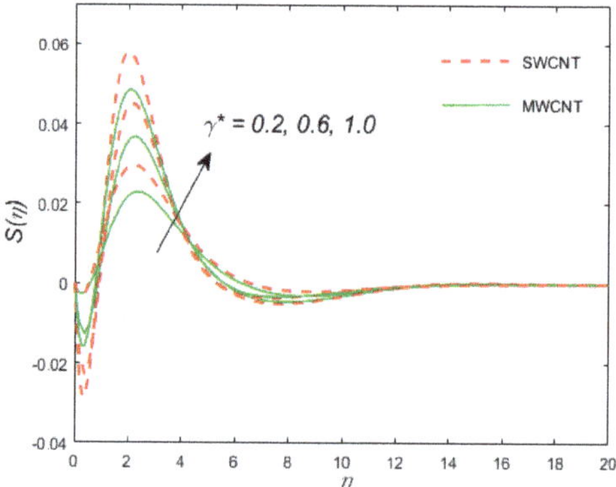

Figure 13. The variation of the micro rotation profile $S(\eta)$ for the case of SWCNT and MWCNT versus the similarity variable for the distinct values of the vortex velocity or material parameter γ^*.

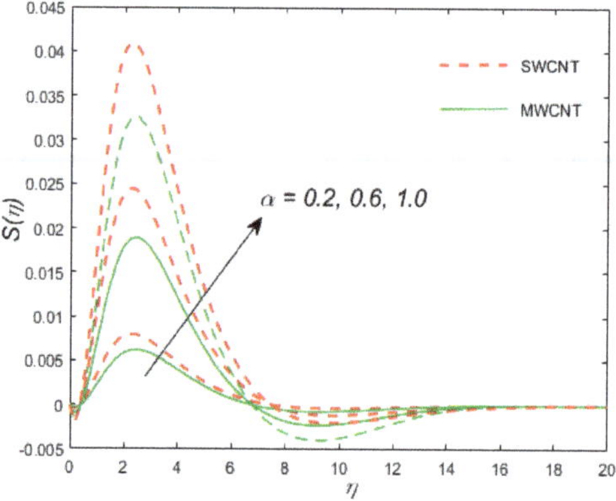

Figure 14. The variation of the micro rotation profile $S(\eta)$ for the case of SWCNT and MWCNT versus the similarity variable for the distinct values of the temperature difference parameter α.

7.3. Concentration

This part presented the discussion and anticipation of the effects of numerous parameters modeled from Equation (4) on $g(\eta)$. The impact of Sc and G_r on concentration profile is discussed, respectively. Figure 15 is illustrated to examine the performance of (Sc) on $g(\eta)$. Higher values of (Sc) the diffusivity of mass declines and thus concentration is deteriorated. The influence of the chemical response factor G_r on the concentration of Maxwell micropolar nanofluid is presented in Figure 16. It is observed that the augmented rate of G_r reduces the concentration of Maxwell micropolar nanofluid.

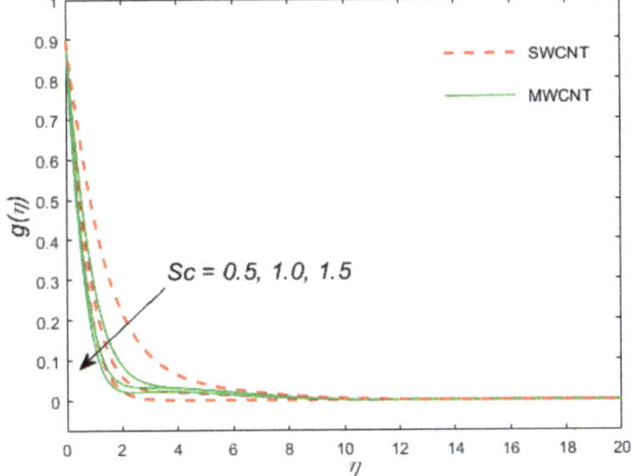

Figure 15. The variation of the concentration profile $g(\eta)$ for the case of SWCNT and MWCNT versus the similarity variable for the distinct values of the Schmidt number Sc.

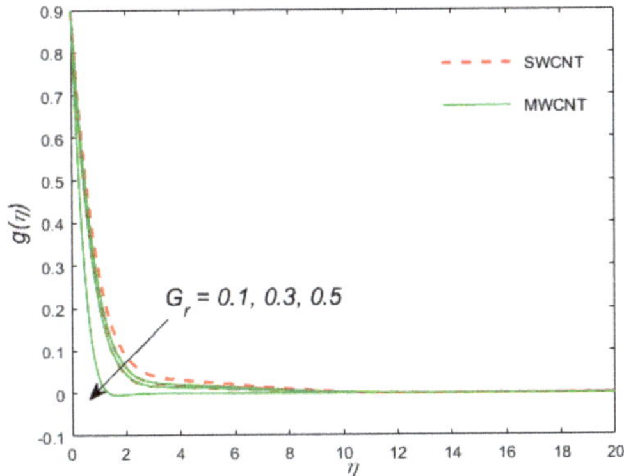

Figure 16. The variation of the concentration profile g(η) for the case of SWCNT and MWCNT versus the similarity variable for the distinct values of the chemical response factor Gr.

7.4. Local Density of Motile Microorganisms

The effects of (δ), (Pe), and (L_b) on Local density of motile microorganism's profile $h(\eta)$ appear in Figures 17–19. The impact of bio-convection constant δ on $h(\eta)$ is shown in Figure 17. The higher value of δ reduces $h(\eta)$ for both Maxwell micropolar nanotubes. The influences of Pe and L_b are presented in Figures 18 and 19. Decreasing behavior are observed for both L_b and Pe.

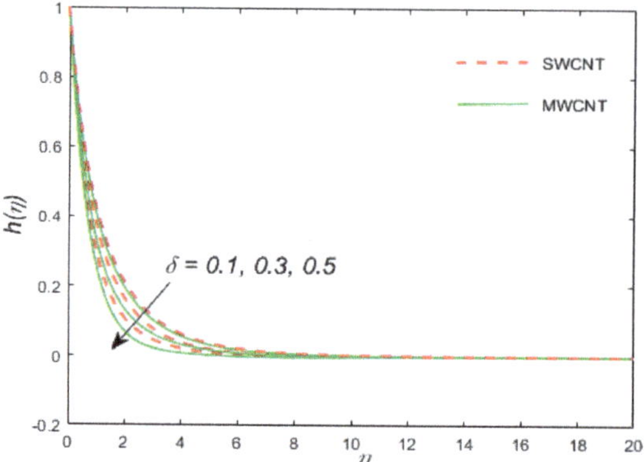

Figure 17. The variation of the local density of motile microorganisms $h(\eta)$ for the case of SWCNT and MWCNT versus the similarity variable for the distinct values of the bio-convection constant δ.

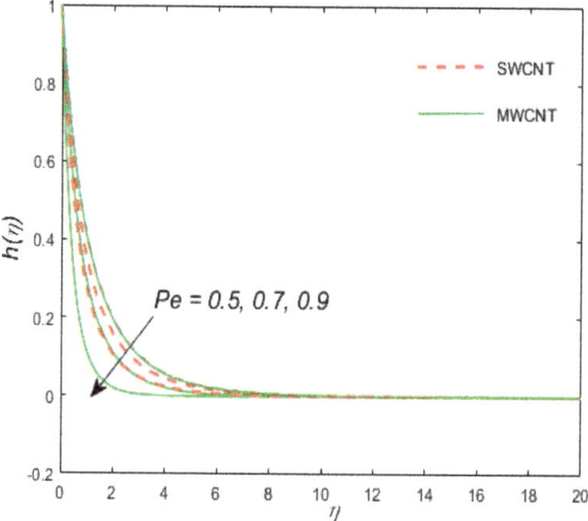

Figure 18. The variation of the local density of motile microorganisms $h(\eta)$ for the case of SWCNT and MWCNT versus the similarity variable for the distinct values of the bio-convection Peclet number *Pe*.

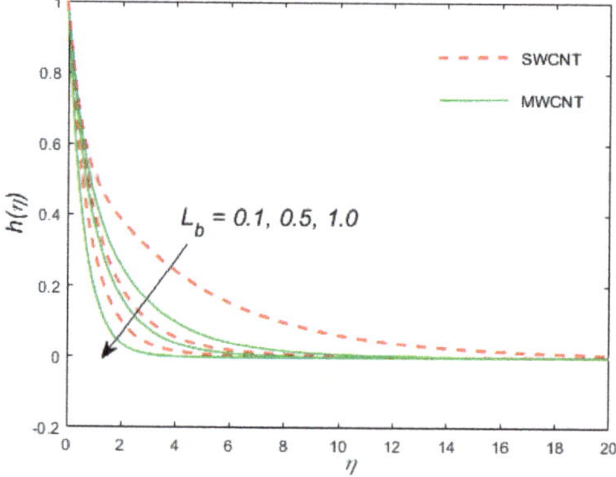

Figure 19. The variation of the local density of motile microorganisms $h(\eta)$ for the case of SWCNT and MWCNT versus the similarity variable for the distinct values of the bio-convection Lewis number L_b.

7.5. Entropy Optimization

The meddled parameters ξ, λ, α, Ra_x, K, and Br on NG are exposed in Figures 20–25. Figures 20 and 21 illustrate the variation of ξ and λ on NG. The increasing value of concentration difference ξ and diffusive constant λ enhances the entropy NG of the nanofluid of SWCNTS and MWCNTS. The Significant effects of α on NG are illustrated in Figure 22. For a higher value of parameter α the entropy NG is found as decreasing function. Figures 23 and 24 show the impact of Reynold number Ra_x and micro rotation parameter K. Entropy optimization of CNTs nanofluid increases with increasing of Ra_x and K. In Figure 25 the impact of Brinkman number Br is introduced. As a matter of fact, Brinkman number is a heat produced source inside the liquid moving district. The heat produced along with the heat moves from the divider and expands the entropy optimization.

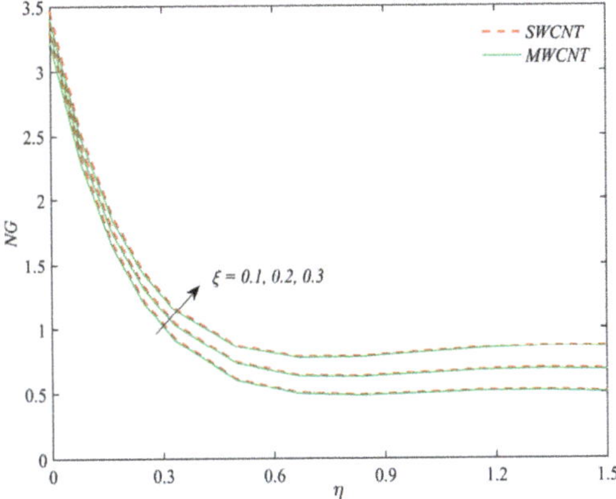

Figure 20. The variation of the entropy optimization NG for the case of SWCNT and MWCNT versus the similarity variable for the distinct values of the concentration difference parameter ξ.

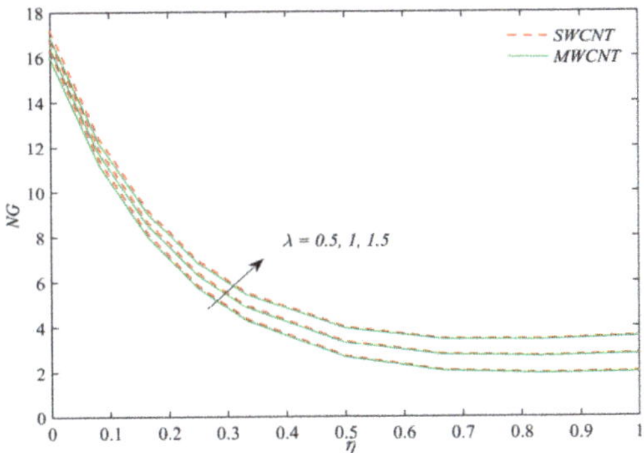

Figure 21. The variation of the entropy optimization NG for the case of SWCNT and MWCNT versus the similarity variable for the distinct values of the diffusive constant parameter λ.

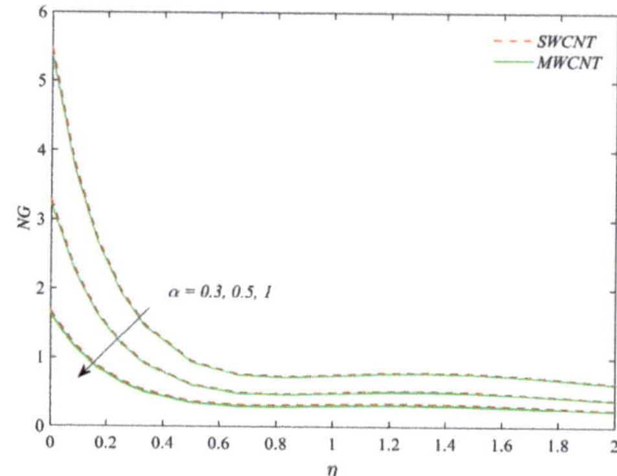

Figure 22. The variation of the entropy optimization *NG* for the case of SWCNT and MWCNT versus the similarity variable for the distinct values of the temperature difference parameter α.

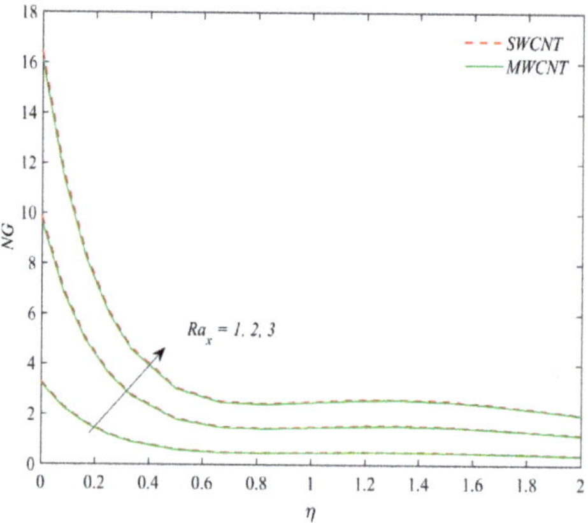

Figure 23. The variation of the entropy optimization *NG* for the case of SWCNT and MWCNT versus the similarity variable for the distinct values of the Reynold number Ra_x.

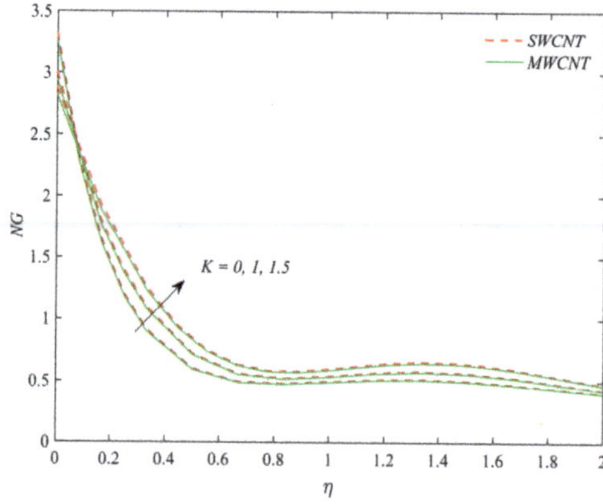

Figure 24. The variation of the entropy optimization *NG* for the case of SWCNT and MWCNT versus the similarity variable for the distinct values of the material parameter *K*.

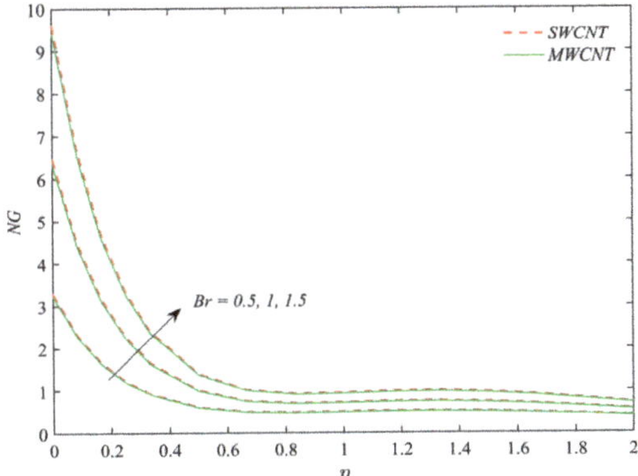

Figure 25. The variation of the entropy optimization *NG* for the case of SWCNT and MWCNT versus the similarity variable for the distinct values of the Brinkman number *Br*.

7.6. Engineering Quantities

Performances of dissimilar engineering parameter on skin friction coefficient C_{Fx}, temperature gradient Nu_x, mass transfer Sh_x, and local density of motile microorganisms N_{nx} are presented in Tables 2–6. Various Thermal-physical properties of carbon nanotubes are shown in Table 2.

Table 2. Thermo-physical properties of base fluid and both type of carbon nanotubes (CNTs) i.e., SWCNTs and MWCNTs.

Material	Water	SWCNT	MWCNT
C_p (j/kgK)	4179	425	796
ρ (kg/m^3)	997.1	2600	1600
k (W/mK)	0.613	6600	3000

Table 3. Estimations of skin friction $f''(0)$ versus different evaluations of various parameters.

Φ	k_1	V_0	R_b	C	$\frac{1}{(1-\phi)^{2.5}}f''(0)$ SWCNTs	MWCNTs
0.01	0.5	1.0	0.1	0.1	1.8355	1.7511
0.03	–	–	–	–	2.2996	1.8007
0.05	–	–	–	–	2.4875	1.8673
–	0.5	–	–	–	1.1596	1.1465
–	0.7	–	–	–	1.4990	1.4987
–	0.9	–	–	–	1.8355	1.8165
–	–	0.5	–	–	2.8463	2.7389
–	–	0.6	–	–	2.5630	2.5314
–	–	0.7	–	–	2.3591	2.3284
–	–	–	0.2	–	2.9477	2.2959
–	–	–	0.3	–	1.9302	1.9281
–	–	–	0.4	–	1.9201	1.8890
–	–	–	–	0.1	2.1976	2.0633
–	–	–	–	0.2	2.0917	2.0528
–	–	–	–	0.3	2.0750	2.0450

Table 4. Estimations of Nusselt number $-\frac{k_{nf}}{k}(1+R_d)\theta\prime(0)$ different evaluations of various parameters.

Φ	R_d	B_1	M	Ec	$-\frac{k_{nf}}{k}(1+R_d)\theta'(0)$	
					SWCNTs	MWCNTs
0.01	0.1	1.0	0.1	0.5	0.0122	0.0138
0.03	–	–	–	–	0.0200	0.0142
0.05	–	–	–	–	0.0222	0.0153
–	0.2	–	–	–	0.0205	0.0323
–	0.3	–	–	–	0.0232	0.0181
–	0.4	–	–	–	0.0290	0.0160
–	–	0.5	–	–	0.0134	0.0133
–	–	0.7	–	–	0.0137	0.0138
–	–	1.0	–	–	0.0139	0.0177
–	–	–	0.1	–	0.0122	0.0119
–	–	–	0.2	–	0.0139	0.0138
–	–	–	0.3	–	0.0142	0.0141
–	–	–	–	0.1	0.0116	0.0115
–	–	–	–	0.5	0.0139	0.0138
–	–	–	–	1.0	0.0159	0.0158

Table 5. Estimations of Sherwood number $-g\prime(0)$ versus different evaluations of various parameters.

Sc	G_r	n	N_r	$-g'(0)$	
				SWCNTs	MWCNTs
0.1	0.1	0.1	0.5	0.3430	0.3428
0.5	–	–	–	0.6290	0.6102
0.9	–	–	–	0.9227	0.8932
–	0.1	–	–	0.6290	0.6102
–	0.2	–	–	0.6977	0.6972
–	0.3	–	–	0.7513	0.7508
–	–	0.0	–	0.6290	0.6274
–	–	0.1	–	0.6147	0.6102
–	–	0.2	–	0.5782	0.6066
–	–	–	0.6	0.6072	0.6187
–	–	–	0.7	0.6005	0.6589
–	–	–	0.8	0.5954	0.6033

Table 6. Values of Motile density number $-h\prime(0)$ versus various estimates of different parameters.

L_b	Pe	R_b	δ	$-h'(0)$	
				SWCNTs	MWCNTs
0.5	0.5	0.1	0.1	0.7525	0.7515
0.6	–	–	–	0.8386	0.8375
0.7	–	–	–	0.9640	0.8806
–	0.1	–	–	0.5504	0.5175
–	0.2	–	–	0.5602	0.5760
–	0.3	–	–	0.6074	0.6779
–	–	0.0	–	0.7493	0.7483
–	–	0.1	–	0.7441	0.7789
–	–	0.2	–	0.7493	0.7847
–	–	–	0.6	0.7792	0.7515
–	–	–	0.7	0.7996	0.8050
–	–	–	0.8	0.8494	0.8231

7.7. Surface Drag Force

Table 3 show the numerical variation in term of skin friction C_{Fx}. Numerical variation of Φ, k_1, V_0, R_b, and M are shown. It is observed that the higher value of Φ, k_1, V_0, and R_b augmented the surface drag force C_{Fx} for both SWCNTs and MWCNTs while the opposite impact is found for M.

7.8. Heat Transfer Rate

Table 4 show numerical variation in Nusselt number Nu_x. Numerical variation of Φ, R_d, B_1, Ec, and M are shown. It is observed that the higher value of Φ, R_d, B_1, Ec, and M augmented the heat transfer rate for both SWCNTs and MWCNTs.

7.9. Mass Transfer Rate

Table 5 show the numerical variation in Sherwood number Sh_x. Numerical variation of Sc, G_r, n, and N_r are shown. It is observed that the higher value of Sc, G_r, n, and N_r augmented the mass transfer rate Sh_x in both SWCNTs and MWCNTs while the opposite impact is found for N_r.

7.10. Local Density of Motile Microorganisms N_{nx}

Table 6 shows the numerical variation in local density of motile microorganisms N_{nx}. Numerical variation of L_b, Pe, R_b, and δ are shown. It is observed that the higher value of L_b, Pe, R_b, and δ augmented the local density of motile microorganisms both SWCNTs and MWCNTs while the opposite impact is found for N_r.

8. Conclusions

In this article we studied the flow of two vertical conical carbon nanotubes (SWCNT and MWCNT) under the action of aqueous nanofluids, accompanied by the movement of swirling microorganisms and the influence of solute stratification in porous media. The investigation is carried out in the presence of chemical reactions, heat generation/absorption, and Joule heating. The noticeable features of the modeled problem are:

- The motion of the nanoparticle increases for enlarging values of solid volume fraction (Φ).
- For a larger value of magnetic parameter (M) the Lorentz forces enhance which raises the forces of resistance of the Maxwell micropolar motion which in turn reduces velocity $f'(\eta)$.
- The augmented N_r reduced the fluid motion.
- The momentum boundary layer reduces with enhances value of k_1.
- Micro rotation velocity $S(\eta)$ augmented with a higher value of K, γ^*, and α.
- Augmentation in the $\theta(\eta)$ with enhancement radiation parameter R_d is observed.
- The higher value of Ec amplified the kinetic energy of CNTs Maxwell micropolar nanofluid molecules, which thus enhanced the heat transmission rate.
- The augmented rate of G_r reduces the concentration of Maxwell micropolar nanofluid.
- As the estimate of the number of Peclets increases, the number of motion densities also increases.
- With the rise in estimations of Pe, $h(\eta)$ are increases.
- For both CNTs, $f'(\eta)$ intensifies against rising values of suction. For rising values of N_r, $\theta(\eta)$ is reducing.
- For these CNTs, $g(\eta)$ is reduced on the increasing of n.
- For the growth estimates of N_r, Sh_x is reduced and raises against numerical values of C_r.
- For solid volume fraction C_f is increased.
- Magnetic force M reduces Nu_x.
- A comparison between the present and previous outcomes for justification is given in Table 1.

Coatings **2020**, *10*, 998

Author Contributions: Z.S. and E.A. modeled and solved the problem. Z.S. wrote the manuscript. M.J. and U.K. contributed in the numerical computations and plotting the graphical results. All the corresponding authors completed the manuscript after its internal evaluation. All authors have read and agreed to the published version of the manuscript.

Funding: This project was funded by the Deanship of Scientific Research (DSR), King Abdulaziz University, Jeddah, under grant No. (DF-143-130-1441).

Acknowledgments: The authors, therefore, gratefully acknowledge DSR for technical and financial support.

Conflicts of Interest: The authors declare that they have no conflict of interest.

Abbreviations

$Pr = \frac{\nu_f}{\alpha}$	Prandtl number
$k_1 = \frac{x^2}{KRa_x^{1/2}}$	Porous parameter
$M = \frac{\sigma \beta_0^2 x^2}{\mu_f Ra_x^{1/2}}$	Magnetic parameter
$Sc = \frac{\alpha}{D_m}$	Schmidt number
$n = \frac{e}{d}$	Solutal stratification
$\gamma* = \frac{kx^2}{\gamma_{nf}^* Ra_x^{1/2}}$	Vortex velocity or material parameter
$\gamma = \frac{Q_0 x^2}{(\rho c_p) Ra_x^{1/2}}$	Heat suction/Injection parameter
$L_b = \frac{\alpha}{D_n}$	Bio-convection Lewis number
$Rd = \frac{16 T_\infty^3 \sigma}{3k*k_{nf}}$	Radiation parameter
$\beta = \frac{\lambda_1 \alpha Ra_x^{1/2}}{x^2}$	Bio-convection Rayleigh number
$N_r = \frac{\beta*(C_w - C_0)}{\beta (T_f - T_0)}$	Buoyancy ratio parameter
$R_b = \frac{\beta*\gamma \Delta p \Delta n_w}{\beta (T_f - T_0)}$	Bio-convection Rayleigh number
$C_r = \frac{K_r x^2}{D_m Ra_x^{1/2}}$	Chemical reaction parameter
$B_1 = \frac{h_f x}{Ra_x^{1/4} k_f}$	Boit number
$Pe = \frac{bW}{D_n}$	Bio-convection Peclet number
$\delta = \frac{n_\infty}{n_w - n_\infty}$	Bio-convection constant
$\alpha = \frac{\Delta T}{T_\infty}$	Temperature difference parameter
$Br = \frac{\mu_f U_w}{k_f \Delta T}$	Brinkman number
$\xi = \frac{\Delta C}{C_\infty}$	concentration difference parameter, and
$\lambda = \frac{RDC_\infty}{k_f}$	diffusive constant parameter
Ra_x	Reynold number

References

1. Choi, S.U.; Eastman, J.A. *Enhancing Thermal Conductivity of Fluids with Nanoparticles*; Argonne National Lab: Lemonte, IL, USA, 1995.
2. Chen, L.; Liu, J.; Fang, X.; Zhang, Z. Reduced graphene oxide dispersed nanofluids with improved photo-thermal conversion performance for direct absorption solar collectors. *Sol. Energy Mater. Sol. Cells* **2017**, *163*, 125–133. [CrossRef]
3. Oudina, F.M. Convective heat transfer of Titania nanofluids of different base fluids in cylindrical annulus with discrete heat source. *Heat Transf. Asian Res.* **2019**, *48*, 135–147. [CrossRef]
4. Oudina, F.M. Numerical modeling of the hydrodynamic stability in vertical annulus with heat source of different lengths. *Eng. Sci. Technol. Int. J.* **2017**, *20*, 1324–1333.
5. Sheikholeslami, M.; Shah, Z.; Shafee, A.; Khan, I.; Tlili, T. Uniform magnetic force impact on water based nanofluid thermal behavior in a porous enclosure with ellipse shaped obstacle. *Sci. Rep.* **2019**, 1196. [CrossRef]
6. Chougule, S.S.; Nirgude, V.V.; Gharge, P.D.; Mayank, M.; Sahu, S.K. Heat Transfer enhancements of low volume concentration CNT/water nanofluid and wire coil inserts in a circular tube. *Energy Procedia* **2016**, *90*, 552–558. [CrossRef]

7. Herrmann-Priesnitz, B.; Calderón-Muñoz, V.R.; Valencia, A.; Soto, R. Thermal design exploration of a swirl flow microchannel heat sink for high heat flux applications based on numerical simulations. *Appl. Therm. Eng.* **2016**, *109*, 22–34. [CrossRef]

8. Ding, Y.; Alias, H.; Wen, D.; Williams, R.A. Heat transfer of aqueous suspensions of carbon nanotubes (CNT nanofluids). *Int. J. Heat Mass Transf.* **2006**, *49*, 240–250. [CrossRef]

9. Pop, I.; Watanabe, T. Free convection with uniform suction or injection from a vertical cone for constant wall heat flux. *Int. Commun. Heat Mass Transf.* **1992**, *19*, 275–283. [CrossRef]

10. Xu, H. Modelling unsteady mixed convection of a nanofluid suspended with multiple kinds of nanoparticles between two rotating disks by generalized hybrid model. *Int. Comm. Heat Mass Tran.* **2019**, *108*, 104275. [CrossRef]

11. Rohsenow, W.M.; Hartnett, J.P.; Cho, Y.I. *Handbook of Heat Transfer*; McGraw-Hill Book Co.: New York, NY, USA, 1985. [CrossRef]

12. Sharma, K.V.; Sundar, L.S.; Sarma, P.K. Estimation of heat transfer coefficient and friction factor in the transition flow with low volume concentration of Al_2O_3 nano fluid flowing in a circular tube and with wisted tape insert. *Int. Commun. Heat Mass Transf.* **2009**, *36*, 503–507. [CrossRef]

13. Zhang, Y. Modified computational methods using effective heat capacity model for the thermal evaluation of PCM outfitted walls. *Int. Commun. Heat Mass Transf.* **2019**, *108*, 104278. [CrossRef]

14. Sun, X.; Zhang, Q.; Medina, M.A.; Lee, K.O. Experimental observations on the heat transfer enhancement caused by natural convection during melting of solid–liquid 17 phase change materials (PCMs). *Appl. Energy* **2016**, *162*, 1453–1461. [CrossRef]

15. Phuoc, T.X.; Massoudi, M.; Chen, R.H. Viscosity and thermal conductivity of nanofluids containing carbon nanotubes stabilized by Chitosan. *Int. J. Therm. Sci.* **2011**, *50*, 12–18. [CrossRef]

16. Harish, S.; Ishikawa, K.; Einarsson, E.; Aikawa, S.; Chiashi, S.; Shiomi, J.; Maruyama, S. Enhanced thermal conductivity of ethylene glycol with single-walled carbon nanotube inclusions. *Int. J. Heat Mass Transf.* **2012**, *55*, 3885–3890. [CrossRef]

17. Hayat, T.; Muhammad, T.; Shehzad, S.A.; Chen, G.Q.; Abbas, I.A. Interaction of magnetic field in flow of Maxwell nanofluid with convective effect. *J. Magn. Magn. Mater.* **2015**, *389*, 48–55. [CrossRef]

18. Bejan, A. A study of entropy generation in fundamentsl convective heat transfer. *J. Heat Trans.* **1979**, *101*, 718–725. [CrossRef]

19. Ellahi, R.; Hassan, M.; Zeeshan, A.; Khan, A.A. The shape effects of nanoparticles suspended in HFE-7100 over wedge with entropy generation and mixed convection. *Appl. Nanosci.* **2016**, *6*, 641–651. [CrossRef]

20. Sheikholeslami, M.; Jafaryar, M.; Hedayat, M.; Shafee, A.; Li, Z.; Khang Nguyen, T.; Bakouri, M. Heat transfer and turbulent simulation of nanomaterial due to compound turbulator including irreversibility analysis. *Int. J. Heat Mass Transf.* **2019**, *137*, 1290–1300. [CrossRef]

21. Ramzan, M.; Mohammad, M.; Howari, F. Magnetized suspended carbon nanotubes based nanofluid flow with bio-convection and entropy generation past a vertical cone. *Sci. Rep.* **2019**, *9*, 12225. [CrossRef] [PubMed]

22. Sarafraz, M.M.; Reza Safaei, M.; Goodarzi, M.; Arjomandi, M. Reforming of methanol with steam in a micro-reactor with Cu–SiO_2 porous catalyst. *Int. J. Hydrogen Energy* **2019**, *44*, 19628–19639. [CrossRef]

Publisher's Note: MDPI stays neutral with regard to jurisdictional claims in published maps and institutional affiliations.

MDPI
St. Alban-Anlage 66
4052 Basel
Switzerland
Tel. +41 61 683 77 34
Fax +41 61 302 89 18
www.mdpi.com

Coatings Editorial Office
E-mail: coatings@mdpi.com
www.mdpi.com/journal/coatings